DEPARTMENT OF THE TREASURY
BUREAU OF ALCOHOL, TOBACCO AND FIREARMS

FEDERAL
FIREARMS REGULATIONS
REFERENCE GUIDE

2000

ATF P 5300.4 (01-00)

Dear Federal Firearms Licensee:

We are pleased to provide you with this 2000 edition of
the Federal Firearms Regulations Reference Guide. With
respect to firearms enforcement, our mission is clear: to
reduce violent crime and protect the public. Toward this
end, our focus is on denying criminals access to firearms
and taking violent offenders off the streets. The efforts
made by licensees in maintaining good records of firearms
transactions and the cooperation shown in assisting us in
tracing crime guns has been invaluable.

Moreover, licensees complying with the interim Brady
law prevented the sale of approximately 320,000 handguns
from March 1994 to November 1998, thanks to the record
checks of handgun purchasers under that law. In most cases,
those would-be handgun buyers who were disqualified from
receiving and possessing firearms were individuals with
prior criminal histories. While it is impossible to know
precisely which future criminal activities were deterred by
the disqualification of these individuals, I have no doubt
that compliance with the law prevented countless crimes and,
indeed, saved many lives.

Effective on November 30, 1998, the interim provisions
of the Brady law expired and were replaced by the permanent
provisions of the Brady law. Now the eligibility of
firearms purchasers is determined at the time of sale by
querying the National Instant Criminal Background Check
System (NICS). You will find in this publication new ATF
and FBI regulations implementing the Brady law, as well as
questions and answers that should help you to understand and
comply with its provisions.

This publication provides you with information designed
to help you comply with all of the laws and regulations
governing the manufacture, importation and distribution of
firearms. The responsible and lawful distribution of
firearms by licensees plays an important role in protecting
the American public from violent firearms crime.

Sincerely yours,

John W. Magaw

FEDERAL FIREARMS REGULATION
REFERENCE GUIDE

TABLE OF CONTENTS

Materials, suggestions and information for future editions may be addressed to:

Bureau of Alcohol, Tobacco and Firearms
Firearms Programs Division
Washington, DC 20226

Editor's Notes

The cross references, bracketed notes, and Editor's notes seen in the regulations are for guidance and assistance only and do not appear in the official Code of Federal Regulations published by the Office of the Federal Register, National Archives and Records Service, General Services Administration.

Throughout this publication reference is made to the "regional director (compliance)", the names of these officers have been changed. Notification of this change will be published in the near future.

THE GUN CONTROL ACT OF 1968, PUBLIC LAW 90-618

TITLE 18, UNITED STATES CODE, CHAPTER 44

An Act to amend title 18, United States Code, to provide for better control of the interstate traffic in firearms.

Be it enacted by the Senate and House of Representatives of the United States of America in Congress assembled, that this Act may be cited as the "Gun Control Act of 1968".

TITLE I - STATE FIREARMS CONTROL ASSISTANCE

PURPOSE

Sec. 101. The Congress hereby declares that the purpose of this title is to provide support to Federal, State, and local law enforcement officials in their fight against crime and violence, and it is not the purpose of this title to place any undue or unnecessary Federal restrictions or burdens on law-abiding citizens with respect to the acquisition, possession, or use of firearms appropriate to the purpose of hunting, trapshooting, target shooting, personal protection, or any other lawful activity, and that this title is not intended to discourage or eliminate the private ownership or use of firearms by law-abiding citizens for lawful purposes, or provide for the imposition by Federal regulations of any procedures or requirements other than those reasonably necessary to implement and effectuate the provisions of this title.

*   *   *   *

Chapter 44 Firearms

§ 921. Definitions

(a) As used in this chapter –

(1) The term **"person"** and the term **"whoever"** include any individual, corporation, company, association, firm, partnership, society, or joint stock company.

(2) The term **"interstate or foreign commerce"** includes commerce between any place in a State and any place outside of that State, or within any possession of the United States (not including the Canal Zone) or the District of Columbia, but such term does not include commerce between places within the same State but through any place outside of that State. The term **"State"** includes the District of Columbia, the Commonwealth of Puerto Rico, and the possessions of the United States (not including

the Canal Zone).

(3) The term **"firearm"** means (A) any weapon (including a starter gun) which will or is designed to or may readily be converted to expel a projectile by the action of an explosive; (B) the frame or receiver of any such weapon; (C) any firearm muffler or firearm silencer; or (D) any destructive device. Such term does not include an antique firearm.

(4) The term **"destructive device"** means -

    (A) any explosive, incendiary, or poison gas -

        (i) bomb,

        (ii) grenade,

        (iii) rocket having a propellant charge of more than four ounces,

        (iv) missile having an explosive or incendiary charge of more than one-quarter ounce,

        (v) mine, or

        (vi) device similar to any of the devices described in the preceding clauses;

    (B) any type of weapon (other than a shotgun or a shotgun shell which the Secretary finds is generally recognized as particularly suitable for sporting purposes) by whatever name known which will, or which may be readily converted to, expel a projectile by the action of an explosive or other propellant, and which has any barrel with a bore of more than one-half inch in diameter; and

    (C) any combination of parts either designed or intended for use in converting any device into any destructive device described in subparagraph (A) or (B) and from which a destructive device may be readily assembled.

The term **"destructive device"** shall not include any device which is neither designed nor redesigned for use as a weapon; any device, although originally designed for use as a weapon, which is redesigned for use as a signaling, pyrotechnic, line throwing, safety, or similar device; surplus ordnance sold, loaned, or given by the Secretary of the Army pursuant to the provisions of section 4684(2), 4685, or 4686 of title 10; or any other device which the Secretary of the Treasury finds is not likely to be used as a weapon, is an antique, or is a rifle which the owner intends to use solely for sporting, recreational or cultural purposes.

(5) The term **"shotgun"** means a weapon designed or redesigned, made or remade, and intended to be fired from the shoulder and designed or redesigned and made or remade to use the energy of an explosive to fire through a smooth bore either a number of ball shot or a single projectile for each single pull of the trigger.

(6) The term **"short-barreled shotgun"** means a shotgun having one or more barrels less than eighteen inches in length and any weapon made from a shotgun (whether by alteration, modification, or otherwise) if such weapon as modified has an overall length of less than twenty-six inches.

(7) The term **"rifle"** means a weapon designed or redesigned, made or remade, and intended to be fired from the shoulder and designed or redesigned and made or remade to use the energy of an explosive to fire only a single projectile through a rifled bore for each single pull of the trigger.

(8) The term **"short-barreled rifle"** means a rifle having one or more barrels less than sixteen inches in length and any weapon made from a rifle (whether by

alteration, modification, or otherwise) if such weapon, as modified, has an overall length of less than twenty-six inches.

(9) The term **"importer"** means any person engaged in the business of importing or bringing firearms or ammunition into the United States for purposes of sale or distribution; and the term "licensed importer" means any such person licensed under the provisions of this chapter.

(10) The term **"manufacturer"** means any person engaged in the business of manufacturing firearms or ammunition for purposes of sale or distribution; and the term **"licensed manufacturer"** means any such person licensed under the provisions of this chapter.

(11) The term **"dealer"** means (A) any person engaged in the business of selling firearms at wholesale or retail, (B) any person engaged in the business of repairing firearms or of making or fitting special barrels, stocks, or trigger mechanisms to firearms, or (C) any person who is a pawnbroker. The term "licensed dealer" means any dealer who is licensed under the provisions of this chapter.

(12) The term **"pawnbroker"** means any person whose business or occupation includes the taking or receiving, by way of pledge or pawn, of any firearm as security for the payment or repayment of money.

(13) The term **"collector"** means any person who acquires, holds, or disposes of firearms as curios or relics, as the Secretary shall by regulation define, and the term **"licensed collector"** means any such person licensed under the provisions of this chapter.

(14) The term **"indictment"** includes an indictment or information in any court under which a crime punishable by imprisonment for a term exceeding one year may be prosecuted.

(15) The term **"fugitive from justice"** means any person who has fled from any State to avoid prosecution for a crime or to avoid giving testimony in any criminal proceeding.

(16) The term **"antique firearm"** means -

(A) any firearm (including any firearm with a matchlock, flintlock, percussion cap, or similar type of ignition system) manufactured in or before 1898; or

(B) any replica of any firearm described in subparagraph (A) if such replica -

(i) is not designed or redesigned for using rimfire or conventional centerfire fixed ammunition, or

(ii) uses rimfire or conventional centerfire fixed ammunition which is no longer manufactured in the United States and which is not readily available in the ordinary channels of commercial trade; or

(C) any muzzle loading rifle, muzzle loading shotgun, or muzzle loading pistol, which is designed to use black powder, or a black powder substitute, and which cannot use fixed ammunition. For purposes of this subparagraph, the term "antique firearm" shall not include any weapon which incorporates a firearm frame or receiver, any firearm which is converted into a muzzle loading weapon, or any muzzle loading weapon which can be readily converted to fire fixed ammunition by replacing the barrel, bolt, breechblock, or any combination thereof.

(17)(A) The term **"ammunition"** means ammunition or cartridge cases, primers, bullets, or propellent powder designed for use in any firearm.

(B) The term **"armor piercing ammunition"** means -

(i) a projectile or projectile core which may be used in a handgun and which is constructed entirely (excluding the presence of traces of other substances) from one or a combination of tungsten alloys, steel, iron, brass, bronze, beryllium copper, or depleted uranium; or

(ii) a full jacketed projectile larger than .22 caliber designed and intended for use in a handgun and whose jacket has a weight of more than 25 percent of the total weight of the projectile.

(C) The term **"armor piercing ammunition"** does not include shotgun shot required by Federal or State environmental or game regulations for hunting purposes, a frangible projectile designed for target shooting, a projectile which the Secretary finds is primarily intended to be used for sporting purposes, or any other projectile or projectile core which the Secretary finds is intended to be used for industrial purposes, including a charge used in an oil and gas well perforating device.

(18) The term **"Secretary"** or **"Secretary of the Treasury"** means the Secretary of the Treasury or his delegate.

(19) The term **"published ordinance"** means a published law of any political subdivision of a State which the Secretary determines to be relevant to the enforcement of this chapter and which is contained on a list compiled by the Secretary, which list shall be published in the Federal Register, revised annually, and furnished to each licensee under this chapter.

(20) The term **"crime punishable by imprisonment for a term exceeding one year"** does not include -

(A) any Federal or State offenses pertaining to antitrust violations, unfair trade practices, restraints of trade, or other similar offenses relating to the regulation of business practices, or

(B) any State offense classified by the laws of the State as a misdemeanor and punishable by a term of imprisonment of two years or less.

What constitutes a conviction of such a crime shall be determined in accordance with the law of the jurisdiction in which the proceedings were held. Any conviction which has been expunged, or set aside or for which a person has been pardoned or has had civil rights restored shall not be considered a conviction for purposes of this chapter, unless such pardon, expungement, or restoration of civil rights expressly provides that the person may not ship, transport, possess, or receive firearms.

(21) The term **"engaged in the business"** means -

(A) as applied to a manufacturer of firearms, a person who devotes time, attention, and labor to manufacturing firearms as a regular course of trade or business with the principal objective of livelihood and profit through the sale or distribution of the firearms manufactured;

(B) as applied to a manufacturer of ammunition, a person who devotes time, attention, and labor to manufacturing ammunition as a regular course of trade or business with the principal objective of livelihood and profit through the sale or distribution of the ammunition manufactured;

(C) as applied to a dealer in firearms, as defined in section 921(a)(11)(A), a person who devotes time, attention, and labor to dealing in firearms as a regular course of trade or business with the principal objective of livelihood and profit through the repetitive purchase and resale of firearms, but such term shall not include a person who makes occasional sales, exchanges, or purchases of firearms for the enhancement of a personal collection or for a hobby, or who sells all or part of his personal collection of firearms;

(D) as applied to a dealer in firearms, as defined in section 921(a)(11)(B), a person who devotes time, attention, and labor to engaging in such activity as a regular course of trade or business with the principal objective of livelihood and profit, but such term shall not include a person who makes occasional repairs of firearms, or who occasionally fits special barrels, stocks, or trigger mechanisms to firearms;

(E) as applied to an importer of firearms, a person who devotes time, attention, and labor to importing firearms as a regular course of trade or business with the principal objective of livelihood and profit through the sale or distribution of the firearms imported; and

(F) as applied to an importer of ammunition, a person who devotes time, attention, and labor to importing ammunition as a regular course of trade or business with the principal objective of livelihood and profit through the sale or distribution of the ammunition imported.

(22) The term **"with the principal objective of livelihood and profit"** means that the intent underlying the sale or disposition of firearms is predominantly one of obtaining livelihood and pecuniary gain, as opposed to other intents, such as improving or liquidating a personal firearms collection: **Provided**, That proof of profit shall not be required as to a person who engages in the regular and repetitive purchase and disposition of firearms for criminal purposes or terrorism. For purposes of this paragraph, the term **"terrorism"** means activity, directed against United States persons, which -

(A) is committed by an individual who is not a national or permanent resident alien of the United States;

(B) involves violent acts or acts dangerous to human life which would be a criminal violation if committed within the jurisdiction of the United States; and

(C) is intended -

(i) to intimidate or coerce a civilian population;

(ii) to influence the policy of a government by intimidation or coercion; or

(iii) to affect the conduct of a government by assassination or kidnapping.

(23) The term **"machinegun"** has the meaning given such term in section 5845(b) of the National Firearms Act (26 U.S.C. 5845(b)).

(24) The terms **"firearm silencer"** and **"firearm muffler"** mean any device for silencing, muffling, or diminishing the report of a portable firearm, including any combination of parts, designed or redesigned, and intended for use in assembling or fabricating a firearm silencer or firearm muffler, and any part intended only for use in such assembly or fabrication.

(25) The term **"school zone"** means -

(A) in, or on the grounds of, a public, parochial or private school; or

(B) within a distance of 1,000 feet from the grounds of a public, parochial or private school.

(26) The term **"school"** means a school which provides elementary or secondary education, as determined under State law.

(27) The term **"motor vehicle"** has the meaning given such term in section 13102 of title 49, United States Code.

(28) The term **"semiautomatic rifle"** means any repeating rifle which utilizes a portion of the energy of a firing cartridge to extract the fired cartridge case and chamber the next round, and which requires a separate pull of the trigger to fire each cartridge.

(29) The term **"handgun"** means –

    (A) a firearm which has a short stock and is designed to be held and fired by the use of a single hand; and

    (B) any combination of parts from which a firearm described in subparagraph (A) can be assembled.

(30) The term **"semiautomatic assault weapon"** means –

    (A) any of the firearms, or copies or duplicates of the firearms in any caliber, known as –

        (i) Norinco, Mitchell, and Poly Technologies Avtomat Kalashnikovs (all models);

        (ii) Action Arms Israeli Military Industries UZI and Galil;

        (iii) Beretta Ar70 (SC-70);

        (iv) Colt AR-15;

        (v) Fabrique National FN/FAL, FN/LAR, and FNC;

        (vi) SWD M-10, M-11, M-11/9, and M-12;

        (vii) Steyr AUG;

        (viii) INTRATEC TEC-9, TEC-DC9 and TEC-22; and

        (ix) revolving cylinder shotguns, such as (or similar to) the Street Sweeper and Striker 12;

    (B) a **semiautomatic rifle** that has an ability to accept a detachable magazine and has at least 2 of –

        (i) a folding or telescoping stock;

        (ii) a pistol grip that protrudes conspicuously beneath the action of the weapon;

        (iii) a bayonet mount;

        (iv) a flash suppressor or threaded barrel designed to accommodate a flash suppressor; and

        (v) a grenade launcher;

    (C) a **semiautomatic pistol** that has an ability to accept a detachable magazine and has at least 2 of –

        (i) an ammunition magazine that attaches to the pistol outside of the pistol grip;

(ii) a threaded barrel capable of accepting a barrel extender, flash suppressor, forward handgrip, or silencer;

(iii) a shroud that is attached to, or partially or completely encircles, the barrel and that permits the shooter to hold the firearm with the nontrigger hand without being burned;

(iv) a manufactured weight of 50 ounces or more when the pistol is unloaded; and

(v) a semiautomatic version of an automatic firearm; and

(D) a **semiautomatic shotgun** that has at least 2 of -

(i) a folding or telescoping stock;

(ii) a pistol grip that protrudes conspicuously beneath the action of the weapon;

(iii) a fixed magazine capacity in excess of 5 rounds; and

(iv) an ability to accept a detachable magazine.

(31) The term **"large capacity ammunition feeding device"** -

(A) means a magazine, belt, drum, feed strip, or similar device manufactured after the date of enactment of the Violent Crime Control and Law Enforcement Act of 1994 that has a capacity of, or that can be readily restored or converted to accept, more than 10 rounds of ammunition; but

(B) does not include an attached tubular device designed to accept, and capable of operating only with, .22 caliber rimfire ammunition.

(32) The term **"intimate partner"** means, with respect to a person, the spouse of the person, a former spouse of the person, an individual who is a parent of a child of the person, and an individual who cohabitates or has cohabited with the person.

(33)(A) Except as provided in subparagraph (C), the term "misdemeanor crime of domestic violence" means an offense that —

(i) is a misdemeanor under Federal or State law; and

(ii) has, as an element, the use or attempted use of physical force, or the threatened use of a deadly weapon, committed by a current or former spouse, parent, or guardian of the victim, by a person with whom the victim shares a child in common, by a person who is cohabiting with or has cohabited with the victim as a spouse, parent, or guardian, or by a person similarly situated to a spouse, parent, or guardian of the victim

(B)(i) A person shall not be considered to have been convicted of such an offense for purposes of this chapter, unless —

(I) the person was represented by counsel in the case, or knowingly and intelligently waived the right to counsel in the case; and

(II) in the case of a prosecution for an offense described in this paragraph for which a person was entitled to a jury trial in the jurisdiction in which the case was tried, either

(aa) the case was tried by a jury, or

(bb) the person knowingly and intelligently waived the right to have the case tried by a jury, by guilty plea or otherwise.

(ii) A person shall not be considered to have been convicted of such an offense for purposes of this chapter if the conviction has been expunged or set aside, or is an offense for which the person has been pardoned or has had civil rights restored (if the law of the applicable jurisdiction provides for the loss of civil rights under such an offense) unless the pardon, expungement, or restoration of civil rights expressly provides that the person may not ship, transport, possess, or receive firearms.

(34) The term **"secure gun storage or safety device"** means –

(A) a device that, when installed on a firearm, is designed to prevent the firearm from being operated without first deactivating the device;

(B) a device incorporated into the design of the firearm that is designed to prevent the operation of the firearm by anyone not having access to the device; or

(C) a safe, gun safe, gun case, lock box, or other device that is designed to be or can be used to store a firearm and that is designed to be unlocked only by means of a key, a combination, or other similar means.

(b) For the purposes of this chapter, a member of the Armed Forces on active duty is a resident of the State in which his permanent duty station is located.

Editor's Note

Section 921(a)(34) was enacted by Public Law 105-277 on October 21, 1998, and became effective on April 19, 1999. No subparagraph (c) was enacted in subsection (a)(33).

(Added Pub.L. 90-351, Title IV, § 902, June 19, 1968, 82 Stat. 226, and amended Pub.L. 90-618, Title I, § 102, Oct. 22, 1968, 82 Stat. 1214; Pub.L. 93-639, § 102, Jan. 4, 1975, 88 Stat. 2217; Pub.L. 99-308, § 101, May 19, 1986, 100 Stat. 449; Pub.L. 99-360, § 1(b), July 8, 1986, 100 Stat. 766; Pub.L. 99-408, § 1, Aug. 28, 1986, 100 Stat. 920; Pub.L. 101-647, Title XVII, § 1702(b)(2), Title XXII, § 2204(a), Nov. 29, 1990, 104 Stat. 4845, 4857; Pub.L. 103-159, Title I, § 102(a)(2), Nov. 30, 1993, 107 Stat. 1539; Pub.L. 103-322, Title XI, §§ 110102(b), 110103(b), 110401(a), 110519, Title XXXIII, § 330021(1), Sept. 13, 1994, 108 Stat. 1997, 1999, 2014, 2020, 2150; Pub.L. 104-88, Title III, § 303(1), Dec. 29, 1995, 109 Stat. 943; Pub.L. 104-208, Div. A, Title I, § 101(f) [Title VI, § 658(a)], Sept. 30, 1996, 110 Stat. 3009-371; Pub.L. 105-277, Div. A, § 101(b) [Title I, § 119(a)], § 101(h) [Title I, § 115], Oct. 21, 1998, 112 Stat. 2681 – ____, 2681 – ____.)

§ 922. Unlawful acts

(a) It shall be unlawful –

(1) for any person –

(A) except a licensed importer, licensed manufacturer, or licensed dealer, to engage in the business of importing, manufacturing, or dealing in firearms, or in the course of such business to ship, transport, or receive any firearm in interstate or foreign commerce; or

(B) except a licensed importer or licensed manufacturer, to engage in the business of importing or manufacturing ammunition, or in the course of such business, to ship, transport, or receive any ammunition in interstate or foreign commerce;

(2) for any importer, manufacturer, dealer, or collector licensed under the provisions of this chapter to ship or transport in interstate or foreign commerce

any firearm to any person other than a licensed importer, licensed manufacturer, licensed dealer, or licensed collector, except that -

(A) this paragraph and subsection (b)(3) shall not be held to preclude a licensed importer, licensed manufacturer, licensed dealer, or licensed collector from returning a firearm or replacement firearm of the same kind and type to a person from whom it was received; and this paragraph shall not be held to preclude an individual from mailing a firearm owned in compliance with Federal, State, and local law to a licensed importer, licensed manufacturer, licensed dealer, or licensed collector;

(B) this paragraph shall not be held to preclude a licensed importer, licensed manufacturer, or licensed dealer from depositing a firearm for conveyance in the mails to any officer, employee, agent, or watchman who, pursuant to the provisions of section 1715 of this title, is eligible to receive through the mails pistols, revolvers, and other firearms capable of being concealed on the person, for use in connection with his official duty; and

(C) nothing in this paragraph shall be construed as applying in any manner in the District of Columbia, the Commonwealth of Puerto Rico, or any possession of the United States differently than it would apply if the District of Columbia, the Commonwealth of Puerto Rico, or the possession were in fact a State of the United States;

(3) for any person, other than a licensed importer, licensed manufacturer, licensed dealer, or licensed collector to transport into or receive in the State where he resides (or if the person is a corporation or other business entity, the State where it maintains a place of business) any firearm purchased or otherwise obtained by such person outside that State, except that this paragraph (A) shall not preclude any person who lawfully acquires a firearm by bequest or intestate succession in a State other than his State of residence from transporting the firearm into or receiving it in that State, if it is lawful for such person to purchase or possess such firearm in that State, (B) shall not apply to the transportation or receipt of a firearm obtained in conformity with subsection (b)(3) of this section, and (C) shall not apply to the transportation of any firearm acquired in any State prior to the effective date of this chapter;

(4) for any person, other than a licensed importer, licensed manufacturer, licensed dealer, or licensed collector, to transport in interstate or foreign commerce any destructive device, machinegun (as defined in section 5845 of the Internal Revenue Code of 1954), short-barreled shotgun, or short-barreled rifle, except as specifically authorized by the Secretary consistent with public safety and necessity;

(5) for any person (other than a licensed importer, licensed manufacturer, licensed dealer, or licensed collector) to transfer, sell, trade, give, transport, or deliver any firearm to any person (other than a licensed importer, licensed manufacturer, licensed dealer, or licensed collector) who the transferor knows or has reasonable cause to believe does not reside in (or if the person is a corporation or other business entity, does not maintain a place of business in) the State in which the transferor resides; except that this paragraph shall not apply to (A) the transfer, transportation, or delivery of a firearm made to carry out a bequest of a firearm to, or an acquisition by intestate succession of a firearm by, a person who is permitted to acquire or possess a firearm under the laws of the State of his residence, and (B) the loan or rental of a firearm to any person for temporary use for lawful sporting purposes;

(6) for any person in connection with the acquisition or attempted acquisition of any firearm or ammunition from a licensed importer, licensed manufacturer, licensed dealer, or licensed collector, knowingly to make any false or fictitious oral or written statement or to furnish or exhibit any false, fictitious, or misrepresented identification, intended or likely to deceive such importer,

manufacturer, dealer, or collector with respect to any fact material to the lawfulness of the sale or other disposition of such firearm or ammunition under the provisions of this chapter;

(7) for any person to manufacture or import armor piercing ammunition, except that this paragraph shall not apply to -

(A) the manufacture or importation of such ammunition for the use of the United States or any department or agency thereof or any State or any department, agency, or political subdivision thereof;

(B) the manufacture of such ammunition for the purpose of exportation; and

(C) any manufacture or importation for the purposes of testing or experimentation authorized by the Secretary;

(8) for any manufacturer or importer to sell or deliver armor piercing ammunition, except that this paragraph shall not apply to -

(A) the sale or delivery by a manufacturer or importer of such ammunition for use of the United States or any department or agency thereof or any State or any department, agency, or political subdivision thereof;

(B) the sale or delivery by a manufacturer or importer of such ammunition for the purpose of exportation;

(C) the sale or delivery by a manufacturer or importer of such ammunition for the purposes of testing or experimenting authorized by the Secretary; and

(9) for any person, other than a licensed importer, licensed manufacturer, licensed dealer, or licensed collector, who does not reside in any State to receive any firearms unless such receipt is for lawful sporting purposes.

(b) It shall be unlawful for any licensed importer, licensed manufacturer, licensed dealer, or licensed collector to sell or deliver -

(1) any firearm or ammunition to any individual who the licensee knows or has reasonable cause to believe is less than eighteen years of age, and, if the firearm, or ammunition is other than a shotgun or rifle, or ammunition for a shotgun or rifle, to any individual who the licensee knows or has reasonable cause to believe is less than twenty-one years of age;

(2) any firearm to any person in any State where the purchase or possession by such person of such firearm would be in violation of any State law or any published ordinance applicable at the place of sale, delivery or other disposition, unless the licensee knows or has reasonable cause to believe that the purchase or possession would not be in violation of such State law or such published ordinance;

(3) any firearm to any person who the licensee knows or has reasonable cause to believe does not reside in (or if the person is a corporation or other business entity, does not maintain a place of business in) the State in which the licensee's place of business is located, except that this paragraph (A) shall not apply to the sale or delivery of any rifle or shotgun to a resident of a State other than a State in which the licensee's place of business is located if the transferee meets in person with the transferor to accomplish the transfer, and the sale, delivery, and receipt fully comply with the legal conditions of sale in both such States (and any licensed manufacturer, importer or dealer shall be presumed, for purposes of this subparagraph, in the absence of evidence to the contrary, to have had actual knowledge of the State laws and published ordinances of both States), and (B) shall not apply to the loan or rental of a firearm to any person for temporary use for lawful sporting purposes;

(4) to any person any destructive device, machinegun (as defined in section 5845 of the Internal Revenue Code of 1954), short-barreled shotgun, or short-barreled rifle, except as specifically authorized by the Secretary consistent with public safety and necessity; and

(5) any firearm or armor-piercing ammunition to any person unless the licensee notes in his records, required to be kept pursuant to section 923 of this chapter, the name, age, and place of residence of such person if the person is an individual, or the identity and principal and local places of business of such person if the person is a corporation or other business entity.

Paragraphs (1), (2), (3), and (4) of this subsection shall not apply to transactions between licensed importers, licensed manufacturers, licensed dealers, and licensed collectors. Paragraph (4) of this subsection shall not apply to a sale or delivery to any research organization designated by the Secretary.

(c) In any case not otherwise prohibited by this chapter, a licensed importer, licensed manufacturer, or licensed dealer may sell a firearm to a person who does not appear in person at the licensee's business premises (other than another licensed importer, manufacturer, or dealer) only if -

(1) the transferee submits to the transferor a sworn statement in the following form:

"Subject to penalties provided by law, I swear that, in the case of any firearm other than a shotgun or a rifle, I am twenty-one years or more of age, or that, in the case of a shotgun or a rifle, I am eighteen years or more of age; that I am not prohibited by the provisions of chapter 44 of title 18, United States Code, from receiving a firearm in interstate or foreign commerce; and that my receipt of this firearm will not be in violation of any statute of the State and published ordinance applicable to the locality in which I reside. Further, the true title, name, and address of the principal law enforcement officer of the locality to which the firearm will be delivered are _____.
Signature _____ Date _____."

and containing blank spaces for the attachment of a true copy of any permit or other information required pursuant to such statute or published ordinance;

(2) the transferor has, prior to the shipment or delivery of the firearm, forwarded by registered or certified mail (return receipt requested) a copy of the sworn statement, together with a description of the firearm, in a form prescribed by the Secretary, to the chief law enforcement officer of the transferee's place of residence, and has received a return receipt evidencing delivery of the statement or has had the statement returned due to the refusal of the named addressee to accept such letter in accordance with United States Post Office Department regulations; and

(3) the transferor has delayed shipment or delivery for a period of at least seven days following receipt of the notification of the acceptance or refusal of delivery of the statement.

A copy of the sworn statement and a copy of the notification to the local law enforcement officer, together with evidence of receipt or rejection of that notification shall be retained by the licensee as a part of the records required to be kept under section 923(g).

(d) It shall be unlawful for any person to sell or otherwise dispose of any firearm or ammunition to any person knowing or having reasonable cause to believe that such person -

(1) is under indictment for, or has been convicted in any court of, a crime

punishable by imprisonment for a term exceeding one year;

(2) is a fugitive from justice;

(3) is an unlawful user of or addicted to any controlled substance (as defined in section 102 of the Controlled Substances Act (21 U.S.C. 802));

(4) has been adjudicated as a mental defective or has been committed to any mental institution;

(5) who, being an alien -

(A) is illegally or unlawfully in the United States; or

(B) except as provided in subsection (y)(2), has been admitted to the United States under a nonimmigrant visa (as that term is defined in section 101(a)(26) of the Immigration and Nationality Act (8 U.S.C. 1101(a)(26)));

(6) who has been discharged from the Armed Forces under dishonorable conditions;

(7) who, having been a citizen of the United States, has renounced his citizenship;

(8) is subject to a court order that restrains such person from harassing, stalking, or threatening an intimate partner of such person or child of such intimate partner or person, or engaging in other conduct that would place an intimate partner in reasonable fear of bodily injury to the partner or child, except that this paragraph shall only apply to a court order that -

(A) was issued after a hearing of which such person received actual notice, and at which such person had the opportunity to participate; and

(B)(i) includes a finding that such person represents a credible threat to the physical safety of such intimate partner or child; or

(ii) by its terms explicitly prohibits the use, attempted use, or threatened use of physical force against such intimate partner or child that would reasonably be expected to cause bodily injury; or

(9) has been convicted in any court of a misdemeanor crime of domestic violence.

This subsection shall not apply with respect to the sale or disposition of a firearm or ammunition to a licensed importer, licensed manufacturer, licensed dealer, or licensed collector who pursuant to subsection (b) of section 925 of this chapter is not precluded from dealing in firearms or ammunition, or to a person who has been granted relief from disabilities pursuant to subsection (c) of section 925 of this chapter.

(e) It shall be unlawful for any person knowingly to deliver or cause to be delivered to any common or contract carrier for transportation or shipment in interstate or foreign commerce, to persons other than licensed importers, licensed manufacturers, licensed dealers, or licensed collectors, any package or other container in which there is any firearm or ammunition without written notice to the carrier that such firearm or ammunition is being transported or shipped; except that any passenger who owns or legally possesses a firearm or ammunition being transported aboard any common or contract carrier for movement with the passenger in interstate or foreign commerce may deliver said firearm or ammunition into the custody of the pilot, captain, conductor or operator of such common or contract carrier for the duration of the trip without violating any of the provisions of this chapter. No common or contract carrier shall require or cause any label, tag, or other written notice to be placed on the outside of any package, luggage, or other container that such package, luggage, or other container contains a firearm.

(f) (1) It shall be unlawful for any common or contract carrier to transport or deliver in interstate or foreign commerce any firearm or ammunition with knowledge or reasonable cause to believe that the shipment, transportation, or receipt thereof would be in violation of the provisions of this chapter.

(2) It shall be unlawful for any common or contract carrier to deliver in interstate or foreign commerce any firearm without obtaining written acknowledgement of receipt from the recipient of the package or other container in which there is a firearm.

(g) It shall be unlawful for any person -

(1) who has been convicted in any court of, a crime punishable by imprisonment for a term exceeding one year;

(2) who is a fugitive from justice;

(3) who is an unlawful user of or addicted to any controlled substance (as defined in section 102 of the Controlled Substances Act (21 U.S.C. 802));

(4) who has been adjudicated as a mental defective or who has been committed to a mental institution;

(5) who, being an alien -

(A) is illegally or unlawfully in the United States; or

(B) except as provided in subsection (y)(2), has been admitted to the United States under a nonimmigrant visa (as that term is defined in section 101(a)(26) of the Immigration and Nationality Act (8 U.S.C. 1101(a)(26)));

(6) who has been discharged from the Armed Forces under dishonorable conditions;

(7) who, having been a citizen of the United States, has renounced his citizenship;

(8) who is subject to a court order that -

(A) was issued after a hearing of which such person received actual notice, and at which such person had an opportunity to participate;

(B) restrains such person from harassing, stalking, or threatening an intimate partner of such person or child of such intimate partner or person, or engaging in other conduct that would place an intimate partner in reasonable fear of bodily injury to the partner or child; and

(C)(i) includes a finding that such person represents a credible threat to the physical safety of such intimate partner or child; or

(ii) by its terms explicitly prohibits the use, attempted use, or threatened use of physical force against such intimate partner or child that would reasonably be expected to cause bodily injury; or

(9) who has been convicted in any court of a misdemeanor crime of domestic violence,

to ship or transport in interstate or foreign commerce, or possess in or affecting commerce, any firearm or ammunition; or to receive any firearm or ammunition which has been shipped or transported in interstate or foreign commerce.

(h) It shall be unlawful for any individual, who to that individual's knowledge and while being employed for any person described in any paragraph of subsection (g) of this section, in the course of such employment -

   (1) to receive, possess, or transport any firearm or ammunition in or affecting interstate or foreign commerce; or

   (2) to receive any firearm or ammunition which has been shipped or transported in interstate or foreign commerce.

(i) It shall be unlawful for any person to transport or ship in interstate or foreign commerce, any stolen firearm or stolen ammunition, knowing or having reasonable cause to believe that the firearm or ammunition was stolen.

(j) It shall be unlawful for any person to receive, possess, conceal, store, barter, sell, or dispose of any stolen firearm or stolen ammunition, or pledge or accept as security for a loan any stolen firearm or stolen ammunition, which is moving as, which is a part of, which constitutes, or which has been shipped or transported in, interstate or foreign commerce, either before or after it was stolen, knowing or having reasonable cause to believe that the firearm or ammunition was stolen.

(k) It shall be unlawful for any person knowingly to transport, ship, or receive, in interstate or foreign commerce, any firearm which has had the importer's or manufacturer's serial number removed, obliterated, or altered or to possess or receive any firearm which has had the importer's or manufacturer's serial number removed, obliterated, or altered and has, at any time, been shipped or transported in interstate or foreign commerce.

(l) Except as provided in section 925(d) of this chapter, it shall be unlawful for any person knowingly to import or bring into the United States or any possession thereof any firearm or ammunition; and it shall be unlawful for any person knowingly to receive any firearm or ammunition which has been imported or brought into the United States or any possession thereof in violation of the provisions of this chapter.

(m) It shall be unlawful for any licensed importer, licensed manufacturer, licensed dealer, or licensed collector knowingly to make any false entry in, to fail to make appropriate entry in, or to fail to properly maintain, any record which he is required to keep pursuant to section 923 of this chapter or regulations promulgated thereunder.

(n) It shall be unlawful for any person who is under indictment for a crime punishable by imprisonment for a term exceeding one year to ship or transport in interstate or foreign commerce any firearm or ammunition or receive any firearm or ammunition which has been shipped or transported in interstate or foreign commerce.

(o) (1) Except as provided in paragraph (2), it shall be unlawful for any person to transfer or possess a machinegun.

   (2) This subsection does not apply with respect to -

      (A) a transfer to or by, or possession by or under the authority of, the United States or any department or agency thereof or a State, or a department, agency, or political subdivision thereof; or

      (B) any lawful transfer or lawful possession of a machinegun that was lawfully possessed before the date this subsection takes effect.

(p) (1) It shall be unlawful for any person to manufacture, import, sell, ship, deliver, possess, transfer, or receive any firearm -

(A) that, after removal of grips, stocks, and magazines, is not as detectable as the Security Exemplar, by walk-through metal detectors calibrated and operated to detect the Security Exemplar; or

(B) any major component of which, when subjected to inspection by the types of x-ray machines commonly used at airports, does not generate an image that accurately depicts the shape of the component. Barium sulfate or other compounds may be used in the fabrication of the component.

(2) For purposes of this subsection -

(A) the term "firearm" does not include the frame or receiver of any such weapon;

(B) the term "major component" means, with respect to a firearm, the barrel, the slide or cylinder, or the frame or receiver of the firearm; and

(C) the term "Security Exemplar" means an object, to be fabricated at the direction of the Secretary, that is -

(i) constructed of, during the 12-month period beginning on the date of the enactment of this subsection, 3.7 ounces of material type 17-4 PH stainless steel in a shape resembling a handgun; and

(ii) suitable for testing and calibrating metal detectors:

Provided, however, That at the close of such 12-month period, and at appropriate times thereafter the Secretary shall promulgate regulations to permit the manufacture, importation, sale, shipment, delivery, possession, transfer, or receipt of firearms previously prohibited under this subparagraph that are as detectable as a "Security Exemplar" which contains 3.7 ounces of material type 17-4 PH stainless steel, in a shape resembling a handgun, or such lesser amount as is detectable in view of advances in state-of-the-art developments in weapons detection technology.

(3) Under such rules and regulations as the Secretary shall prescribe, this subsection shall not apply to the manufacture, possession, transfer, receipt, shipment, or delivery of a firearm by a licensed manufacturer or any person acting pursuant to a contract with a licensed manufacturer, for the purpose of examining and testing such firearm to determine whether paragraph (1) applies to such firearm. The Secretary shall ensure that rules and regulations adopted pursuant to this paragraph do not impair the manufacture of prototype firearms or the development of new technology.

(4) The Secretary shall permit the conditional importation of a firearm by a licensed importer or licensed manufacturer, for examination and testing to determine whether or not the unconditional importation of such firearm would violate this subsection.

(5) This subsection shall not apply to any firearm which -

(A) has been certified by the Secretary of Defense or the Director of Central Intelligence, after consultation with the Secretary and the Administrator of the Federal Aviation Administration, as necessary for military or intelligence applications; and

(B) is manufactured for and sold exclusively to military or intelligence agencies of the United States.

(6) This subsection shall not apply with respect to any firearm manufactured in, imported into, or possessed in the United States before the date of the enactment of the Undetectable Firearms Act of 1988.

(q) (1) The Congress finds and declares that -

(A) crime, particularly crime involving drugs and guns, is a pervasive, nationwide problem;

(B) crime at the local level is exacerbated by the interstate movement of drugs, guns, and criminal gangs;

(C) firearms and ammunition move easily in interstate commerce and have been found in increasing numbers in and around schools, as documented in numerous hearings in both the Committee n the Judiciary of the House of Representatives and the Committee on the Judiciary of the Senate;

(D) in fact, even before the sale of a firearm, the gun, its component parts, ammunition, and the raw materials from which they are made have considerably moved in interstate commerce;

(E) while criminals freely move from State to State, ordinary citizens and foreign visitors may fear to travel to or through certain parts of the country due to concern about violent crime and gun violence, and parents may decline to send their children to school for the same reason;

(F) the occurrence of violent crime in school zones has resulted in a decline in the quality of education in our country;

(G) this decline in the quality of education has an adverse impact on interstate commerce and the foreign commerce of the United States;

(H) States, localities, and school systems find it almost impossible to handle gun-related crime by themselves - even States, localities, and school systems that have made strong efforts to prevent, detect, and punish gun-related crime find their efforts unavailing due in part to the failure or inability of other States or localities to take strong measures; and

(I) Congress has the power, under the interstate commerce clause and other provisions of the Constitution, to enact measures to ensure the integrity and safety of the Nation's schools by enactment of this subsection.

(2)(A) It shall be unlawful for any individual knowingly to possess a firearm that has moved in or that otherwise affects interstate or foreign commerce at a place that the individual knows, or has reasonable cause to believe, is a school zone.

(B) Subparagraph (A) does not apply to the possession of a firearm -

(i) on private property not part of school grounds;

(ii) if the individual possessing the firearm is licensed to do so by the State in which the school zone is located or a political subdivision of the State, and the law of the State or political subdivision requires that, before an individual obtains such a license, the law enforcement authorities of the State or political subdivision verify that the individual is qualified under law to receive the license;

(iii) that is -

(I) not loaded; and

(II) in a locked container, or a locked firearms rack that is on a motor vehicle;

(iv) by an individual for use in a program approved by a school in the school zone;

(v) by an individual in accordance with a contract entered into between a school in the school zone and the individual or an employer of the individual;

(vi) by a law enforcement officer acting in his or her official capacity; or

(vii) that is unloaded and is possessed by an individual while traversing school premises for the purpose of gaining access to public or private lands open to hunting, if the entry on school premises is authorized by school authorities.

(3)(A) Except as provided in subparagraph (B), it shall be unlawful for any person, knowingly or with reckless disregard for the safety of another, to discharge or attempt to discharge a firearm that has moved in or that otherwise affects interstate or foreign commerce at a place that the person knows is a school zone.

(B) Subparagraph (A) does not apply to the discharge of a firearm -

(i) on private property not part of school grounds;

(ii) as part of a program approved by a school in the school zone, by an individual who is participating in the program;

(iii) by an individual in accordance with a contract entered into between a school in a school zone and the individual or an employer of the individual; or

(iv) by a law enforcement officer acting in his or her official capacity.

(4) Nothing in this subsection shall be construed as preempting or preventing a State or local government from enacting a statute establishing gun free school zones as provided in this subsection.

(r) It shall be unlawful for any person to assemble from imported parts any semiautomatic rifle or any shotgun which is identical to any rifle or shotgun prohibited from importation under section 925(d)(3) of this chapter as not being particularly suitable for or readily adaptable to sporting purposes except that this subsection shall not apply to -

(1) the assembly of any such rifle or shotgun for sale or distribution by a licensed manufacturer to the United States or any department or agency thereof or to any State or any department, agency, or political subdivision thereof; or

(2) the assembly of any such rifle or shotgun for the purposes of testing or experimentation authorized by the Secretary.

(s)(1) Beginning on the date that is 90 days after the date of enactment of this subsection and ending on the day before the date that is 60 months after such date of enactment, it shall be unlawful for any licensed importer, licensed manufacturer, or licensed dealer to sell, deliver, or transfer a handgun (other than the return of a handgun to the person from whom it was received) to an individual who is not licensed under section 923, unless -

(A) after the most recent proposal of such transfer by the transferee -

(i) the transferor has -

(I) received from the transferee a statement of the transferee containing the information described in paragraph (3);

(II) verified the identity of the transferee by examining the identification document presented;

(III) within 1 day after the transferee furnishes the statement, provided notice of the contents of the statement to the chief law enforcement officer of the place of residence of the transferee; and

    (IV) within 1 day after the transferee furnishes the statement, transmitted a copy of the statement to the chief law enforcement officer of the place of residence of the transferee; and

    (ii)(I) 5 business days (meaning days on which State offices are open) have elapsed from the date the transferor furnished notice of the contents of the statement to the chief law enforcement officer, during which period the transferor has not received information from the chief law enforcement officer that receipt or possession of the handgun by the transferee would be in violation of Federal, State, or local law; or

    (II) the transferor has received notice from the chief law enforcement officer that the officer has no information indicating that receipt or possession of the handgun by the transferee would violate Federal, State, or local law;

(B) the transferee has presented to the transferor a written statement, issued by the chief law enforcement officer of the place of residence of the transferee during the 10-day period ending on the date of the most recent proposal of such transfer by the transferee, stating that the transferee requires access to a handgun because of a threat to the life of the transferee or of any member of the household of the transferee;

(C)(i) the transferee has presented to the transferor a permit that -

    (I) allows the transferee to possess or acquire a handgun; and

    (II) was issued not more than 5 years earlier by the State in which the transfer is to take place; and

    (ii) the law of the State provides that such a permit is to be issued only after an authorized government official has verified that the information available to such official does not indicate that possession of a handgun by the transferee would be in violation of the law;

(D) the law of the State requires that, before any licensed importer, licensed manufacturer, or licensed dealer completes the transfer of a handgun to an individual who is not licensed under section 923, an authorized government official verify that the information available to such official does not indicate that possession of a handgun by the transferee would be in violation of law;

(E) the Secretary has approved the transfer under section 5812 of the Internal Revenue Code of 1986; or

(F) on application of the transferor, the Secretary has certified that compliance with subparagraph (A)(i)(III) is impracticable because -

    (i) the ratio of the number of law enforcement officers of the State in which the transfer is to occur to the number of square miles of land area of the State does not exceed 0.0025;

    (ii) the business premises of the transferor at which the transfer is to occur are extremely remote in relation to the chief law enforcement officer; and

    (iii) there is an absence of telecommunications facilities in the geographical area in which the business premises are located.

(2) A chief law enforcement officer to whom a transferor has provided notice pursuant to paragraph (1)(A)(i)(III) shall make a reasonable effort to ascertain

within 5 business days whether receipt or possession would be in violation of the law, including research in whatever State and local recordkeeping systems are available and in a national system designated by the Attorney General.

(3) The statement referred to in paragraph (1)(A)(i)(I) shall contain only -

(A) the name, address, and date of birth appearing on a valid identification document (as defined in section 1028(d)(1)) of the transferee containing a photograph of the transferee and a description of the identification used;

(B) a statement that the transferee -

(i) is not under indictment for, and has not been convicted in any court of, a crime punishable by imprisonment for a term exceeding 1 year and has not been convicted in any court of a misdemeanor crime of domestic violence;

(ii) is not a fugitive from justice;

(iii) is not an unlawful user of or addicted to any controlled substance (as defined in section 102 of the Controlled Substances Act);

(iv) has not been adjudicated as a mental defective or been committed to a mental institution;

(v) is not an alien who -

(I) is illegally or unlawfully in the United States; or

(II) subject to subsection (y)(2), has been admitted to the United States under a nonimmigrant visa (as that term is defined in section 101(a)(26) of the Immigration and Nationality Act (8 U.S.C. 1101(a)(26)));

(vi) has not been discharged from the Armed Forces under dishonorable conditions; and

(vii) is not a person who, having been a citizen of the United States, has renounced such citizenship;

(C) the date the statement is made; and

(D) notice that the transferee intends to obtain a handgun from the transferor.

(4) Any transferor of a handgun who, after such transfer, receives a report from a chief law enforcement officer containing information that receipt or possession of the handgun by the transferee violates Federal, State, or local law shall, within 1 business day after receipt of such request, communicate any information related to the transfer that the transferor has about the transfer and the transferee to -

(A) the chief law enforcement officer of the place of business of the transferor; and

(B) the chief law enforcement officer of the place of residence of the transferee.

(5) Any transferor who receives information, not otherwise available to the public, in a report under this subsection shall not disclose such information except to the transferee, to law enforcement authorities, or pursuant to the direction of a court of law.

(6)(A) Any transferor who sells, delivers, or otherwise transfers a handgun to a transferee shall retain the copy of the statement of the transferee with respect to the handgun transaction, and shall retain evidence that the transferor has complied with subclauses (III) and (IV) of paragraph (1)(A)(i) with respect to the statement.

(B) Unless the chief law enforcement officer to whom a statement is transmitted under paragraph (1)(A)(i)(IV) determines that a transaction would violate Federal, State, or local law -

      (i) the officer shall, within 20 business days after the date the transferee made the statement on the basis of which the notice was provided, destroy the statement, any record containing information derived from the statement, and any record created as a result of the notice required by paragraph (1)(A)(i)(III);

      (ii) the information contained in the statement shall not be conveyed to any person except a person who has a need to know in order to carry out this subsection; and

      (iii) the information contained in the statement shall not be used for any purpose other than to carry out this subsection.

      (C) If a chief law enforcement officer determines that an individual is ineligible to receive a handgun and the individual requests the officer to provide the reason for such determination, the officer shall provide such reasons to the individual in writing within 20 business days after receipt of the request.

      (7) A chief law enforcement officer or other person responsible for providing criminal history background information pursuant to this subsection shall not be liable in an action at law for damages -

      (A) for failure to prevent the sale or transfer of a handgun to a person whose receipt or possession of the handgun is unlawful under this section; or

      (B) for preventing such a sale or transfer to a person who may lawfully receive or possess a handgun.

      (8) For purposes of this subsection, the term "chief law enforcement officer" means the chief of police, the sheriff, or an equivalent officer or the designee of any such individual.

      (9) The Secretary shall take necessary actions to ensure that the provisions of this subsection are published and disseminated to licensed dealers, law enforcement officials, and the public.

      (t) (1) Beginning on the date that is 30 days after the Attorney General notifies licensees under section 103(d) of the Brady Handgun Violence Prevention Act that the national instant criminal background check system is established, a licensed importer, licensed manufacturer, or licensed dealer shall not transfer a firearm to any other person who is not licensed under this chapter, unless -

      (A) before the completion of the transfer, the licensee contacts the national instant criminal background check system established under section 103 of that Act;

      (B)(i) the system provides the licensee with a unique identification number; or

      (ii) 3 business days (meaning a day on which State offices are open) have elapsed since the licensee contacted the system, and the system has not notified the licensee that the receipt of a firearm by such other person would violate subsection (g) or (n) of this section; and

      (C) the transferor has verified the identity of the transferee by examining a valid identification document (as defined in section 1028(d)(1) of this title) of the transferee containing a photograph of the transferee.

      (2) If receipt of a firearm would not violate subsection 922(g) or (n) or State

law, the system shall -

    (A) assign a unique identification number to the transfer;

    (B) provide the licensee with the number; and

    (C) destroy all records of the system with respect to the call (other than the identifying number and the date the number was assigned) and all records of the system relating to the person or the transfer.

    (3) Paragraph (1) shall not apply to a firearm transfer between a licensee and another person if -

    (A)(i) such other person has presented to the licensee a permit that -

        (I) allows such other person to possess or acquire a firearm; and

        (II) was issued not more than 5 years earlier by the State in which the transfer is to take place; and

    (ii) the law of the State provides that such a permit is to be issued only after an authorized government official has verified that the information available to such official does not indicate that possession of a firearm by such other person would be in violation of law;

    (B) the Secretary has approved the transfer under section 5812 of the Internal Revenue Code of 1986; or

    (C) on application of the transferor, the Secretary has certified that compliance with paragraph (1)(A) is impracticable because -

    (i) the ratio of the number of law enforcement officers of the State in which the transfer is to occur to the number of square miles of land area of the State does not exceed 0.0025;

    (ii) the business premises of the licensee at which the transfer is to occur are extremely remote in relation to the chief law enforcement officer as defined in subsection (s)(8)); and

    (iii) there is an absence of telecommunications facilities in the geographical area in which the business premises are located.

    (4) If the national instant criminal background check system notifies the licensee that the information available to the system does not demonstrate that the receipt of a firearm by such other person would violate subsection (g) or (n) or State law, and the licensee transfers a firearm to such other person, the licensee shall include in the record of the transfer the unique identification number provided by the system with respect to the transfer.

    (5) If the licensee knowingly transfers a firearm to such other person and knowingly fails to comply with paragraph (1) of this subsection with respect to the transfer and, at the time such other person most recently proposed the transfer, the national instant criminal background check system was operating and information was available to the system demonstrating that receipt of a firearm by such other person would violate subsection (g) or (n) of this section or State law, the Secretary may, after notice and opportunity for a hearing, suspend for not more than 6 months or revoke any license issued to the licensee under section 923, and may impose on the licensee a civil fine of not more than $5,000.

    (6) Neither a local government nor an employee of the Federal Government or of any State or local government, responsible for providing information to the national instant criminal background check system shall be liable in an action at law for damages -

(A) for failure to prevent the sale or transfer of a firearm to a person whose receipt or possession of the firearm is unlawful under this section; or

(B) for preventing such a sale or transfer to a person who may lawfully receive or possess a firearm.

(u) It shall be unlawful for a person to steal or unlawfully take or carry away from the person or the premises of a person who is licensed to engage in the business of importing, manufacturing, or dealing in firearms, any firearm in the licensee's business inventory that has been shipped or transported in interstate or foreign commerce.

(v)(1) It shall be unlawful for a person to manufacture, transfer, or possess a semiautomatic assault weapon.

(2) Paragraph (1) shall not apply to the possession or transfer of any semiautomatic assault weapon otherwise lawfully possessed under Federal law on the date of the enactment of this subsection.

(3) Paragraph (1) shall not apply to -

(A) any of the firearms, or replicas or duplicates of the firearms, specified in Appendix A to this section, as such firearms were manufactured on October 1, 1993;

(B) any firearm that -

(i) is manually operated by bolt, pump, lever, or slide action;

(ii) has been rendered permanently inoperable; or

(iii) is an antique firearm;

(C) any semiautomatic rifle that cannot accept a detachable magazine that holds more than 5 rounds of ammunition; or

(D) any semiautomatic shotgun that cannot hold more than 5 rounds of ammunition in a fixed or detachable magazine.

The fact that a firearm is not listed in Appendix A shall not be construed to mean that paragraph (1) applies to such firearm. No firearm exempted by this subsection may be deleted from Appendix A so long as this subsection is in effect.

(4) Paragraph (1) shall not apply to -

(A) the manufacture for, transfer to, or possession by the United States or a department or agency of the United States or a State or a department, agency, or political subdivision of a State, or a transfer to or possession by a law enforcement officer employed by such an entity for purposes of law enforcement (whether on or off duty);

(B) the transfer to a licensee under title I of the Atomic Energy Act of 1954 for purposes of establishing and maintaining an on-site physical protection system and security organization required by Federal law, or possession by an employee or contractor of such licensee on-site for such purposes or off-site for purposes of licensee-authorized training or transportation of nuclear materials;

(C) the possession, by an individual who is retired from service with a law enforcement agency and is not otherwise prohibited from receiving a firearm, of a semiautomatic assault weapon transferred to the individual by the agency upon such retirement; or

(D) the manufacture, transfer, or possession of a semiautomatic assault weapon by a licensed manufacturer or licensed importer for the purposes of testing or experimentation authorized by the Secretary.

(w) (1) Except as provided in paragraph (2), it shall be unlawful for a person to transfer or possess a large capacity ammunition feeding device.

(2) Paragraph (1) shall not apply to the possession or transfer of any large capacity ammunition feeding device otherwise lawfully possessed on or before the date of the enactment of this subsection.

(3) This subsection shall not apply to -

(A) the manufacture for, transfer to, or possession by the United States or a department or agency of the United States or a State or a department, agency, or political subdivision of a State, or a transfer to or possession by a law enforcement officer employed by such an entity for purposes of law enforcement (whether on or off duty);

(B) the transfer to a licensee under title I of the Atomic Energy Act of 1954 for purposes of establishing and maintaining an on-site physical protection system and security organization required by Federal law, or possession by an employee or contractor of such licensee on-site for such purposes or off-site for purposes of licensee-authorized training or transportation of nuclear materials;

(C) the possession, by an individual who is retired from service with a law enforcement agency and is not otherwise prohibited from receiving ammunition, of a large capacity ammunition feeding device transferred to the individual by the agency upon such retirement; or

(D) the manufacture, transfer, or possession of any large capacity ammunition feeding device by a licensed manufacturer or licensed importer for the purposes of testing or experimentation authorized by the Secretary.

(4) If a person charged with violating paragraph (1) asserts that paragraph (1) does not apply to such person because of paragraph (2) or (3), the Government shall have the burden of proof to show that such paragraph (1) applies to such person. The lack of a serial number as described in section 923(i) of this title shall be a presumption that the large capacity ammunition feeding device is not subject to the prohibition of possession in paragraph (1).

(x) (1) It shall be unlawful for a person to sell, deliver, or otherwise transfer to a person who the transferor knows or has reasonable cause to believe is a juvenile-

(A) a handgun; or

(B) ammunition that is suitable for use only in a handgun.

(2) It shall be unlawful for any person who is a juvenile to knowingly possess

(A) a handgun; or

(B) ammunition that is suitable for use only in a handgun.

(3) This subsection does not apply to -

(A) a temporary transfer of a handgun or ammunition to a juvenile or to the possession or use of a handgun or ammunition by a juvenile if the handgun and ammunition are possessed and used by the juvenile -

(i) in the course of employment, in the course of ranching or farming related to activities at the residence of the juvenile (or on property used for ranching or farming at which the juvenile, with the permission of the property owner or lessee, is performing activities related to the operation of the farm or ranch), target practice, hunting, or a course of instruction in the safe and lawful use of a handgun;

(ii) with the prior written consent of the juvenile's parent or guardian who is not prohibited by Federal, State, or local law from possessing a firearm, except -

(I) during transportation by the juvenile of an unloaded handgun in a locked container directly from the place of transfer to a place at which an activity described in clause (i) is to take place and transportation by the juvenile of that handgun, unloaded and in a locked container, directly from the place at which such an activity took place to the transferor; or

(II) with respect to ranching or farming activities as described in clause (i), a juvenile may possess and use a handgun or ammunition with the prior written approval of the juvenile's parent or legal guardian and at the direction of an adult who is not prohibited by Federal, State or local law from possessing a firearm;

(iii) the juvenile has the prior written consent in the juvenile's possession at all times when a handgun is in the possession of the juvenile; and

(iv) in accordance with State and local law;

(B) a juvenile who is a member of the Armed Forces of the United States or the National Guard who possesses or is armed with a handgun in the line of duty;

(C) a transfer by inheritance of title (but not possession) of a handgun or ammunition to a juvenile; or

(D) the possession of a handgun or ammunition by a juvenile taken in defense of the juvenile or other persons against an intruder into the residence of the juvenile or a residence in which the juvenile is an invited guest.

(4) A handgun or ammunition, the possession of which is transferred to a juvenile in circumstances in which the transferor is not in violation of this subsection shall not be subject to permanent confiscation by the Government if its possession by the juvenile subsequently becomes unlawful because of the conduct of the juvenile, but shall be returned to the lawful owner when such handgun or ammunition is no longer required by the Government for the purposes of investigation or prosecution.

(5) For purposes of this subsection, the term "juvenile" means a person who is less than 18 years of age.

(6)(A) In a prosecution of a violation of this subsection, the court shall require the presence of a juvenile defendant's parent or legal guardian at all proceedings.

(B) The court may use the contempt power to enforce subparagraph (A).

(C) The court may excuse attendance of a parent or legal guardian of a juvenile defendant at a proceeding in a prosecution of a violation of this subsection for good cause shown.

(y) Provisions relating to aliens admitted under nonimmigrant visas. -

(1) Definitions. - In this subsection -

   (A) the term "alien" has the same meaning as in section 101(a)(3) of the Immigration and Nationality Act (8 U.S.C. 1101(a)(3)); and

   (B) the term "nonimmigrant visa" has the same meaning as in section 101(a)(26) of the Immigration and Nationality Act (8 U.S.C. 1101(a)(26)).

   (2) Exceptions. - Subsections (d)(5)(B), (g)(5)(B), and (s)(3)(B)(v)(II) do not apply to any alien who has been lawfully admitted to the United States under a nonimmigrant visa, if that alien is -

   (A) admitted to the United States for lawful hunting or sporting purposes or is in possession of a hunting license or permit lawfully issued in the United States;

   (B) an official representative of a foreign government who is -

      (i) accredited to the United States Government or the Government's mission to an international organization having its headquarters in the United States; or

      (ii) en route to or from another country to which that alien is accredited;

   (C) an official of a foreign government or a distinguished foreign visitor who has been so designated by the Department of State; or

   (D) a foreign law enforcement officer of a friendly foreign government entering the United States on official law enforcement business.

   (3) Waiver. -

   (A) Conditions for waiver. - Any individual who has been admitted to the United States under a nonimmigrant visa may receive a waiver from the requirements of subsection (g)(5), if -

      (i) the individual submits to the Attorney General a petition that meets the requirements of subparagraph (C); and

      (ii) the Attorney General approves the petition.

   (B) Petition. - Each petition under subparagraph (B) shall -

      (i) demonstrate that the petitioner has resided in the United States for a continuous period of not less than 180 days before the date on which the petition is submitted under this paragraph; and

      (ii) include a written statement from the embassy or consulate of the petitioner, authorizing the petitioner to acquire a firearm or ammunition and certifying that the alien would not, absent the application of subsection (g)(5)(B), otherwise be prohibited from such acquisition under subsection (g).

   (C) Approval of petition. - The Attorney General shall approve a petition submitted in accordance with this paragraph, if the Attorney General determines that waiving the requirements of subsection (g)(5)(B) with respect to the petitioner -

      (i) would be in the interests of justice; and

      (ii) would not jeopardize the public safety.

Centerfire Rifles - Autoloaders

Browning BAR Mark II Safari Semi-Auto Rifle
Browning BAR Mark II Safari Magnum Rifle
Browning High-Power Rifle
Heckler & Koch Model 300 Rifle
Iver Johnson M-1 Carbine
Iver Johnson 50th Anniversary M-1 Carbine
Marlin Model 9 Camp Carbine
Marlin Model 45 Carbine
Remington Nylon 66 Auto-Loading Rifle
Remington Model 7400 Auto Rifle
Remington Model 7400 Rifle
Remington Model 7400 Special Purpose Auto Rifle
Ruger Mini-14 Autoloading Rifle (w/o folding stock)
Ruger Mini Thirty Rifle

Centerfire Rifles - Lever & Slide

Browning Model 81 BLR Lever-Action Rifle
Browning Model 81 Long Action BLR
Browning Model 1886 Lever-Action Carbine
Browning Model 1886 High Grade Carbine
Cimarron 1860 Henry Replica
Cimarron 1866 Winchester Replicas
Cimarron 1873 Short Rifle
Cimarron 1873 Sporting Rifle
Cimarron 1873 30" Express Rifle
Dixie Engraved 1873 Rifle
E.M.F. 1866 Yellowboy Lever Actions
E.M.F. 1860 Henry Rifle
E.M.F. Model 73 Lever-Action Rifle
Marlin Model 336CS Lever-Action Carbine
Marlin Model 30AS Lever-Action Carbine
Marlin Model 444SS Lever-Action Sporter
Marlin Model 1894S Lever-Action Carbine
Marlin Model 1894CS Carbine
Marlin Model 1894CL Classic
Marlin Model 1895SS Lever-Action Rifle
Mitchell 1858 Henry Replica
Mitchell 1866 Winchester Replica
Mitchell 1873 Winchester Replica
Navy Arms Military Henry Rifle
Navy Arms Henry Trapper
Navy Arms Iron Frame Henry
Navy Arms Henry Carbine
Navy Arms 1866 Yellowboy Rifle
Navy Arms 1873 Winchester-Style Rifle
Navy Arms 1873 Sporting Rifle
Remington 7600 Slide Action
Remington Model 7600 Special Purpose Slide Action
Rossi M92 SRC Saddle-Ring Carbine
Rossi M92 SRS Short Carbine
Savage 99C Lever-Action Rifle
Uberti Henry Rifle
Uberti 1866 Sporting Rifle
Uberti 1873 Sporting Rifle
Winchester Model 94 Side Eject Lever-Action Rifle
Winchester Model 94 Trapper Side Eject
Winchester Model 94 Big Bore Side Eject
Winchester Model 94 Ranger Side Eject Lever-Action Rifle

Winchester Model 94 Wrangler Side Eject

Centerfire Rifles - Bolt Action

Alpine Bolt-Action Rifle
A-Square Caesar Bolt-Action Rifle
A-Square Hannibal Bolt-Action Rifle
Anschutz 1700D Classic Rifles
Anschutz 1700D Custom Rifles
Anschutz 1700D Bavarian Bolt-Action Rifle
Anschutz 1733D Mannlicher Rifle
Barret Model 90 Bolt-Action Rifle
Beeman/HW 60J Bolt-Action Rifle
Blaser R84 Bolt-Action Rifle
BRNO 537 Sporter Bolt-Action Rifle
BRNO ZKB 527 Fox Bolt-Action Rifle
BRNO ZKK 600, 601, 602 Bolt-Action Rifles
Browning A-Bolt Rifle
Browning A-Bolt Stainless Stalker
Browning A-Bolt Left Hand
Browning A-Bolt Short Action
Browning Euro-Bolt Rifle
Browning A-Bolt Gold Medallion
Browning A-Bolt Micro Medallion
Century Centurion 14 Sporter
Century Enfield Sporter #4
Century Swedish Sporter #38
Century Mauser 98 Sporter
Cooper Model 38 Centerfire Sporter
Dakota 22 Sporter Bolt-Action Rifle
Dakota 76 Classic Bolt-Action Rifle
Dakota 76 Short Action Rifles
Dakota 76 Safari Bolt-Action Rifle
Dakota 416 Rigby African
E.A.A./Sabatti Rover 870 Bolt-Action Rifle
Auguste Francotte Bolt-Action Rifles
Carl Gustaf 2000 Bolt-Action Rifle
Heym Magnum Express Series Rifle
Howa Lightning Bolt-Action Rifle
Howa Realtree Camo Rifle
Interarms Mark X Viscount Bolt-Action Rifle
Interarms Mini-Mark X Rifle
Interarms Mark X Whitworth Bolt-Action Rifle
Interarms Whitworth Express Rifle
Iver Johnson Model 5100A1 Long-Range Rifle
KDF K15 American Bolt-Action Rifle
Krico Model 600 Bolt-Action Rifle
Krico Model 700 Bolt-Action Rifles
Mauser Model 66 Bolt-Action Rifle
Mauser Model 99 Bolt-Action Rifle
McMillan Signature Classic Sporter
McMillan Signature Super Varminter
McMillan Signature Alaskan
McMillan Signature Titanium Mountain Rifle
McMillan Classic Stainless Sporter
McMillan Talon Safari Rifle
McMillan Talon Sporter Rifle
Midland 1500S Survivor Rifle
Navy Arms TU-33/40 Carbine
Parker-Hale Model 81 Classic Rifle
Parker-Hale Model 81 Classic African Rifle
Parker-Hale Model 1000 Rifle
Parker-Hale Model 1100M African Magnum

Parker-Hale Model 1100 Lightweight Rifle
Parker-Hale Model 1200 Super Rifle
Parker-Hale Model 1200 Super Clip Rifle
Parker-Hale Model 1300C Scout Rifle
Parker-Hale Model 2100 Midland Rifle
Parker-Hale Model 2700 Lightweight Rifle
Parker-Hale Model 2800 Midland Rifle
Remington Model Seven Bolt-Action Rifle
Remington Model Seven Youth Rifle
Remington Model Seven Custom KS
Remington Model Seven Custom MS Rifle
Remington 700 ADL Bolt-Action Rifle
Remington 700 BDL Bolt-Action Rifle
Remington 700 BDL Varmint Special
Remington 700 BDL European Bolt-Action Rifle
Remington 700 Varmint Synthetic Rifle
Remington 700 BDL SS Rifle
Remington 700 Stainless Synthetic Rifle
Remington 700 MTRSS Rifle
Remington 700 BDL Left Hand
Remington 700 Camo Synthetic Rifle
Remington 700 Safari
Remington 700 Mountain Rifle
Remington 700 Custom KS Mountain Rifle
Remington 700 Classic Rifle
Ruger M77 Mark II Rifle
Ruger M77 Mark II Magnum Rifle
Ruger M77RL Ultra Light
Ruger M77 Mark II All-Weather Stainless Rifle
Ruger M77 RSI International Carbine
Ruger M77 Mark II Express Rifle
Ruger M77VT Target Rifle
Sako Hunter Rifle
Sako FiberClass Sporter
Sako Safari Grade Bolt Action
Sako Hunter Left-Hand Rifle
Sako Classic Bolt Action
Sako Hunter LS Rifle
Sako Deluxe Lightweight
Sako Super Deluxe Sporter
Sako Mannlicher-Style Carbine
Sako Varmint Heavy Barrel
Sako TRG-S Bolt-Action Rifle
Sauer 90 Bolt-Action Rifle
Savage 110G Bolt-Action Rifle
Savage 110CY Youth/Ladies Rifle
Savage 110WLE One of One Thousand Limited Edition Rifle
Savage 110GXP3 Bolt-Action Rifle
Savage 110F Bolt-Action Rifle
Savage 110FXP3 Bolt-Action Rifle
Savage 110GV Varmint Rifle
Savage 112FV Varmint Rifle
Savage Model 112FVS Varmint Rifle
Savage Model 112BV Heavy Barrel Varmint Rifle
Savage 116FSS Bolt-Action Rifle
Savage Model 116FSK Kodiak Rifle
Savage 110FP Police Rifle
Steyr-Mannlicher Sporter Models SL, L, M, S, S/T
Steyr-Mannlicher Luxus Model L, M, S
Steyr-Mannlicher Model M Professional Rifle
Tikka Bolt-Action Rifle
Tikka Premium Grade Rifles
Tikka Varmint/Continental Rifle

Tikka Whitetail/Battue Rifle
Ultra Light Arms Model 20 Rifle
Ultra Light Arms Model 28, Model 40 Rifles
Voere VEC 91 Lightning Bolt-Action Rifle
Voere Model 2165 Bolt-Action Rifle
Voere Model 2155, 2150 Bolt-Action Rifles
Weatherby Mark V Deluxe Bolt-Action Rifle
Weatherby Lasermark V Rifle
Weatherby Mark V Crown Custom Rifles
Weatherby Mark V Sporter Rifle
Weatherby Mark V Safari Grade Custom Rifles
Weatherby Weathermark Rifle
Weatherby Weathermark Alaskan Rifle
Weatherby Classicmark No. 1 Rifle
Weatherby Weatherguard Alaskan Rifle
Weatherby Vanguard VGX Deluxe Rifle
Weatherby Vanguard Classic Rifle
Weatherby Vanguard Classic No. 1 Rifle
Weatherby Vanguard Weatherguard Rifle
Wichita Classic Rifle
Wichita Varmint Rifle
Winchester Model 70 Sporter
Winchester Model 70 Sporter WinTuff
Winchester Model 70 SM Sporter
Winchester Model 70 Stainless Rifle
Winchester Model 70 Varmint
Winchester Model 70 Synthetic Heavy Varmint Rifle
Winchester Model 70 DBM Rifle
Winchester Model 70 DBM-S Rifle
Winchester Model 70 Featherweight
Winchester Model 70 Featherweight WinTuff
Winchester Model 70 Featherweight Classic
Winchester Model 70 Lightweight Rifle
Winchester Ranger Rifle
Winchester Model 70 Super Express Magnum
Winchester Model 70 Super Grade
Winchester Model 70 Custom Sharpshooter
Winchester Model 70 Custom Sporting Sharpshooter Rifle

Centerfire Rifles - Single Shot

Armsport 1866 Sharps Rifle, Carbine
Brown Model One Single Shot Rifle
Browning Model 1885 Single Shot Rifle
Dakota Single Shot Rifle
Desert Industries G-90 Single Shot Rifle
Harrington & Richardson Ultra Varmint Rifle
Model 1885 High Wall Rifle
Navy Arms Rolling Block Buffalo Rifle
Navy Arms #2 Creedmoor Rifle
Navy Arms Sharps Cavalry Carbine
Navy Arms Sharps Plains Rifle
New England Firearms Handi-Rifle
Red Willow Armory Ballard No. 5 Pacific
Red Willow Armory Ballard No. 1.5 Hunting Rifle
Red Willow Armory Ballard No. 8 Union Hill Rifle
Red Willow Armory Ballard No. 4.5 Target Rifle
Remington-Style Rolling Block Carbine
Ruger No. 1B Single Shot
Ruger No. 1A Light Sporter
Ruger No. 1H Tropical Rifle
Ruger No. 1S Medium Sporter
Ruger No. 1 RSI International

Ruger No. 1V Special Varminter
C.Sharps Arms New Model 1874 Old Reliable
C.Sharps Arms New Model 1875 Rifle
C.Sharps Arms 1875 Classic Sharps
C.Sharps Arms New Model 1875 Target & Long Range
Shiloh Sharps 1874 Long Range Express
Shiloh Sharps 1874 Montana Roughrider
Shiloh Sharps 1874 Military Carbine
Shiloh Sharps 1874 Business Rifle
Shiloh Sharps 1874 Military Rifle
Sharps 1874 Old Reliable
Thompson/Center Contender Carbine
Thompson/Center Stainless Contender Carbine
Thompson/Center Contender Carbine Survival System
Thompson/Center Contender Carbine Youth Model
Thompson/Center TCR '87 Single Shot Rifle
Uberti Rolling Block Baby Carbine

Drillings, Combination Guns, Double Rifles

Beretta Express SSO O/U Double Rifles
Beretta Model 455 SxS Express Rifle
Chapuis RGExpress Double Rifle
Auguste Francotte Sidelock Double Rifles
Auguste Francotte Boxlock Double Rifle
Heym Model 55B O/U Double Rifle
Heym Model 55FW O/U Combo Gun
Heym Model 88b Side-by-Side Double Rifle
Kodiak Mk. IV Double Rifle
Kreighoff Teck O/U Combination Gun
Kreighoff Trumpf Drilling
Merkel Over/Under Combination Guns
Merkel Drillings
Merkel Model 160 Side-by-Side Double Rifles
Merkel Over/Under Double Rifles
Savage 24F O/U Combination Gun
Savage 24F-12T Turkey Gun
Springfield Inc. M6 Scout Rifle/Shotgun
Tikka Model 412s Combination Gun
Tikka Model 412S Double Fire
A. Zoli Rifle-Shotgun O/U Combo

Rimfire Rifles - Autoloaders

AMT Lightning 25/22 Rifle
AMT Lightning Small-Game Hunting Rifle II
AMT Magnum Hunter Auto Rifle
Anschutz 525 Deluxe Auto
Armscor Model 20P Auto Rifle
Browning Auto-22 Rifle
Browning Auto-22 Grade VI
Krico Model 260 Auto Rifle
Lakefield Arms Model 64B Auto Rifle
Marlin Model 60 Self-Loading Rifle
Marlin Model 60ss Self-Loading Rifle
Marlin Model 70 HC Auto
Marlin Model 990l Self-Loading Rifle
Marlin Model 70P Papoose
Marlin Model 922 Magnum Self-Loading Rifle
Marlin Model 995 Self-Loading Rifle
Norinco Model 22 ATD Rifle
Remington Model 522 Viper Autoloading Rifle
Remington 552BDL Speedmaster Rifle

Ruger 10/22 Autoloading Carbine (w/o folding stock)
Survival Arms AR-7 Explorer Rifle
Texas Remington Revolving Carbine
Voere Model 2115 Auto Rifle

Rimfire Rifles - Lever & Slide Action

Browning BL-22 Lever-Action Rifle
Marlin 39TDS Carbine
Marlin Model 39AS Golden Lever-Action Rifle
Remington 572BDL Fieldmaster Pump Rifle
Norinco EM-321 Pump Rifle
Rossi Model 62 SA Pump Rifle
Rossi Model 62 SAC Carbine
Winchester Model 9422 Lever-Action Rifle
Winchester Model 9422 Magnum Lever-Action Rifle

Rimfire Rifles - Bolt Actions & Single Shots

Anschutz Achiever Bolt-Action Rifle
Anschutz 1416D/1516D Classic Rifles
Anschutz 1418D/1518D Mannlicher Rifles
Anschutz 1700D Classic Rifles
Anschutz 1700D Custom Rifles
Anschutz 1700 FWT Bolt-Action Rifle
Anschutz 1700D Graphite Custom Rifle
Anschutz 1700D Bavarian Bolt-Action Rifle
Armscor Model 14P Bolt-Action Rifle
Armscor Model 1500 Rifle
BRNO ZKM-452 Deluxe Bolt-Action Rifle
BRNO ZKM 452 Deluxe
Beeman/HW 60-J-ST Bolt-Action Rifle
Browning A-Bolt 22 Bolt-Action Rifle
Browning A-Bolt Gold Medallion
Cabanas Phaser Rifle
Cabanas Master Bolt-Action Rifle
Cabanas Espronceda IV Bolt-Action Rifle
Cabanas Leyre Bolt-Action Rifle
Chipmunk Single Shot Rifle
Cooper Arms Model 36S Sporter Rifle
Dakota 22 Sporter Bolt-Action Rifle
Krico Model 300 Bolt-Action Rifles
Lakefield Arms Mark II Bolt-Action Rifle
Lakefield Arms Mark I Bolt-Action Rifle
Magtech Model MT-22C Bolt-Action Rifle
Marlin Model 880 Bolt-Action Rifle
Marlin Model 881 Bolt-Action Rifle
Marlin Model 882 Bolt-Action Rifle
Marlin Model 883 Bolt-Action Rifle
Marlin Model 883SS Bolt-Action Rifle
Marlin Model 25MN Bolt-Action Rifle
Marlin Model 25N Bolt-Action Repeater
Marlin Model 15YN "Little Buckaroo"
Mauser Model 107 Bolt-Action Rifle
Mauser Model 201 Bolt-Action Rifle
Navy Arms TU-KKW Training Rifle
Navy Arms TU-33/40 Carbine
Navy Arms TU-KKW Sniper Trainer
Norinco JW-27 Bolt-Action Rifle
Norinco JW-15 Bolt-Action Rifle
Remington 541-T
Remington 40-XR Rimfire Custom Sporter
Remington 541-T HB Bolt-Action Rifle

Remington 581-S Sportsman Rifle
Ruger 77/22 Rimfire Bolt-Action Rifle
Ruger K77/22 Varmint Rifle
Ultra Light Arms Model 20 RF Bolt-Action Rifle
Winchester Model 52B Sporting Rifle

Competition Rifles - Centerfire & Rimfire

Anschutz 64-MS Left Silhouette
Anschutz 1808D RT Super Match 54 Target
Anschutz 1827B Biathlon Rifle
Anschutz 1903D Match Rifle
Anschutz 1803D Intermediate Match
Anschutz 1911 Match Rifle
Anschutz 54.18MS REP Deluxe Silhouette Rifle
Anschutz 1913 Super Match Rifle
Anschutz 1907 Match Rifle
Anschutz 1910 Super Match II
Anschutz 54.18MS Silhouette Rifle
Anschutz Super Match 54 Target Model 2013
Anschutz Super Match 54 Target Model 2007
Beeman/Feinwerkbau 2600 Target Rifle
Cooper Arms Model TRP-1 ISU Standard Rifle
E.A.A./Weihrauch HW 60 Target Rifle
E.A.A./HW 660 Match Rifle
Finnish Lion Standard Target Rifle
Krico Model 360 S2 Biathlon Rifle
Krico Model 400 Match Rifle
Krico Model 360S Biathlon Rifle
Krico Model 500 Kricotronic Match Rifle
Krico Model 600 Sniper Rifle
Krico Model 600 Match Rifle
Lakefield Arms Model 90B Target Rifle
Lakefield Arms Model 91T Target Rifle
Lakefield Arms Model 92S Silhouette Rifle
Marlin Model 2000 Target Rifle
Mauser Model 86-SR Specialty Rifle
McMillan M-86 Sniper Rifle
McMillan Combo M-87/M-88 50-Caliber Rifle
McMillan 300 Phoenix Long Range Rifle
McMillan M-89 Sniper Rifle
McMillan National Match Rifle
McMillan Long Range Rifle
Parker-Hale M-87 Target Rifle
Parker-Hale M-85 Sniper Rifle
Remington 40-XB Rangemaster Target Centerfire
Remington 40-XR KS Rimfire Position Rifle
Remington 40-XBBR KS
Remington 40-XC KS National Match Course Rifle
Sako TRG-21 Bolt-Action Rifle
Steyr-Mannlicher Match SPG-UIT Rifle
Steyr-Mannlicher SSG P-I Rifle
Steyr-Mannlicher SSG P-III Rifle
Steyr-Mannlicher SSG P-IV Rifle
Tanner Standard UIT Rifle
Tanner 50 Meter Free Rifle
Tanner 300 Meter Free Rifle
Wichita Silhouette Rifle

Shotguns - Autoloaders

American Arms/Franchi Black Magic 48/AL
Benelli Super Black Eagle Shotgun

Benelli Super Black Eagle Slug Gun
Benelli M1 Super 90 Field Auto Shotgun
Benelli Montefeltro Super 90 20-Gauge Shotgun
Benelli Montefeltro Super 90 Shotgun
Benelli M1 Sporting Special Auto Shotgun
Benelli Black Eagle Competition Auto Shotgun
Beretta A-303 Auto Shotgun
Beretta 390 Field Auto Shotgun
Beretta 390 Super Trap, Super Skeet Shotguns
Beretta Vittoria Auto Shotgun
Beretta Model 1201F Auto Shotgun
Browning BSA 10 Auto Shotgun
Browning BSA 10 Stalker Auto Shotgun
Browning A-500R Auto Shotgun
Browning A-500G Auto Shotgun
Browning A-500G Sporting Clays
Browning Auto-5 Light 12 and 20
Browning Auto-5 Stalker
Browning Auto-5 Magnum 20
Browning Auto-5 Magnum 12
Churchill Turkey Automatic Shotgun
Cosmi Automatic Shotgun
Maverick Model 60 Auto Shotgun
Mossberg Model 5500 Shotgun
Mossberg Model 9200 Regal Semi-Auto Shotgun
Mossberg Model 9200 USST Auto Shotgun
Mossberg Model 9200 Camo Shotgun
Mossberg Model 6000 Auto Shotgun
Remington Model 1100 Shotgun
Remington 11-87 Premier Shotgun
Remington 11-87 Sporting Clays
Remington 11-87 Premier Skeet
Remington 11-87 Premier Trap
Remington 11-87 Special Purpose Magnum
Remington 11-87 SPS-T Camo Auto Shotgun
Remington 11-87 Special Purpose Deer Gun
Remington 11-87 SPS-BG-Camo Deer/Turkey Shotgun
Remington 11-87 SPS-Deer Shotgun
Remington 11-87 Special Purpose Synthetic Camo
Remington SP-10 Magnum-Camo Auto Shotgun
Remington SP-10 Magnum Auto Shotgun
Remington SP-10 Magnum Turkey Combo
Remington 1100 LT-20 Auto
Remington 1100 Special Field
Remington 1100 20-Gauge Deer Gun
Remington 1100 LT-20 Tournament Skeet
Winchester Model 1400 Semi-Auto Shotgun

Shotguns - Slide Actions

Browning Model 42 Pump Shotgun
Browning BPS Pump Shotgun
Browning BPS Stalker Pump Shotgun
Browning BPS Pigeon Grade Pump Shotgun
Browning BPS Pump Shotgun (Ladies and Youth Model)
Browning BPS Game Gun Turkey Special
Browning BPS Game Gun Deer Special
Ithaca Model 87 Supreme Pump Shotgun
Ithaca Model 87 Deerslayer Shotgun
Ithaca Deerslayer II Rifled Shotgun
Ithaca Model 87 Turkey Gun
Ithaca Model 87 Deluxe Pump Shotgun
Magtech Model 586-VR Pump Shotgun

Maverick Models 88, 91 Pump Shotguns
Mossberg Model 500 Sporting Pump
Mossberg Model 500 Camo Pump
Mossberg Model 500 Muzzleloader Combo
Mossberg Model 500 Trophy Slugster
Mossberg Turkey Model 500 Pump
Mossberg Model 500 Bantam Pump
Mossberg Field Grade Model 835 Pump Shotgun
Mossberg Model 835 Regal Ulti-Mag Pump
Remington 870 Wingmaster
Remington 870 Special Purpose Deer Gun
Remington 870 SPS-BG-Camo Deer/Turkey Shotgun
Remington 870 SPS-Deer Shotgun
Remington 870 Marine Magnum
Remington 870 TC Trap
Remington 870 Special Purpose Synthetic Camo
Remington 870 Wingmaster Small Gauges
Remington 870 Express Rifle Sighted Deer Gun
Remington 879 SPS Special Purpose Magnum
Remington 870 SPS-T Camo Pump Shotgun
Remington 870 Special Field
Remington 870 Express Turkey
Remington 870 High Grades
Remington 870 Express
Remington Model 870 Express Youth Gun
Winchester Model 12 Pump Shotgun
Winchester Model 42 High Grade Shotgun
Winchester Model 1300 Walnut Pump
Winchester Model 1300 Slug Hunter Deer Gun
Winchester Model 1300 Ranger Pump Gun Combo & Deer Gun
Winchester Model 1300 Turkey Gun
Winchester Model 1300 Ranger Pump Gun

Shotguns - Over/Unders

American Arms/Franchi Falconet 2000 O/U
American Arms Silver I O/U
American Arms Silver II Shotgun
American Arms Silver Skeet O/U
American Arms/Franchi Sporting 2000 O/U
American Arms Silver Sporting O/U
American Arms Silver Trap O/U
American Arms WS/OU 12, TS/OU 12 Shotguns
American Arms WT/OU 10 Shotgun
Armsport 2700 O/U Goose Gun
Armsport 2700 Series O/U
Armsport 2900 Tri-Barrel Shotgun
Baby Bretton Over/Under Shotgun
Beretta Model 686 Ultralight O/U
Beretta ASE 90 Competition O/U Shotgun
Beretta Over/Under Field Shotguns
Beretta Onyx Hunter Sport O/U Shotgun
Beretta Model SO5, SO6, SO9 Shotguns
Beretta Sporting Clay Shotguns
Beretta 687EL Sporting O/U
Beretta 682 Super Sporting O/U
Beretta Series 682 Competition Over/Unders
Browning Citori O/U Shotgun
Browning Superlight Citori Over/Under
Browning Lightning Sporting Clays
Browning Micro Citori Lightning
Browning Citori Plus Trap Combo
Browning Citori Plus Trap Gun

Browning Citori O/U Skeet Models
Browning Citori O/U Trap Models
Browning Special Sporting Clays
Browning Citori GTI Sporting Clays
Browning 325 Sporting Clays
Centurion Over/Under Shotgun
Chapuis Over/Under Shotgun
Connecticut Valley Classics Classic Sporter O/U
Connecticut Valley Classics Classic Field Waterfowler
Charles Daly Field Grade O/U
Charles Daly Lux Over/Under
E.A.A./Sabatti Sporting Clays Pro-Gold O/U
E.A.A/Sabatti Falcon-Mon Over/Under
Kassnar Grade I O/U Shotgun
Krieghoff K-80 Sporting Clays O/U
Krieghoff K-80 Skeet Shotgun
Krieghoff K-80 International Skeet
Krieghoff K-80 Four-Barrel Skeet Set
Krieghoff K-80/RT Shotguns
Krieghoff K-80 O/U Trap Shotgun
Laurona Silhouette 300 Sporting Clays
Laurona Silhouette 300 Trap
Laurona Super Model Over/Unders
Ljutic LM-6 Deluxe O/U Shotgun
Marocchi Conquista Over/Under Shotgun
Marocchi Avanza O/U Shotgun
Merkel Model 200E O/U Shotgun
Merkel Model 200E Skeet, Trap Over/Unders
Merkel Model 203E, 303E Over/Under Shotguns
Perazzi Mirage Special Sporting O/U
Perazzi Mirage Special Four-Gauge Skeet
Perazzi Sporting Classic O/U
Perazzi MX7 Over/Under Shotguns
Perazzi Mirage Special Skeet Over/Under
Perazzi MX8/MX8 Special Trap, Skeet
Perazzi MX 8/20 Over/Under Shotgun
Perazzi MX9 Single Over/Under Shotguns
Perazzi MX12 Hunting Over/Under
Perazzi MX28, MX410 Game O/U Shotguns
Perazzi MX20 Hunting Over/Under
Piotti Boss Over/Under Shotgun
Remington Peerless Over/Under Shotgun
Ruger Red Label O/U Shotgun
Ruger Sporting Clays O/U Shotgun
San Marco 12-Ga. Wildflower Shotgun
San Marco Field Special O/U Shotgun
San Marco 10-Ga. O/U Shotgun
SKB Model 505 Deluxe Over/Under Shotgun
SKB Model 685 Over/Under Shotgun
SKB Model 885 Over/Under Trap, Skeet, Sporting Clays
Stoeger/IGA Condor I O/U Shotgun
Stoeger/IGA ERA 2000 Over/Under Shotgun
Techni-Mec Model 610 Over/Under
Tikka Model 412S Field Grade Over/Under
Weatherby Athena Grade IV O/U Shotguns
Weatherby Athena Grade V Classic Field O/U
Weatherby Orion O/U Shotguns
Weatherby II, III Classic Field O/Us
Weatherby Orion II Classic Sporting Clays O/U
Weatherby Orion II Sporting Clays O/U
Winchester Model 1001 O/U Shotgun
Winchester Model 1001 Sporting Clays O/U
Pietro Zanoletti Model 2000 Field O/U

Shotguns - Side by Sides

American Arms Brittany Shotgun
American Arms Gentry Double Shotgun
American Arms Derby Side-by-Side
American Arms Grulla #2 Double Shotgun
American Arms WS/SS 10
American Arms TS/SS 10 Double Shotgun
American Arms TS/SS 12 Side-by-Side
Arrieta Sidelock Double Shotguns
Armsport 1050 Series Double Shotguns
Arizaga Model 31 Double Shotgun
AYA Boxlock Shotguns
AYA Sidelock Double Shotguns
Beretta Model 452 Sidelock Shotgun
Beretta Side-by-Side Field Shotguns
Crucelegui Hermanos Model 150 Double
Chapuis Side-by-Side Shotgun
E.A.A./Sabatti Saba-Mon Double Shotgun
Charles Daly Model Dss Double
Ferlib Model F VII Double Shotgun
Auguste Francotte Boxlock Shotgun
Auguste Francotte Sidelock Shotgun
Garbi Model 100 Double
Garbi Model 101 Side-by-Side
Garbi Model 103A, B Side-by-Side
Garbi Model 200 Side-by-Side
Bill Hanus Birdgun Doubles
Hatfield Uplander Shotgun
Merkel Model 8, 47E Side-by-Side Shotguns
Merkel Model 47LSC Sporting Clays Double
Merkel Model 47S, 147S Side-by-Sides
Parker Reproductions Side-by-Side
Piotti King No.1 Side-by-Side
Piotti Lunik Side-by-Side
Piotti King Extra Side-by-Side
Piotti Piuma Side-by-Side
Precision Sports Model 600 Series Doubles
Rizzini Boxlock Side-by-Side
Rizzini Sidelock Side-by-Side
Stoeger/IGA Uplander Side-by-Side Shotgun
Ugartechea 10-Ga. Magnum Shotgun

Shotguns - Bolt Actions & Single Shots

Armsport Single Barrel Shotgun
Browning BT-99 Competition Trap Special
Browning BT-99 Plus Trap Gun
Browning BT-99 Plus Micro
Browning Recoilless Trap Shotgun
Browning Micro Recoilless Trap Shotgun
Desert Industries Big Twenty Shotgun
Harrington & Richardson Topper Model 098
Harrington & Richardson Topper Classic Youth Shotgun
Harrington & Richardson N.W.T.F. Turkey Mag
Harrington & Richardson Topper Deluxe Model 098
Krieghoff KS-5 Trap Gun
Krieghoff KS-5 Special
Krieghoff K-80 Single Barrel Trap Gun
Ljutic Mono Gun Single Barrel
Ljutic LTX Super Deluxe Mono Gun
Ljutic Recoilless Space Gun Shotgun

Marlin Model 55 Goose Gun Bolt Action
New England Firearms Turkey and Goose Gun
New England Firearms N.W.T.F. Shotgun
New England Firearms Tracker Slug Gun
New England Firearms Standard Pardner
New England Firearms Survival Gun
Perazzi TM1 Special Single Trap
Remington 90-T Super Single Shotgun
Snake Charmer II Shotgun
Stoeger/IGA Reuna Single Barrel Shotgun
Thompson/Center TCR '87 Hunter Shotgun.

(Added Pub.L. 90-351, Title IV, § 902, June 19, 1968, 82 Stat. 228, and amended Pub.L. 90-618, Title I, § 102, Oct. 22, 1968, 82 Stat. 1216; Pub.L. 97-377, Title I, § 165(a), Dec. 21, 1982, 96 Stat. 1923; Pub.L. 99-308, § 102, May 19, 1986, 100 Stat. 451; Pub.L. 99-408, § 2, Aug. 28, 1986, 100 Stat. 920; Pub.L. 100-649, § 2(a), Nov. 10, 1988, 102 Stat. 3816, 3818; Pub.L. 100-690, Title VII, § 7060(c), Nov. 18, 1988, 102 Stat. 4404; Pub.L. 101-647, Title XVII, § 1702(b)(1), Title XXII, §§ 2201, 2202, 2204(b), Title XXXV, § 3524, Nov. 29, 1990, 104 Stat. 4844, 4856, 4857, 4924; Pub.L. 103-159, Title I, § 102(a)(1),(b), Title III, § 302(a) to (c), Nov. 30, 1993, 107 Stat. 1536, 1539, 1545; Pub.L. 103-322, Title XI, §§ 110102(a), 110103(a), 110106, 110201(a), 110401(b), (c), 110511, 110514, Title XXXII, §§ 320904, 320927, Title XXXIII, § 330011(i), Sept. 13, 1994, 108 Stat. 1996, 1998, 2000, 2010, 2014, 2019, 2125, 2131, 2145; Pub.L. 104-208, Div. A, Title I, § 101(f) [Title VI, §§ 657, 658(b)], Sept. 30, 1996, 110 Stat. 3009-369, 3009-372; Pub.L. 104-294, Title VI, § 603(b), (c)(1), (d) to (f)(1), (g), Oct. 11, 1996, 110 Stat. 3503, 3504; Pub.L. 105-277, Div. A, § 101(b) [Title I, § 121], Oct. 21, 1998, 112 Stat. 2681 - ____.)

§ 923. Licensing

(a) No person shall engage in the business of importing, manufacturing, or dealing in firearms, or importing or manufacturing ammunition, until he has filed an application with and received a license to do so from the Secretary. The application shall be in such form and contain only that information necessary to determine eligibility for licensing as the Secretary shall by regulation prescribe and shall include a photograph and fingerprints of the applicant. Each applicant shall pay a fee for obtaining such a license, a separate fee being required for each place in which the applicant is to do business, as follows:

(1) If the applicant is a manufacturer -

(A) of destructive devices, ammunition for destructive devices or armor piercing ammunition, a fee of $1,000 per year;

(B) of firearms other than destructive devices, a fee of $50 per year; or

(C) of ammunition for firearms, other than ammunition for destructive devices or armor piercing ammunition, a fee of $10 per year.

(2) If the applicant is an importer -

(A) of destructive devices, ammunition for destructive devices or armor piercing ammunition, a fee of $1,000 per year; or

(B) of firearms other than destructive devices or ammunition for firearms other than destructive devices, or ammunition other than armor piercing ammunition, a fee of $50 per year.

(3) If the applicant is a dealer -

(A) in destructive devices or ammunition for destructive devices, a fee of $1,000 per year; or

(B) who is not a dealer in destructive devices, a fee of $200 for 3 years, except that the fee for renewal of a valid license shall be $90 for 3 years.

[(C) Repealed. Pub.L. 103-159, Title III, § 303(4), Nov. 30, 1993, 107 Stat. 1546.]

(b) Any person desiring to be licensed as a collector shall file an application for such license with the Secretary. The application shall be in such form and contain only that information necessary to determine eligibility as the Secretary shall by regulation prescribe. The fee for such license shall be $10 per year. Any license granted under this subsection shall only apply to transactions in curios and relics.

(c) Upon the filing of a proper application and payment of the prescribed fee, the Secretary shall issue to a qualified applicant the appropriate license which, subject to the provisions of this chapter and other applicable provisions of law, shall entitle the licensee to transport, ship, and receive firearms and ammunition covered by such license in interstate or foreign commerce during the period stated in the license. Nothing in this chapter shall be construed to prohibit a licensed manufacturer, importer, or dealer from maintaining and disposing of a personal collection of firearms, subject only to such restrictions as apply in this chapter to dispositions by a person other than a licensed manufacturer, importer, or dealer. If any firearm is so disposed of by a licensee within one year after its transfer from his business inventory into such licensee's personal collection or if such disposition or any other acquisition is made for the purpose of willfully evading the restrictions placed upon licensees by this chapter, then such firearm shall be deemed part of such licensee's business inventory, except that any licensed manufacturer, importer, or dealer who has maintained a firearm as part of a personal collection for one year and who sells or otherwise disposes of such firearm shall record the description of the firearm in a bound volume, containing the name and place of residence and date of birth of the transferee if the transferee is an individual, or the identity and principal and local places of business of the transferee if the transferee is a corporation or other business entity: **Provided,** That no other recordkeeping shall be required.

(d) (1) Any application submitted under subsection (a) or (b) of this section shall be approved if -

(A) the applicant is twenty-one years of age or over;

(B) the applicant (including, in the case of a corporation, partnership, or association, any individual possessing, directly or indirectly, the power to direct or cause the direction of the management and policies of the corporation, partnership, or association) is not prohibited from transporting, shipping, or receiving firearms or ammunition in interstate or foreign commerce under section 922(g) and (n) of this chapter;

(C) the applicant has not willfully violated any of the provisions of this chapter or regulations issued thereunder;

(D) the applicant has not willfully failed to disclose any material information required, or has not made any false statement as to any material fact, in connection with his application;

(E) the applicant has in a State (i) premises from which he conducts business subject to license under this chapter or from which he intends to conduct such business within a reasonable period of time, or (ii) in the case of a collector, premises from which he conducts his collecting subject to license under this chapter or from which he intends to conduct such collecting within a reasonable period of time;

(F) the applicant certifies that -

(i) the business to be conducted under the license is not prohibited by State or local law in the place where the licensed premise is located;

(ii)(I) within 30 days after the application is approved the business will comply with the requirements of State and local law applicable to the conduct of the business; and

(II) the business will not be conducted under the license until the requirements of State and local law applicable to the business have been met; and

(iii) that the applicant has sent or delivered a form to be prescribed by the Secretary, to the chief law enforcement officer of the locality in which the premises are located, which indicates that the applicant intends to apply for a Federal firearms license; and

(G) in the case of an application to be licensed as a dealer, the applicant certifies that secure gun storage or safety devices will be available at any place in which firearms are sold under the license to persons who are not licensees (subject to the exception that in any case in which a secure gun storage or safety device is temporarily unavailable because of theft, casualty loss, consumer sales, backorders from a manufacturer, or any other similar reason beyond the control of the licensee, the dealer shall not be considered to be in violation of the requirement under this subparagraph to make available such a device).

Editor's Note

Section 923(d)(1)(G) was enacted by Public Law 105-277 on October 21, 1998, and became effective April 19, 1999.

(2) The Secretary must approve or deny an application for a license within the 60-day period beginning on the date it is received. If the Secretary fails to act within such period, the applicant may file an action under section 1361 of title 28 to compel the Secretary to act. If the Secretary approves an applicant's application, such applicant shall be issued a license upon the payment of the prescribed fee.

(e) The Secretary may, after notice and opportunity for hearing, revoke any license issued under this section if the holder of such license has willfully violated any provision of this chapter or any rule or regulation prescribed by the Secretary under this chapter or fails to have secure gun storage or safety devices available at any place in which firearms are sold under the license to persons who are not licensees (except that in any case in which a secure gun storage or safety device is temporarily unavailable because of theft, casualty loss, consumer sales, backorders from a manufacturer, or any other similar reason beyond the control of the licensee, the dealer shall not be considered to be in violation of the requirement to make available such a device). The Secretary may, after notice and opportunity for hearing, revoke the license of a dealer who willfully transfers armor piercing ammunition. The Secretary's action under this subsection may be reviewed only as provided in subsection (f) of this section.

Editor's Note

The provision in Section 923(e) for revoking a license if the holder fails to have secure gun storage or safety devices available where firearms are sold was enacted by Public Law 105-277 on October 21, 1998 and became effective April 19, 1999.

(f) (1) Any person whose application for a license is denied and any holder of a license which is revoked shall receive a written notice from the Secretary stating specifically the grounds upon which the application was denied or upon which the license was revoked. Any notice of a revocation of a license shall be given to the holder of such license before the effective date of the revocation.

(2) If the Secretary denies an application for, or revokes, a license, he shall, upon request by the aggrieved party, promptly hold a hearing to review his denial or revocation. In the case of a revocation of a license, the Secretary shall upon the request of the holder of the license stay the effective date of the revocation. A hearing held under this paragraph shall be held at a location convenient to the aggrieved party.

(3) If after a hearing held under paragraph (2) the Secretary decides not to reverse his decision to deny an application or revoke a license, the Secretary shall give notice of his decision to the aggrieved party. The aggrieved party may at any time within sixty days after the date notice was given under this paragraph file a petition with the United States district court for the district in which he resides or has his principal place of business for a de novo judicial review of such denial or revocation. In a proceeding conducted under this subsection, the court may consider any evidence submitted by the parties to the proceeding whether or not such evidence was considered at the hearing held under paragraph (2). If the court decides that the Secretary was not authorized to deny the application or to revoke the license, the court shall order the Secretary to take such action as may be necessary to comply with the judgment of the court.

(4) If criminal proceedings are instituted against a licensee alleging any violation of this chapter or of rules or regulations prescribed under this chapter, and the licensee is acquitted of such charges, or such proceedings are terminated, other than upon motion of the Government before trial upon such charges, the Secretary shall be absolutely barred from denying or revoking any license granted under this chapter where such denial or revocation is based in whole or in part on the facts which form the basis of such criminal charges. No proceedings for the revocation of a license shall be instituted by the Secretary more than one year after the filing of the indictment or information.

(g) (1) (A) Each licensed importer, licensed manufacturer, and licensed dealer shall maintain such records of importation, production, shipment, receipt, sale, or other disposition of firearms at his place of business for such period, and in such form, as the Secretary may by regulations prescribe. Such importers, manufacturers, and dealers shall not be required to submit to the Secretary reports and information with respect to such records and the contents thereof, except as expressly required by this section. The Secretary, when he has reasonable cause to believe a violation of this chapter has occurred and that evidence thereof may be found on such premises, may, upon demonstrating such cause before a Federal magistrate and securing from such magistrate a warrant authorizing entry, enter during business hours the premises (including places of storage) of any licensed firearms importer, licensed manufacturer, licensed dealer, licensed collector, or any licensed importer or manufacturer of ammunition, for the purpose of inspecting or examining -

(i) any records or documents required to be kept by such licensed importer, licensed manufacturer, licensed dealer, or licensed collector under this chapter or rules or regulations under this chapter, and

(ii) any firearms or ammunition kept or stored by such licensed importer, licensed manufacturer, licensed dealer, or licensed collector, at such premises.

(B) The Secretary may inspect or examine the inventory and records of a licensed importer, licensed manufacturer, or licensed dealer without such reasonable cause or warrant -

(i) in the course of a reasonable inquiry during the course of a criminal investigation of a person or persons other than the licensee;

(ii) for ensuring compliance with the record keeping requirements of this chapter -

(I) not more than once during any 12-month period; or

(II) at any time with respect to records relating to a firearm involved in a criminal investigation that is traced to the licensee; or

(iii) when such inspection or examination may be required for determining the disposition of one or more particular firearms in the course of a bona fide criminal investigation.

(C) The Secretary may inspect the inventory and records of a licensed collector without such reasonable cause or warrant –

(i) for ensuring compliance with the record keeping requirements of this chapter not more than once during any twelve-month period; or

(ii) when such inspection or examination may be required for determining the disposition of one or more particular firearms in the course of a bona fide criminal investigation.

(D) At the election of a licensed collector, the annual inspection of records and inventory permitted under this paragraph shall be performed at the office of the Secretary designated for such inspections which is located in closest proximity to the premises where the inventory and records of such licensed collector are maintained. The inspection and examination authorized by this paragraph shall not be construed as authorizing the Secretary to seize any records or other documents other than those records or documents constituting material evidence of a violation of law. If the Secretary seizes such records or documents, copies shall be provided the licensee within a reasonable time. The Secretary may make available to any Federal, State, or local law enforcement agency any information which he may obtain by reason of this chapter with respect to the identification of persons prohibited from purchasing or receiving firearms or ammunition who have purchased or received firearms or ammunition, together with a description of such firearms or ammunition, and he may provide information to the extent such information may be contained in the records required to be maintained by this chapter, when so requested by any Federal, State, or local law enforcement agency.

(2) Each licensed collector shall maintain in a bound volume the nature of which the Secretary may by regulations prescribe, records of the receipt, sale, or other disposition of firearms. Such records shall include the name and address of any person to whom the collector sells or otherwise disposes of a firearm. Such collector shall not be required to submit to the Secretary reports and information with respect to such records and the contents thereof, except as expressly required by this section.

(3)(A) Each licensee shall prepare a report of multiple sales or other dispositions whenever the licensee sells or otherwise disposes of, at one time or during any five consecutive business days, two or more pistols, or revolvers, or any combination of pistols and revolvers totalling two or more, to an unlicensed person. The report shall be prepared on a form specified by the Secretary and forwarded to the office specified thereon and to the department of State police or State law enforcement agency of the State or local law enforcement agency of the local jurisdiction in which the sale or other disposition took place, not later than the close of business on the day that the multiple sale or other disposition occurs.

(B) Except in the case of forms and contents thereof regarding a purchaser who is prohibited by subsection (g) or (n) of section 922 of this title from receipt of a firearm, the department of State police or State law enforcement agency or local law enforcement agency of the local jurisdiction shall not disclose any such form or the contents thereof to any person or entity, and shall destroy each such form and any record of the contents thereof no more than 20 days from the date such form is received. No later than the date that is 6 months after the effective date of this subparagraph, and at the end of each 6-month period thereafter, the department of State police or State law enforcement agency or local law enforcement agency of the local jurisdiction shall certify to the Attorney General of the United States that no disclosure contrary to this subparagraph has been made and that all forms and any record of the contents thereof have been destroyed as provided in this subparagraph.

(4) Where a firearms or ammunition business is discontinued and succeeded by a new licensee, the records required to be kept by this chapter shall appropriately reflect such facts and shall be delivered to the successor. Where discontinuance of the business is absolute, such records shall be delivered within thirty days after the business discontinuance to the Secretary. However, where State law or local ordinance requires the delivery of records to other responsible authority, the Secretary may arrange for the delivery of such records to such other responsible authority.

(5)(A) Each licensee shall, when required by letter issued by the Secretary, and until notified to the contrary in writing by the Secretary, submit on a form specified by the Secretary, for periods and at the times specified in such letter, all record information required to be kept by this chapter or such lesser record information as the Secretary in such letter may specify.

(B) The Secretary may authorize such record information to be submitted in a manner other than that prescribed in subparagraph (A) of this paragraph when it is shown by a licensee that an alternate method of reporting is reasonably necessary and will not unduly hinder the effective administration of this chapter. A licensee may use an alternate method of reporting if the licensee describes the proposed alternate method of reporting and the need therefor in a letter application submitted to the Secretary, and the Secretary approves such alternate method of reporting.

(6) Each licensee shall report the theft or loss of a firearm from the licensee's inventory or collection, within 48 hours after the theft or loss is discovered, to the Secretary and to the appropriate local authorities.

(7) Each licensee shall respond immediately to, and in no event later than 24 hours after the receipt of, a request by the Secretary for information contained in the records required to be kept by this chapter as may be required for determining the disposition of 1 or more firearms in the course of a bona fide criminal investigation. The requested information shall be provided orally or in writing, as the Secretary may require. The Secretary shall implement a system whereby the licensee can positively identify and establish that an individual requesting information via telephone is employed by and authorized by the agency to request such information.

(h) Licenses issued under the provisions of subsection (c) of this section shall be kept posted and kept available for inspection on the premises covered by the license.

(i) Licensed importers and licensed manufacturers shall identify, by means of a serial number engraved or cast on the receiver or frame of the weapon, in such manner as the Secretary shall by regulations prescribe, each firearm imported or manufactured by such importer or manufacturer. The serial number of any semiautomatic assault weapon manufactured after the date of the enactment of this sentence shall clearly show the date on which the weapon was manufactured. A large capacity ammunition feeding device manufactured after the date of the enactment of this sentence shall be identified by a serial number that clearly shows that the device was manufactured or imported after the effective date of this subsection, and such other identification as the Secretary may by regulation prescribe.

(j) A licensed importer, licensed manufacturer, or licensed dealer may, under rules or regulations prescribed by the Secretary, conduct business temporarily at a location other than the location specified on the license if such temporary location is the location for a gun show or event sponsored by any national, State, or local organization, or any affiliate of any such organization devoted to the collection, competitive use, or other sporting use of firearms in the community, and such location is in the State which is specified on the license. Records of receipt and disposition of firearms transactions conducted at such temporary location shall include the location of the sale or other disposition and shall be entered in the permanent records of the licensee and retained on the location specified on the license. Nothing in this subsection shall authorize any licensee to conduct business in or from any motorized or towed vehicle. Notwithstanding the provisions of subsection (a) of this section, a separate fee shall not be required of a licensee with respect to business conducted under this subsection. Any inspection or examination of inventory or records under this chapter by the Secretary at such temporary location shall be limited to inventory consisting of, or records relating to, firearms held or disposed at such temporary location. Nothing in this subsection shall be construed to authorize the Secretary to inspect or examine the inventory or records of a licensed importer, licensed manufacturer, or licensed dealer at any location other than the location specified on the license. Nothing in this subsection shall be construed to diminish in any manner any right to display, sell, or otherwise dispose of firearms or ammunition, which is in effect before the date of the enactment of the Firearms Owners' Protection Act, including the right of a licensee to conduct "curios or relics" firearms transfers and business away from their business premises with another licensee without regard as to whether the location of where the business is conducted is located in the State specified on the license of either licensee.

(k) Licensed importers and licensed manufacturers shall mark all armor piercing projectiles and packages containing such projectiles for distribution in the manner prescribed by the Secretary by regulation. The Secretary shall furnish information to each dealer licensed under this chapter defining which projectiles are considered armor piercing ammunition as defined in section 921(a)(17)(B).

(l) The Secretary of the Treasury shall notify the chief law enforcement officer in the appropriate State and local jurisdictions of the names and addresses of all persons in the State to whom a firearms license is issued.

(Added Pub.L. 90-351, Title IV, § 902, June 19, 1968, 82 Stat. 231, and amended Pub.L. 90-618, Title I, § 102, Oct. 22, 1968, 82 Stat. 1221; Pub.L. 97-377, Title I, § 165(b), Dec. 21, 1982, 96 Stat. 1923; Pub.L. 99-308, § 103, May 19, 1986, 100 Stat. 453; Pub.L. 99-360, § 1(c), July 8, 1986, 100 Stat. 766; Pub.L. 99-408, §§ 3 to 7, Aug. 28, 1986, 100 Stat. 921; Pub.L. 100-690, Title VII, § 7060(d), Nov. 18, 1988, 102 Stat. 4404; Pub.L. 101-647, Title XXII, § 2203(a), Title XXXV, § 3525, Nov. 29, 1990, 104 Stat. 4857, 4924; Pub.L. 103-159, Title II, § 201, Title III, § 303, Nov. 30, 1993, 107 Stat. 1544, 1545; Pub.L. 103-322, Title XI, §§ 110102(d), 110103(d), 110301(a), 110302 to 110307, Title XXXIII, § 330011(i), Sept. 13, 1994, 108 Stat. 1998, 1999, 2012, 2013, 2014, 2145; Pub.L. 104-208, Div. A, Title I, § 101(f) [Title I, § 118], Sept. 30, 1996, 110 Stat. 3009-326; Pub.L. 104-294, Title VI, § 603(j)(1), (k), (l), Oct. 11, 1996, 110 Stat. 3504, 3405; Pub. L. 105-277, Oct. 21, 1998, 112 Stat. 2681 - ____, 2681 - ____).

§ 924. Penalties

(a)(1) Except as otherwise provided in this subsection, subsection (b), (c), or (f) of this section, or in section 929, whoever -

(A) knowingly makes any false statement or representation with respect to the information required by this chapter to be kept in the records of a person licensed under this chapter or in applying for any license or exemption or relief from disability under the provisions of this chapter;

(B) knowingly violates subsection (a)(4), (f), (k), (r), (v), or (w) of section 922;

(C) knowingly imports or brings into the United States or any possession thereof any firearm or ammunition in violation of section 922(l); or

(D) willfully violates any other provision of this chapter,

shall be fined under this title, imprisoned not more than five years, or both.

(2) Whoever knowingly violates subsection (a)(6), (d), (g), (h), (i), (j), or (o) of section 922 shall be fined as provided in this title, imprisoned not more than 10 years, or both.

(3) Any licensed dealer, licensed importer, licensed manufacturer, or licensed collector who knowingly -

(A) makes any false statement or representation with respect to the information required by the provisions of this chapter to be kept in the records of a person licensed under this chapter, or

(B) violates subsection (m) of section 922,

shall be fined under this title, imprisoned not more than one year, or both.

(4) Whoever violates section 922(q) shall be fined under this title, imprisoned for not more than 5 years, or both. Notwithstanding any other provision of law, the term of imprisonment imposed under this paragraph shall not run concurrently with any other term of imprisonment imposed under any other provision of law. Except for the authorization of a term of imprisonment of not more than 5 years made in this paragraph, for the purpose of any other law a violation of section 922(q) shall be deemed to be a misdemeanor.

(5) Whoever knowingly violates subsection (s) or (t) of section 922 shall be fined under this title, imprisoned for not more than 1 year, or both.

(6)(A)(i) A juvenile who violates section 922(x) shall be fined under this title, imprisoned not more than 1 year, or both, except that a juvenile described in clause (ii) shall be sentenced to probation on appropriate conditions and shall not be incarcerated unless the juvenile fails to comply with a condition of probation.

(ii) A juvenile is described in this clause if -

(I) the offense of which the juvenile is charged is possession of a handgun or ammunition in violation of section 922(x)(2); and

(II) the juvenile has not been convicted in any court of an offense (including an offense under section 922(x) or a similar State law, but not including any other offense consisting of conduct that if engaged in by an adult would not constitute an offense) or adjudicated as a juvenile delinquent for conduct that if engaged in by an adult would constitute an offense.

(B) A person other than a juvenile who knowingly violates section 922(x) -

(i) shall be fined under this title, imprisoned not more than 1 year, or both; and

(ii) if the person sold, delivered, or otherwise transferred a handgun or ammunition to a juvenile knowing or having reasonable cause to know that the juvenile intended to carry or otherwise possess or discharge or otherwise use the handgun or ammunition in the commission of a crime of violence, shall be fined under this title, imprisoned not more than 10 years, or both.

(b) Whoever, with intent to commit therewith an offense punishable by imprisonment for a term exceeding one year, or with knowledge or reasonable cause to believe that an offense punishable by imprisonment for a term exceeding one year is to be committed therewith, ships, transports, or receives a firearm or any ammunition in interstate or foreign commerce shall be fined under this title, or imprisoned not more than ten years, or both.

(c) (1) (A) Except to the extent that a greater minimum sentence is otherwise provided by this subsection or by any other provision of law, any person who, during and in relation to any crime of violence or drug trafficking crime (including a crime of violence or drug trafficking crime that provides for an enhanced punishment if committed by the use of a deadly or dangerous weapon or device) for which the person may be prosecuted in a court of the United States, uses or carries a firearm, or who, in furtherance of any such crime, possesses a firearm, shall, in addition to the punishment provided for such crime of violence or drug trafficking crime -

(i) be sentenced to a term of imprisonment of not less than 5 years;

(ii) if the firearm is brandished, be sentenced to a term of imprisonment of not less than 7 years; and

(iii) if the firearm is discharged, be sentenced to a term of imprisonment of not less than 10 years.

(B) If the firearm possessed by a person convicted of a violation of this subsection -

(i) is a short-barreled rifle, short-barreled shotgun, or semiautomatic assault weapon, the person shall be sentenced to a term of imprisonment of not less than 10 years; or

(ii) is a machinegun or a destructive device, or is equipped with a firearm silencer or firearm muffler, the person shall be sentenced to a term of imprisonment of not less than 30 years.

(C) In the case of a second or subsequent conviction under this subsection, the person shall -

(i) be sentenced to a term of imprisonment of not less than 25 years; and

(ii) if the firearm involved is a machinegun or a destructive device, or is equipped with a firearm silencer or firearm muffler, be sentenced to imprisonment for life.

(D) Notwithstanding any other provision of law -

(i) a court shall not place on probation any person convicted of a violation of this subsection; and

(ii) no term of imprisonment imposed on a person under this subsection shall run concurrently with any other term of imprisonment imposed on the person, including any term of imprisonment imposed for the crime of violence or drug trafficking crime during which the firearm was used, carried, or possessed.

(2) For purposes of this subsection, the term **"drug trafficking crime"** means any felony punishable under the Controlled Substances Act (21 U.S.C. 801 et seq.), the Controlled Substances Import and Export Act (21 U.S.C. 951 et seq.), or the Maritime Drug Law Enforcement Act (46 U.S.C. App. 1901 et seq.).

(3) For purposes of this subsection the term **"crime of violence"** means an offense that is a felony and -

(A) has as an element the use, attempted use, or threatened use of physical force against the person or property of another, or

(B) that by its nature, involves a substantial risk that physical force against the person or property of another may be used in the course of committing the offense.

(4) For purposes of this subsection, the term **"brandish"** means, with respect to a firearm, to display all or part of the firearm, or otherwise make the presence of the firearm known to another person, in order to intimidate that person, regardless of whether the firearm is directly visible to that person.

(d) (1) Any firearm or ammunition involved in or used in any knowing violation of subsection (a)(4), (a)(6), (f), (g), (h), (i), (j), or (k) of section 922, or knowing importation or bringing into the United States or any possession thereof any firearm or ammunition in violation of section 922(l), or knowing violation of section 924, or willful violation of any other provision of this chapter or any rule or regulation promulgated thereunder, or any violation of any other criminal law of the United States, or any firearm or ammunition intended to be used in any offense referred to in paragraph (3) of this subsection, where such intent is demonstrated by clear and convincing evidence, shall be subject to seizure and forfeiture, and all provisions of the Internal Revenue Code of 1954 relating to the seizure, forfeiture, and disposition of firearms, as defined in section 5845(a) of that Code, shall, so far as applicable, extend to seizures and forfeitures under the provisions of this chapter: **Provided,** That upon acquittal of the owner or possessor, or dismissal of the charges against him other than upon motion of the Government prior to trial, or lapse of or court termination of the restraining order to which he is subject, the seized or relinquished firearms or ammunition shall be returned forthwith to the owner or possessor or to a person delegated by the owner or possessor unless the return of the firearms or ammunition would place the owner or possessor or his delegate in violation of law. Any action or proceeding for the forfeiture of firearms or ammunition shall be commenced within one hundred and twenty days of such seizure.

(2)(A) In any action or proceeding for the return of firearms or ammunition seized under the provisions of this chapter, the court shall allow the prevailing party, other than the United States, a reasonable attorney's fee, and the United States shall be liable therefor.

(B) In any other action or proceeding under the provisions of this chapter, the court, when it finds that such action was without foundation, or was initiated vexatiously, frivolously, or in bad faith, shall allow the prevailing party, other than the United States, a reasonable attorney's fee, and the United States shall be liable therefor.

(C) Only those firearms or quantities of ammunition particularly named and individually identified as involved in or used in any violation of the provisions of this chapter or any rule or regulation issued thereunder, or any other criminal law of the United States or as intended to be used in any offense referred to in paragraph (3) of this subsection, where such intent is demonstrated by clear and convincing evidence, shall be subject to seizure, forfeiture, and disposition.

(D) The United States shall be liable for attorneys' fees under this paragraph only to the extent provided in advance by appropriation Acts.

(3) The offenses referred to in paragraphs (1) and (2)(C) of this subsection are -

(A) any crime of violence, as that term is defined in section 924(c)(3) of this title;

(B) any offense punishable under the Controlled Substances Act (21 U.S.C. 801 et seq.) or the Controlled Substances Import and Export Act (21 U.S.C. 951 et seq.);

(C) any offense described in section 922(a)(1), 922(a)(3), 922(a)(5), or 922(b)(3) of this title, where the firearm or ammunition intended to be used in any such offense is involved in a pattern of activities which includes a violation of any offense described in section 922(a)(1), 922(a)(3), 922(a)(5), or 922(b)(3) of this title;

(D) any offense described in section 922(d) of this title where the firearm or ammunition is intended to be used in such offense by the transferor of such firearm or ammunition;

(E) any offense described in section 922(i), 922(j), 922(l), 922(n), or 924(b) of this title; and

(F) any offense which may be prosecuted in a court of the United States which involves the exportation of firearms or ammunition.

(e) (1) In the case of a person who violates section 922(g) of this title and has three previous convictions by any court referred to in section 922(g)(1) of this title for a violent felony or a serious drug offense, or both, committed on occasions different from one another, such person shall be fined not more than $25,000 and imprisoned not less than fifteen years, and, notwithstanding any other provision of law, the court shall not suspend the sentence of, or grant a probationary sentence to, such person with respect to the conviction under section 922(g).

(2) As used in this subsection -

(A) the term **"serious drug offense"** means -

(i) an offense under the Controlled Substances Act (21 U.S.C. 801 et seq.), the Controlled Substances Import and Export Act (21 U.S.C. 951 et seq.), or the Maritime Drug Law Enforcement Act (46 U.S.C. App. 1901 et seq.), for which a maximum term of imprisonment of ten years or more is prescribed by law; or

(ii) an offense under State law, involving manufacturing, distributing, or possessing with intent to manufacture or distribute, a controlled substance (as defined in section 102 of the Controlled Substances Act (21 U.S.C. 802)), for which a maximum term of imprisonment of ten years or more is prescribed by law;

(B) the term **"violent felony"** means any crime punishable by imprisonment for a term exceeding one year, or any act of juvenile delinquency involving the use or carrying of a firearm, knife, or destructive device that would be punishable by imprisonment for such term if committed by an adult, that -

(i) has as an element the use, attempted use, or threatened use of physical force against the person of another; or

(ii) is burglary, arson, or extortion, involves use of explosives, or otherwise involves conduct that presents a serious potential risk of physical injury to another; and

(C) the term **"conviction"** includes a finding that a person has committed an act of juvenile delinquency involving a violent felony.

(f) In the case of a person who knowingly violates section 922(p), such person shall be fined under this title, or imprisoned not more than 5 years, or both.

(g) Whoever, with the intent to engage in conduct which -

(1) constitutes an offense listed in section 1961(1),

(2) is punishable under the Controlled Substances Act (21 U.S.C. 802 et seq.), the Controlled Substances Import and Export Act (21 U.S.C. 951 et seq.), or the Maritime Drug Law Enforcement Act (46 U.S.C.App. 1901 et seq.),

(3) violates any State law relating to any controlled substance (as defined in section 102(6) of the Controlled Substances Act (21 U.S.C. 802(6))), or

(4) constitutes a crime of violence (as defined in subsection (c)(3)),

travels from any State or foreign country into any other State and acquires, transfers, or attempts to acquire or transfer, a firearm in such other State in furtherance of such purpose, shall be imprisoned not more than 10 years, fined in accordance with this title, or both.

(h) Whoever knowingly transfers a firearm, knowing that such firearm will be used to commit a crime of violence (as defined in subsection (c)(3)) or drug trafficking crime (as defined in subsection (c)(2)) shall be imprisoned not more than 10 years, fined in accordance with this title, or both.

(i)(1) A person who knowingly violates section 922(u) shall be fined under this title, imprisoned not more than 10 years, or both.

(2) Nothing contained in this subsection shall be construed as indicating an intent on the part of Congress to occupy the field in which provisions of this subsection operate to the exclusion of State laws on the same subject matter, nor shall any provision of this subsection be construed as invalidating any provision of State law unless such provision is inconsistent with any of the purposes of this subsection.

(j) A person who, in the course of a violation of subsection (c), causes the death of a person through the use of a firearm, shall -

(1) if the killing is a murder (as defined in section 1111), be punished by death or by imprisonment for any term of years or for life; and

(2) if the killing is manslaughter (as defined in section 1112), be punished as provided in that section.

(k) A person who, with intent to engage in or to promote conduct that -

(1) is punishable under the Controlled Substances Act (21 U.S.C. 801 et seq.), the Controlled Substances Import and Export Act (21 U.S.C. 951 et seq.), or the Maritime Drug Law Enforcement Act (46 U.S.C. App. 1901 et seq.);

(2) violates any law of a State relating to any controlled substance (as defined in section 102 of the Controlled Substances Act, 21 U.S.C. 802); or

(3) constitutes a crime of violence (as defined in subsection (c)(3)),

smuggles or knowingly brings into the United States a firearm, or attempts to do so, shall be imprisoned not more than 10 years, fined under this title, or both.

(l) A person who steals any firearm which is moving as, or is a part of, or which has moved in, interstate or foreign commerce shall be imprisoned for not more than 10 years, fined under this title, or both.

(m) A person who steals any firearm from a licensed importer, licensed manufacturer, licensed dealer, or licensed collector shall be fined under this title, imprisoned not more than 10 years, or both.

(n) A person who, with the intent to engage in conduct that constitutes a violation of section 922(a)(1)(A), travels from any State or foreign country into any other State and acquires, or attempts to acquire, a firearm in such other State in furtherance of such purpose shall be imprisoned for not more than 10 years.

(o) A person who conspires to commit an offense under subsection (c) shall be imprisoned for not more than 20 years, fined under this title, or both; and if the firearm is a machinegun or destructive device, or is equipped with a firearm silencer or muffler, shall be imprisoned for any term of years or life.

(Added Pub.L. 90-351, Title IV, § 902, June 19, 1968, 82 Stat. 233, and amended Pub.L. 90-618, Title I, § 102, Oct. 22, 1968, 82 Stat. 1223; Pub.L. 91-644, Title II, § 13, Jan. 2, 1971, 84 Stat. 1889; Pub.L. 98-473, Title II, §§ 223(a), 1005(a), Oct. 12, 1984, 98 Stat. 2028, 2138; Pub.L. 99-308, § 104(a), May 19, 1986, 100 Stat. 456; Pub.L. 99-570, Title I, § 1402, Oct. 27, 1986, 100 Stat. 3207-39; Pub.L. 100-649, § 2(b), (f)(2)(B), (E), Nov. 10, 1988, 102 Stat. 3817, 3818; Pub.L. 100-690, Title VI, §§ 6211, 6212, 6451, 6460, 6462, Title VII, §§ 7056, 7060(a), Nov. 18, 1988, 102 Stat. 4359, 4360, 4371, 4373, 4374, 4402, 4403; Pub.L. 101-647, Title XI, § 1101, Title XVII, § 1702(b)(3), Title XXII, §§ 2203(d), 2204(c), Title XXXV, §§ 3526, 3527, 3528, 3529, Nov. 29, 1990, 104 Stat. 4829, 4845, 4857, 4924; Pub.L. 103-159, Title I, § 102(c), Title III, § 302(d), Nov. 30, 1993, 107 Stat. 1541, 1545; Pub.L. 103-322, Title VI, § 60013, Title XI, §§ 110102(c), 110103(c), 110201(b), 110401(e), 110503, 110504(a), 110507, 110510, 110515(a), 110517, 110518(a), Title XXXIII, §§ 330002(h), 330003(f)(2), 330011(i), (j), 330016(1)(H), (K), (L), Sept. 13, 1994, 108 Stat. 1973, 1998, 1999, 2011, 2015, 2016, 2018, 2019, 2020, 2140, 2141, 2145, 2147; Pub.L. 104-294, Title VI, § 603(m)(1), (n) to (p)(1), (q) to (s), Oct. 11, 1996, 110 Stat. 3505; Pub.L. 105-386, § 1(a), Nov. 13, 1998, 112 Stat. 3469.)

§ 925. Exceptions: Relief from disabilities

(a) (1) The provisions of this chapter, except for sections 922(d)(9) and 922(g)(9) and provisions relating to firearms subject to the prohibitions of section 922(p), shall not apply with respect to the transportation, shipment, receipt, possession, or importation of any firearm or ammunition imported for, sold or shipped to, or issued for the use of, the United States or any department or agency thereof or any State or any department, agency, or political subdivision thereof.

(2) The provisions of this chapter shall not apply with respect to (A) the shipment or receipt of firearms or ammunition when sold or issued by the Secretary of the Army pursuant to section 4308 of title 10, before the repeal of such section by section 1624(a) of the Corporation for the Promotion of Rifle Practice and Firearms Safety Act and (B) the transportation of any such firearm or ammunition carried out to enable a person, who lawfully received such firearm or ammunition from the Secretary of the Army, to engage in military training or in competitions.

(3) Unless otherwise prohibited by this chapter or any other Federal law, a licensed importer, licensed manufacturer, or licensed dealer may ship to a member of the United States Armed Forces on active duty outside the United States or to clubs, recognized by the Department of Defense, whose entire membership is composed of such members, and such members or clubs may receive a firearm or ammunition determined by the Secretary of the Treasury to be generally recognized as particularly suitable for sporting purposes and intended for the personal use of such member or club.

(4) When established to the satisfaction of the Secretary to be consistent with the provisions of this chapter and other applicable Federal and State laws and published ordinances, the Secretary may authorize the transportation, shipment, receipt, or importation into the United States to the place of residence of any member of the United States Armed Forces who is on active duty outside the United States (or who has been on active duty outside the United States within the sixty day period immediately preceding the transportation, shipment, receipt, or importation), of any firearm or ammunition which is (A) determined by the Secretary to be generally recognized as particularly suitable for sporting purposes, or determined by the Department of Defense to be a type of firearm normally classified as a war souvenir, and (B) intended for the personal use of such member.

(5) For the purpose of paragraph (3) of this subsection, the term **"United States"** means each of the several States and the District of Columbia.

(b) A licensed importer, licensed manufacturer, licensed dealer, or licensed collector who is indicted for a crime punishable by imprisonment for a term exceeding one year, may, notwithstanding any other provision of this chapter, continue operation pursuant to his existing license (if prior to the expiration of the term of the existing license timely application is made for a new license) during the term of such indictment and until any conviction pursuant to the indictment becomes final.

(c) A person who is prohibited from possessing, shipping, transporting, or receiving firearms or ammunition may make application to the Secretary for relief from the disabilities imposed by Federal laws with respect to the acquisition, receipt, transfer, shipment, transportation, or possession of firearms, and the Secretary may grant such relief if it is established to his satisfaction that the circumstances regarding the disability, and the applicant's record and reputation, are such that the applicant will not be likely to act in a manner dangerous to public safety and that the granting of the relief would not be contrary to the public interest. Any person whose application for relief from disabilities is denied by the Secretary may file a petition with the United States district court for the district in which he resides for a judicial review of such denial. The court may in its discretion admit additional evidence where failure to do so would result in a miscarriage of justice. A licensed importer, licensed manufacturer, licensed dealer, or licensed collector conducting operations under this chapter, who makes application for relief from the disabilities incurred under this chapter, shall not be barred by such disability from further operations under his license pending final action on an application for relief filed pursuant to this section. Whenever the Secretary grants relief to any person pursuant to this section he shall promptly publish in the Federal Register notice of such action, together with the reasons therefor.

(d) The Secretary shall authorize a firearm or ammunition to be imported or brought into the United States or any possession thereof if the firearm or ammunition-

(1) is being imported or brought in for scientific or research purposes, or is for use in connection with competition or training pursuant to chapter 401 of title 10;

(2) is an unserviceable firearm, other than a machinegun as defined in section 5845(b) of the Internal Revenue Code of 1954 (not readily restorable to firing condition), imported or brought in as a curio or museum piece;

(3) is of a type that does not fall within the definition of a firearm as defined in section 5845(a) of the Internal Revenue Code of 1954 and is generally recognized as particularly suitable for or readily adaptable to sporting purposes, excluding surplus military firearms, except in any case where the Secretary has not authorized the importation of the firearm pursuant to this paragraph, it shall be unlawful to import any frame, receiver, or barrel of such firearm which would be prohibited if assembled; or

(4) was previously taken out of the United States or a possession by the person who is bringing in the firearm or ammunition.

The Secretary shall permit the conditional importation or bringing in of a firearm or ammunition for examination and testing in connection with the making of a determination as to whether the importation or bringing in of such firearm or ammunition will be allowed under this subsection.

(e) Notwithstanding any other provision of this title, the Secretary shall authorize the importation of, by any licensed importer, the following:

(1) All rifles and shotguns listed as curios or relics by the Secretary pursuant to section 921(a)(13), and

(2) All handguns, listed as curios or relics by the Secretary pursuant to section 921(a)(13), provided that such handguns are generally recognized as particularly suitable for or readily adaptable to sporting purposes.

(f) In the case of a person who knowingly violates section 922(p), such person shall be fined under this title, or imprisoned not more than 5 years, or both.

(Added Pub.L. 90-351, Title IV, § 902, June 19, 1968, 82 Stat. 233, and amended Pub.L. 90-618, Title I, § 102, Oct. 22, 1968, 82 Stat. 1224; Pub.L. 98-573, Title II, § 233, Oct. 30, 1984, 98 Stat. 2991; Pub.L. 99-308, § 105, May 19, 1986, 100 Stat. 459; Pub.L. 100-649, § 2(c)(f)(2)(C), (E), Nov. 10, 1988, 102 Stat. 3817; Pub.L. 101-647, Title XXII, § 2203(b), (c), Nov. 29, 1990, 104 Stat. 4857; Pub.L. 104-106, Div. A, Title XVI, § 1624(b)(3), Feb. 10, 1996, 110 Stat. 522; Pub.L. 104-208, Div. A, Title I, § 101(f) [Title VI, § 658(c)], Sept. 30, 1996, 110 Stat. 3009-372; Pub.L. 104-294, Title VI, § 607(c), Oct. 11, 1996, 110 Stat. 3511.)

§ 925A. Remedy for erroneous denial of firearm

Any person denied a firearm pursuant to subsection (s) or (t) of section 922 -

(1) due to the provision of erroneous information relating to the person by any State or political subdivision thereof, or by the national instant criminal background check system established under section 103 of the Brady Handgun Violence Prevention Act; or

(2) who was not prohibited from receipt of a firearm pursuant to subsection (g) or (n) of section 922,

may bring an action against the State or political subdivision responsible for providing the erroneous information, or responsible for denying the transfer, or against the United States, as the case may be, for an order directing that the erroneous information be corrected or that the transfer be approved, as the case may be. In any action under this section, the court, in its discretion, may allow the prevailing party a reasonable attorney's fee as part of the costs.

(Added Pub.L. 103-159, Title I, § 104(a), Nov. 30, 1993, 107 Stat. 1543.)

§ 926. Rules and regulations

(a) The Secretary may prescribe only such rules and regulations as are necessary to carry out the provisions of this chapter, including -

(1) regulations providing that a person licensed under this chapter, when dealing with another person so licensed, shall provide such other licensed person a certified copy of this license;

(2) regulations providing for the issuance, at a reasonable cost, to a person licensed under this chapter, of certified copies of his license for use as provided under regulations issued under paragraph (1) of this subsection; and

(3) regulations providing for effective receipt and secure storage of firearms relinquished by or seized from persons described in subsection (d)(8) or (g)(8) of section 922.

No such rule or regulation prescribed after the date of the enactment of the Firearms Owners' Protection Act may require that records required to be maintained under this chapter or any portion of the contents of such records, be recorded at or transferred to a facility owned, managed, or controlled by the United States or any State or any political subdivision thereof, nor that any system of registration of firearms, firearms owners, or firearms transactions or dispositions be established. Nothing in this section expands or restricts the Secretary's authority to inquire into the disposition of any firearm in the course of a criminal investigation.

(b) The Secretary shall give not less than ninety days public notice, and shall afford interested parties opportunity for hearing, before prescribing such rules and regulations.

(c) The Secretary shall not prescribe rules or regulations that require purchasers of black powder under the exemption provided in section 845(a)(5) of this title to complete affidavits or forms attesting to that exemption.

(Added Pub.L. 90-351, Title IV, § 902, June 19, 1968, 82 Stat. 234, and amended Pub.L. 90-618, Title I, § 102, Oct. 22, 1968, 82 Stat. 1226; Pub.L. 99-308, § 106, May 19, 1986, 100 Stat. 459; Pub.L. 103-322, Title XI, § 110401(d), Sept. 13, 1994, 108 Stat. 2015.)

§ 926A. Interstate transportation of firearms

Notwithstanding any other provision of any law or any rule or regulation of a State or any political subdivision thereof, any person who is not otherwise prohibited by this chapter from transporting, shipping, or receiving a firearm shall be entitled to transport a firearm for any lawful purpose from any place where he may lawfully possess and carry such firearm to any other place where he may lawfully possess and carry such firearm if, during such transportation the firearm is unloaded, and neither the firearm nor any ammunition being transported is readily accessible or is directly accessible from the passenger compartment of such transporting vehicle: **Provided**, That in the case of a vehicle without a compartment separate from the driver's compartment the firearm or ammunition shall be contained in a locked container other than the glove compartment or console.

(Added Pub.L. 99-360, § 1(a), July 8, 1986, 100 Stat. 766.)

§ 927. Effect on State law

No provision of this chapter shall be construed as indicating an intent on the part of the Congress to occupy the field in which such provision operates to the exclusion of the law of any State on the same subject matter, unless there is a direct and positive conflict between such provision and the law of the State so that the two cannot be reconciled or consistently stand together.

(Added Pub.L. 90-351, Title IV, § 902, June 19, 1968, 82 Stat. 234, and amended Pub.L. 90-618, Title I, § 102, Oct. 22, 1968, 82 Stat. 1226.)

§ 928. Separability

If any provision of this chapter or the application thereof to any person or circumstance is held invalid, the remainder of the chapter and the application of such provision to other persons not similarly situated or to other circumstances shall not be affected thereby.

(Added Pub.L. 90-351, Title IV, § 902, June 19, 1968, 82 Stat. 234, and amended Pub.L. 90-618, Title I, § 102, Oct. 22, 1968, 82 Stat. 1226.)

§ 929. Use of restricted ammunition

(a)(1) Whoever, during and in relation to the commission of a crime of violence or drug trafficking crime (including a crime of violence or drug trafficking crime which provides for an enhanced punishment if committed by the use of a deadly or dangerous weapon or device) for which he may be prosecuted in a court of the United States, uses or carries a firearm and is in possession of armor piercing ammunition capable of being fired in that firearm, shall, in addition to the punishment provided for the commission of such crime of violence or drug trafficking crime be sentenced to a term of imprisonment for not less than five years.

(2) For purposes of this subsection, the term **"drug trafficking crime"** means any felony punishable under the Controlled Substances Act (21 U.S.C. 801 et seq.), the Controlled Substances Import and Export Act (21 U.S.C. 951 et seq.), or the

Maritime Drug Law Enforcement Act (46 U.S.C. App. 1901 et seq.).

(b) Notwithstanding any other provision of law, the court shall not suspend the sentence of any person convicted of a violation of this section, nor place the person on probation, nor shall the terms of imprisonment run concurrently with any other terms of imprisonment, including that imposed for the crime in which the armor piercing ammunition was used or possessed. No person sentenced under this section shall be eligible for parole during the term of imprisonment imposed herein.

(Added Pub.L. 98-473, Title II, § 1006(a), Oct. 12, 1984, 98 Stat. 2139; and amended Pub.L. 99-308, § 108, May 19, 1986, 100 Stat. 460; Pub.L. 99-408, § 8, Aug. 28, 1986, 100 Stat. 921; Pub.L. 100-690, Title VI, § 6212, Title VII, § 7060(b), Nov. 18, 1988, 102 Stat. 4360, 4404.)

§ 930. Possession of firearms and dangerous weapons in Federal facilities

(a) Except as provided in subsection (d), whoever knowingly possesses or causes to be present a firearm or other dangerous weapon in a Federal facility (other than a Federal court facility), or attempts to do so, shall be fined under this title or imprisoned not more than 1 year, or both.

(b) Whoever, with intent that a firearm or other dangerous weapon be used in the commission of a crime, knowingly possesses or causes to be present such firearm or dangerous weapon in a Federal facility, or attempts to do so, shall be fined under this title or imprisoned not more than 5 years, or both.

(c) A person who kills or attempts to kill any person in the course of a violation of subsection (a) or (b), or in the course of an attack on a Federal facility involving the use of a firearm or other dangerous weapon, shall be punished as provided in sections 1111, 1112, and 1113.

(d) Subsection (a) shall not apply to -

(1) the lawful performance of official duties by an officer, agent, or employee of the United States, a State, or a political subdivision thereof, who is authorized by law to engage in or supervise the prevention, detection, investigation, or prosecution of any violation of law;

(2) the possession of a firearm or other dangerous weapon by a Federal official or a member of the Armed Forces if such possession is authorized by law; or

(3) the lawful carrying of firearms or other dangerous weapons in a Federal facility incident to hunting or other lawful purposes.

(e) (1) Except as provided in paragraph (2), whoever knowingly possesses or causes to be present a firearm in a Federal court facility, or attempts to do so, shall be fined under this title, imprisoned not more than 2 years, or both.

(2) Paragraph (1) shall not apply to conduct which is described in paragraph (1) or (2) of subsection (d).

(f) Nothing in this section limits the power of a court of the United States to punish for contempt or to promulgate rules or orders regulating, restricting, or prohibiting the possession of weapons within any building housing such court or any of its proceedings, or upon any grounds appurtenant to such building.

(g) As used in this section:

(1) The term **"Federal facility"** means a building or part thereof owned or leased by the Federal Government, where Federal employees are regularly present for the purpose of performing their official duties.

(2) The term **"dangerous weapon"** means a weapon, device, instrument, material, or substance, animate or inanimate, that is used for, or is readily capable of, causing death or serious bodily injury, except that such term does not include a pocket knife with a blade of less than 2 1/2 inches in length.

(3) The term **"Federal court facility"** means the courtroom, judges' chambers, witness rooms, jury deliberation rooms, attorney conference rooms, prisoner holding cells, offices of the court clerks, the United States attorney, and the United States marshal, probation and parole offices, and adjoining corridors of any court of the United States.

(h) Notice of the provisions of subsections (a) and (b) shall be posted conspicuously at each public entrance to each Federal facility, and notice of subsection (e) shall be posted conspicuously at each public entrance to each Federal court facility, and no person shall be convicted of an offense under subsection (a) or (e) with respect to a Federal facility if such notice is not so posted at such facility, unless such person had actual notice of subsection (a) or (e), as the case may be.

(Added Pub.L. 100-690, Title VI, § 6215(a), Nov. 18, 1988, 102 Stat. 4361, and amended Pub.L. 101-647, Title XXII, § 2205(a), Nov. 29, 1990, 104 Stat. 4857; Pub.L. 103-322, Title VI, § 60014, Sept. 13, 1994, 108 Stat. 1973; Pub.L. 104-294, Title VI, § 603(t), (u), Oct. 11, 1996, 110 Stat. 3506.)

Editor's Note

The **Secretary of the Treasury** has the responsibility to administer and enforce the provisions of the Gun Control Act. See section 103 of Pub. L. 90-618 and section 903 of Pub. L. 90-351.

THE NATIONAL FIREARMS ACT

TITLE 26, UNITED STATES CODE, CHAPTER 53
INTERNAL REVENUE CODE

CHAPTER 53 - MACHINE GUNS, DESTRUCTIVE DEVICES, AND CERTAIN OTHER
FIREARMS

SUBCHAPTER
A. Taxes.
B. General provisions and exemptions.
C. Prohibited acts.
D. Penalties and forfeitures.

SUBCHAPTER A - TAXES

PART
I.   Special (occupational) taxes.
II. Tax on transferring firearms.
III. Tax on making firearms.

PART I - SPECIAL (OCCUPATIONAL) TAXES

SEC.
5801. Imposition of tax.
5802. Registration of importers, manufacturers, and dealers.

§ 5801. Imposition of tax

   (a) General rule. - On first engaging in business and thereafter on or before July 1 of each year, every importer, manufacturer, and dealer in firearms shall pay a special (occupational) tax for each place of business at the following rates:

      (1) Importers and manufacturers: $1,000 a year or fraction thereof.

      (2) Dealers: $500 a year or fraction thereof.

   (b) Reduced rates of tax for small importers and manufacturers. -

      (1) **In general.** - Paragraph (1) of subsection (a) shall be applied by substituting "$500" for "$1,000" with respect to any taxpayer the gross receipts of which (for the most recent taxable year ending before the 1st day of the taxable period to which the tax imposed by subsection (a) relates) are less than $500,000.

      (2) **Controlled group rules.** - All persons treated as 1 taxpayer under section 5061(e)(3) shall be treated as 1 taxpayer for purposes of paragraph (1).

      (3) **Certain rules to apply.** - For purposes of paragraph (1), rules similar to the rules of subparagraphs (B) and (C) of section 448(c)(3) shall apply.

   (Added Pub.L. 90-618, Title II, § 201, Oct. 22, 1968, 82 Stat. 1227, and amended Pub.L. 180 - 203, Title X, § 10512 (g)(1), Dec. 22, 1987, 101 Stat. 1330 - 449.)

§ 5802. Registration of importers, manufacturers, and dealers

   On first engaging in business and thereafter on or before the first day of July of each year, each importer, manufacturer, and dealer in firearms shall register with the Secretary in each internal revenue district in which such business is to be carried on, his name, including any trade name, and the address of each location in the district where he will conduct such business. An individual required to register under this section shall include a photograph and fingerprints of the individual with the initial application. Where there is a change during the taxable year in the location of, or the trade name used in, such business, the importer, manufacturer, or dealer shall file an application with the Secretary to amend his registration. Firearms operations of an importer, manufacturer, or dealer may not be commenced at the new location or under a new trade name prior to approval by the Secretary of the application.

(Added Pub.L. 90-618, Title II, § 201, Oct. 22, 1968, 82 Stat. 1227, and amended Pub.L. 94-455, Title XIX, § 1906(b)(13)(A), Oct. 4, 1976, 90 Stat. 1834; Pub.L. 103-322, Title XI, § 110301(b), Sept. 13, 1994, 108 Stat. 2012.)

PART II - TAX ON TRANSFERRING FIREARMS

SEC.
5811. Transfer tax.
5812. Transfers.

§ 5811. Transfer tax

(a) Rate. - There shall be levied, collected, and paid on firearms transferred a tax at the rate of $200 for each firearm transferred, except, the transfer tax on any firearm classified as any other weapon under section 5845(e) shall be at the rate of $5 for each such firearm transferred.

(b) By whom paid. - The tax imposed by subsection (a) of this section shall be paid by the transferor.

(c) Payment. - The tax imposed by subsection (a) of this section shall be payable by the appropriate stamps prescribed for payment by the Secretary.

(Added Pub.L. 90-618, Title II, § 201, Oct. 22, 1968, 82 Stat. 1228, and amended Pub.L. 94-455, Title XIX, § 1906(b)(13)(A), Oct. 4, 1976, 90 Stat. 1834.)

§ 5812. Transfers

(a) Application. - A firearm shall not be transferred unless (1) the transferor of the firearm has filed with the Secretary a written application, in duplicate, for the transfer and registration of the firearm to the transferee on the application form prescribed by the Secretary; (2) any tax payable on the transfer is paid as evidenced by the proper stamp affixed to the original application form; (3) the transferee is identified in the application form in such manner as the Secretary may by regulations prescribe, except that, if such person is an individual, the identification must include his fingerprints and his photograph; (4) the transferor of the firearm is identified in the application form in such manner as the Secretary may by regulations prescribe; (5) the firearm is identified in the application form in such manner as the Secretary may by regulations prescribe; and (6) the application form shows that the Secretary has approved the transfer and the registration of the firearm to the transferee. Applications shall be denied if the transfer, receipt, or possession of the firearm would place the transferee in violation of law.

(b) Transfer of possession. - The transferee of a firearm shall not take possession of the firearm unless the Secretary has approved the transfer and registration of the firearm to the transferee as required by subsection (a) of this section.

(Added Pub.L. 90-618, Title II, § 201, Oct. 22, 1968, 82 Stat. 1228, and amended Pub.L. 94-455, Title XIX, § 1906(b)(13)(A), Oct. 4, 1976, 90 Stat. 1834.)

PART III - TAX ON MAKING FIREARMS

§ 5821. Making tax

(a) Rate. - There shall be levied, collected, and paid upon the making of a firearm a tax at the rate of $200 for each firearm made.

(b) By whom paid. - The tax imposed by subsection (a) of this section shall be paid by the person making the firearm.

(c) Payment. - The tax imposed by subsection (a) of this section shall be payable by the stamp prescribed for payment by the Secretary.

(Added Pub.L. 90-618, Title II, § 201, Oct. 22, 1968, 82 Stat. 1228, and amended Pub.L. 94-455, Title XIX, § 1906(b)(13)(A), Oct. 4, 1976, 90 Stat. 1834.)

§ 5822. Making

No person shall make a firearm unless he has (a) filed with the Secretary a written application, in duplicate, to make and register the firearm on the form prescribed by the Secretary; (b) paid any tax payable on the making and such payment is evidenced by the proper stamp affixed to the original application form; (c) identified the firearm to be made in the application form in such manner as the Secretary may by regulations prescribe; (d) identified himself in the application form in such manner as the Secretary may by regulations prescribe, except that, if such person is an individual, the identification must include his fingerprints and his photograph; and (e) obtained the approval of the Secretary to make and register the firearm and the application form shows such approval. Applications shall be denied if the making or possession of the firearm would place the person making the firearm in violation of law.

(Added Pub.L. 90-618, Title II, § 201, Oct. 22, 1968, 82 Stat. 1228, and amended Pub.L. 94-455, Title XIX, § 1906(b)(13)(A), Oct. 4, 1976, 90 Stat. 1834.)

SUBCHAPTER B - GENERAL PROVISIONS AND EXEMPTIONS

PART I - GENERAL PROVISIONS

§ 5841. Registration of firearms

(a) Central registry. - The Secretary shall maintain a central registry of all firearms in the United States which are not in the possession or under the control of the United States. This registry shall be known as the National Firearms Registration and Transfer Record. The registry shall include -

(1) identification of the firearm;

(2) date of registration; and

(3) identification and address of person entitled to possession of the firearm.

(b) By whom registered. - Each manufacturer, importer, and maker shall register each firearm he manufactures, imports, or makes. Each firearm transferred shall be registered to the transferee by the transferor.

(c) How registered. - Each manufacturer shall notify the Secretary of the manufacture of a firearm in such manner as may by regulations be prescribed and such notification shall effect the registration of the firearm required by this section. Each importer, maker, and transferor of a firearm shall, prior to importing, making, or transferring a firearm, obtain authorization in such manner as required by this chapter or regulations issued thereunder to import, make, or transfer the firearm, and such authorization shall effect the registration of the firearm required by this section.

(d) Firearms registered on effective date of this act. - A person shown as possessing a firearm by the records maintained by the Secretary pursuant to the National Firearms Act in force on the day immediately prior to the effective date of the National Firearms Act of 1968 shall be considered to have registered under this section the firearms in his possession which are disclosed by that record as being in his possession.

(e) Proof of registration. - A person possessing a firearm registered as required by this section shall retain proof of registration which shall be made available to the Secretary upon request.

(Added Pub.L. 90-618, Title II, § 201, Oct. 22, 1968, 82 Stat. 1229, and amended Pub.L. 94-455, Title XIX, § 1906(b)(13)(A), Oct. 4, 1976, 90 Stat. 1834.)

§ 5842. Identification of firearms

(a) Identification of firearms other than destructive devices. - Each manufacturer and importer and anyone making a firearm shall identify each firearm, other than a destructive device, manufactured, imported, or made by a serial number which may not be readily removed, obliterated, or altered, the name of the manufacturer, importer, or maker, and such other identification as the Secretary may by regulations prescribe.

(b) Firearms without serial number. - Any person who possesses a firearm, other than a destructive device, which does not bear the serial number and other information required by subsection (a) of this section shall identify the firearm with a serial number assigned by the Secretary and any other information the Secretary may by regulations prescribe.

(c) Identification of destructive device. - Any firearm classified as a destructive device shall be identified in such manner as the Secretary may by regulations prescribe.

(Added Pub.L. 90-618, Title II, § 201, Oct. 22, 1968, 82 Stat. 1230, and amended Pub.L. 94-455, Title XIX, § 1906(b)(13)(A), Oct. 4, 1976, 90 Stat. 1834.)

§ 5843. Records and returns

Importers, manufacturers, and dealers shall keep such records of, and render such returns in relation to, the importation, manufacture, making, receipt, and sale, or other disposition, of firearms as the Secretary may by regulations prescribe.

(Added Pub.L. 90-618, Title II, § 201, Oct. 22, 1968, 82 Stat. 1230, and amended Pub.L. 94-455, Title XIX, § 1906(b)(13)(A), Oct. 4, 1976, 90 Stat. 1834.)

§ 5844. Importation

No firearm shall be imported or brought into the United States or any territory under its control or jurisdiction unless the importer establishes, under regulations as may be prescribed by the Secretary, that the firearm to be imported or brought in is -

(1) being imported or brought in for the use of the United States or any department, independent establishment, or agency thereof or any State or possession or any political subdivision thereof; or

(2) being imported or brought in for scientific or research purposes; or

(3) being imported or brought in solely for testing or use as a model by a registered manufacturer or solely for use as a sample by a registered importer or registered dealer;

except that, the Secretary may permit the conditional importation or bringing in of a firearm for examination and testing in connection with classifying the firearm.

(Added Pub.L. 90-618, Title II, § 201, Oct. 22, 1968, 82 Stat. 1230, and amended Pub.L. 94-455, Title XIX, § 1906(b)(13)(A), Oct. 4, 1976, 90 Stat. 1834.)

§ 5845. Definitions

For the purpose of this chapter -

(a) **Firearm**. - The term **"firearm"** means (1) a shotgun having a barrel or barrels of less than 18 inches in length; (2) a weapon made from a shotgun if such weapon as modified has an overall length of less than 26 inches or a barrel or barrels of less than 18 inches in length; (3) a rifle having a barrel or barrels of less than 16 inches in length; (4) a weapon made from a rifle if such weapon as modified has an overall length of less than 26 inches or a barrel or barrels of less than 16 inches in length; (5) any other weapon, as defined in subsection (e); (6) a machinegun; (7) any silencer (as defined in section 921 of title 18, United States Code); and (8) a destructive device. The term **"firearm"** shall not include an antique firearm or any device (other than a machinegun or destructive device) which, although designed as a weapon, the Secretary finds by reason of the date of its manufacture, value, design, and other characteristics is primarily a collector's item and is not likely to be used as a weapon.

(b) **Machinegun**. - The term **"machinegun"** means any weapon which shoots, is designed to shoot, or can be readily restored to shoot, automatically more than one shot, without manual reloading, by a single function of the trigger. The term shall also include the frame or receiver of any such weapon, any part designed and intended solely and exclusively, or combination of parts designed and intended, for use in converting a weapon into a machinegun, and any combination of parts from which a machinegun can be assembled if such parts are in the possession or under the control of a person.

(c) **Rifle**. - The term **"rifle"** means a weapon designed or redesigned, made or remade, and intended to be fired from the shoulder and designed or redesigned and made or remade to use the energy of the explosive in a fixed cartridge to fire only a single projectile through a rifled bore for each single pull of the trigger, and shall include any such weapon which may be readily restored to fire a fixed cartridge.

(d) Shotgun. - The term **"shotgun"** means a weapon designed or redesigned, made or remade, and intended to be fired from the shoulder and designed or redesigned and made or remade to use the energy of the explosive in a fixed shotgun shell to fire through a smooth bore either a number of projectiles (ball shot) or a single projectile for each pull of the trigger, and shall include any such weapon which may be readily restored to fire a fixed shotgun shell.

(e) Any other weapon. - The term **"any other weapon"** means any weapon or device capable of being concealed on the person from which a shot can be discharged through the energy of an explosive, a pistol or revolver having a barrel with a smooth bore designed or redesigned to fire a fixed shotgun shell, weapons with combination shotgun and rifle barrels 12 inches or more, less than 18 inches in length, from which only a single discharge can be made from either barrel without manual reloading, and shall include any such weapon which may be readily restored to fire. Such term shall not include a pistol or a revolver having a rifled bore, or rifled bores, or weapons designed, made, or intended to be fired from the shoulder and not capable of firing fixed ammunition.

(f) Destructive device. - The term **"destructive device"** means (1) any explosive, incendiary, or poison gas (A) bomb, (B) grenade, (C) rocket having a propellant charge of more than four ounces, (D) missile having an explosive or incendiary charge of more than one-quarter ounce, (E) mine, or (F) similar device; (2) any type of weapon by whatever name known which will, or which may be readily converted to, expel a projectile by the action of an explosive or other propellant, the barrel or barrels of which have a bore of more than one-half inch in diameter, except a shotgun or shotgun shell which the Secretary finds is generally recognized as particularly suitable for sporting purposes; and (3) any combination of parts either designed or intended for use in converting any device into a destructive device as defined in subparagraphs (1) and (2) and from which a destructive device may be readily assembled. The term **"destructive device"** shall not include any device which is neither designed nor redesigned for use as a weapon; any device, although originally designed for use as a weapon, which is redesigned for use as a signaling, pyrotechnic, line throwing, safety, or similar device; surplus ordnance sold, loaned, or given by the Secretary of the Army pursuant to the provisions of section 4684(2), 4685, or 4686 of title 10 of the United States Code; or any other device which the Secretary finds is not likely to be used as a weapon, or is an antique or is a rifle which the owner intends to use solely for sporting purposes.

(g) Antique firearm. - The term **"antique firearm"** means any firearm not designed or redesigned for using rim fire or conventional center fire ignition with fixed ammunition and manufactured in or before 1898 (including any matchlock, flintlock, percussion cap, or similar type of ignition system or replica thereof, whether actually manufactured before or after the year 1898) and also any firearm using fixed ammunition manufactured in or before 1898, for which ammunition is no longer manufactured in the United States and is not readily available in the ordinary channels of commercial trade.

(h) Unserviceable firearm. - The term **"unserviceable firearm"** means a firearm which is incapable of discharging a shot by means of an explosive and incapable of being readily restored to a firing condition.

(i) Make. - The term **"make"**, and the various derivatives of such word, shall include manufacturing (other than by one qualified to engage in such business under this chapter), putting together, altering, any combination of these, or otherwise producing a firearm.

(j) Transfer. - The term **"transfer"** and the various derivatives of such word, shall include selling, assigning, pledging, leasing, loaning, giving away, or otherwise disposing of.

(k) Dealer. - The term **"dealer"** means any person, not a manufacturer or importer, engaged in the business of selling, renting, leasing, or loaning firearms and shall include pawnbrokers who accept firearms as collateral for loans.

(l) Importer. - The term **"importer"** means any person who is engaged in the business of importing or bringing firearms into the United States.

(m) Manufacturer. - The term **"manufacturer"** means any person who is engaged in the business of manufacturing firearms.

(Added Pub.L. 90-618, Title II, § 201, Oct. 22, 1968, 82 Stat. 1230, and amended Pub.L. 94-455, Title XIX, § 1906(b)(13)(A), (J), Oct. 4, 1976, 90 Stat. 1834, 1835; Pub.L. 99-308, § 109, May 19, 1986, 100 Stat. 460.)

§ 5846. Other laws applicable

All provisions of law relating to special taxes imposed by chapter 51 and to engraving, issuance, sale, accountability, cancellation, and distribution of stamps for tax payment shall, insofar as not inconsistent with the provisions of this chapter, be applicable with respect to the taxes imposed by sections 5801, 5811, and 5821.

(Added Pub.L. 90-618, Title II, § 201, Oct. 22, 1968, 82 Stat. 1232.)

§ 5847. Effect on other laws

Nothing in this chapter shall be construed as modifying or affecting the requirements of section 414 of the Mutual Security Act of 1954, as amended, with respect to the manufacture, exportation, and importation of arms, ammunition, and implements of war.

(Added Pub.L. 90-618, Title II, § 201, Oct. 22, 1968, 82 Stat. 1232.)

§ 5848. Restrictive use of information

(a) General rule. - No information or evidence obtained from an application, registration, or records required to be submitted or retained by a natural person in order to comply with any provision of this chapter or regulations issued thereunder, shall, except as provided in subsection (b) of this section, be used, directly or indirectly, as evidence against that person in a criminal proceeding with respect to a violation of law occurring prior to or concurrently with the filing of the application or registration, or the compiling of the records containing the information or evidence.

(b) Furnishing false information. - Subsection (a) of this section shall not preclude the use of any such information or evidence in a prosecution or other action under any applicable provision of law with respect to the furnishing of false information.

(Added Pub.L. 90-618, Title II, § 201, Oct. 22, 1968, 82 Stat. 1232.)

§ 5849. Citation of chapter

This chapter may be cited as the "National Firearms Act" and any reference in any other provision of law to the "National Firearms Act" shall be held to refer to the provisions of this chapter.

(Added Pub.L. 90-618, Title II, § 201, Oct. 22, 1968, 82 Stat. 1232.)

PART II - EXEMPTIONS

§ 5851. Special (occupational) tax exemption

(a) Business with United States. - Any person required to pay special (occupational) tax under section 5801 shall be relieved from payment of that tax if he establishes to the satisfaction of the Secretary that his business is conducted exclusively with, or on behalf of, the United States or any department, independent establishment, or agency thereof. The Secretary may relieve any person manufacturing firearms for, or on behalf of, the United States from compliance with any provision of this chapter in the conduct of such business.

(b) Application. - The exemption provided for in subsection (a) of this section may be obtained by filing with the Secretary an application on such form and containing such information as may by regulations be prescribed. The exemptions must thereafter be renewed on or before July 1 of each year. Approval of the application by the Secretary shall entitle the applicant to the exemptions stated on the approved application.

(Added Pub.L. 90-618, Title II, § 201, Oct. 22, 1968, 82 Stat. 1233, and amended Pub.L. 94-455, Title XIX, § 1906(b)(13)(A), Oct. 4, 1976, 90 Stat. 1834.)

§ 5852. General transfer and making tax exemption

(a) Transfer. - Any firearm may be transferred to the United States or any department, independent establishment, or agency thereof, without payment of the transfer tax imposed by section 5811.

(b) Making by a person other than a qualified manufacturer. - Any firearm may be made by, or on behalf of, the United States, or any department, independent establishment, or agency thereof, without payment of the making tax imposed by section 5821.

(c) Making by a qualified manufacturer. - A manufacturer qualified under this chapter to engage in such business may make the type of firearm which he is qualified to manufacture without payment of the making tax imposed by section 5821.

(d) Transfers between special (occupational) taxpayers. - A firearm registered to a person qualified under this chapter to engage in business as an importer, manufacturer, or dealer may be transferred by that person without payment of the transfer tax imposed by section 5811 to any other person qualified under this chapter to manufacture, import, or deal in that type of firearm.

(e) Unserviceable firearm. - An unserviceable firearm may be transferred as a curio or ornament without payment of the transfer tax imposed by section 5811, under such requirements as the Secretary may by regulations prescribe.

(f) Right to exemption. - No firearm may be transferred or made exempt from tax under the provisions of this section unless the transfer or making is performed pursuant to an application in such form and manner as the Secretary may by regulations prescribe.

(Added Pub.L. 90-618, Title II, § 201, Oct. 22, 1968, 82 Stat. 1233, and amended Pub.L. 94-455, Title XIX, § 1906(b)(13)(A), Oct. 4, 1976, 90 Stat. 1834.)

§ 5853. Transfer and making tax exemption available to certain governmental entities

(a) Transfer. - A firearm may be transferred without the payment of the transfer tax imposed by section 5811 to any State, possession of the United States, any political subdivision thereof, or any official police organization of such a government entity engaged in criminal investigations.

(b) Making. - A firearm may be made without payment of the making tax imposed by section 5821 by, or on behalf of, any State, or possession of the United States, any political subdivision thereof, or any official police organization of such a government entity engaged in criminal investigations.

(c) Right to exemption. - No firearm may be transferred or made exempt from tax under this section unless the transfer or making is performed pursuant to an application in such form and manner as the Secretary may by regulations prescribe.

(Added Pub.L. 90-618, Title II, § 201, Oct. 22, 1968, 82 Stat. 1233, and amended Pub.L. 94-455, Title XIX, § 1906(b)(13)(A), Oct. 4, 1976, 90 Stat. 1834.)

§ 5854. Exportation of firearms exempt from transfer tax

A firearm may be exported without payment of the transfer tax imposed under section 5811 provided that proof of the exportation is furnished in such form and manner as the Secretary may by regulations prescribe.

(Added Pub.L. 90-618, Title II, § 201, Oct. 22, 1968, 82 Stat. 1234, and amended Pub.L. 94-455, Title XIX, § 1906(b)(13)(A), Oct. 4, 1976, 90 Stat. 1834.)

SUBCHAPTER C - PROHIBITED ACTS

SEC.
5861. Prohibited acts.

§ 5861. Prohibited acts

It shall be unlawful for any person -

(a) to engage in business as a manufacturer or importer of, or dealer in, firearms without having paid the special (occupational) tax required by section 5801 for his business or having registered as required by section 5802; or

(b) to receive or possess a firearm transferred to him in violation of the provisions of this chapter; or

(c) to receive or possess a firearm made in violation of the provisions of this chapter; or

(d) to receive or possess a firearm which is not registered to him in the National Firearms Registration and Transfer Record; or

(e) to transfer a firearm in violation of the provisions of this chapter; or

(f) to make a firearm in violation of the provisions of this chapter; or

(g) to obliterate, remove, change, or alter the serial number or other identification of a firearm required by this chapter; or

(h) to receive or possess a firearm having the serial number or other identification required by this chapter obliterated, removed, changed, or altered; or

(i) to receive or possess a firearm which is not identified by a serial number as required by this chapter; or

(j) to transport, deliver, or receive any firearm in interstate commerce which has not been registered as required by this chapter; or

(k) to receive or possess a firearm which has been imported or brought into the United States in violation of section 5844; or

(l) to make, or cause the making of, a false entry on any application, return, or record required by this chapter, knowing such entry to be false.

(Added Pub.L. 90-618, Title II, § 201, Oct. 22, 1968, 82 Stat. 1234.)

SUBCHAPTER D – PENALTIES AND FORFEITURES

SEC.
5871. Penalties.
5872. Forfeitures.

§ 5871. Penalties

Any person who violates or fails to comply with any provision of this chapter shall, upon conviction, be fined not more than $10,000, or be imprisoned not more than ten years, or both.

(Added Pub.L. 90-618, Title II, § 201, Oct. 22, 1968, 82 Stat. 1234, and amended Pub.L. 98-473, Title II, § 227, Oct. 12, 1984, 98 Stat. 2030.)

§ 5872. Forfeitures

(a) Laws applicable. - Any firearm involved in any violation of the provisions of this chapter shall be subject to seizure and forfeiture, and (except as provided in subsection (b)) all the provisions of internal revenue laws relating to searches, seizures, and forfeitures of unstamped articles are extended to and made to apply to the articles taxed under this chapter, and the persons to whom this chapter applies.

(b) Disposal. - In the case of the forfeiture of any firearm by reason of a violation of this chapter, no notice of public sale shall be required; no such firearm shall be sold at public sale; if such firearm is forfeited for a violation of this chapter and there is no remission or mitigation of forfeiture thereof, it shall be delivered by the Secretary to the Administrator of General Services, General Services Administration, who may order such firearm destroyed or may sell it to any State, or possession, or political subdivision thereof, or at the request of the Secretary, may authorize its retention for official use of the Treasury Department, or may transfer it without charge to any executive department or independent establishment of the Government for use by it.

(Added Pub.L. 90-618, Title II, § 201, Oct. 22, 1968, 82 Stat. 1235, and amended Pub.L. 94-455, Title XIX, § 1906(b)(13)(A), Oct. 4, 1976, 90 Stat. 1834.)

Editor's Note

Section 7801, Title 26, U.S.C., provides that the administration and enforcement of the Internal Revenue Code, which includes the National Firearms Act, shall be performed by or under the supervision of the Secretary of the Treasury.

ARMS EXPORT CONTROL ACT OF 1976, Sec. 38, AS AMENDED

TITLE 22, UNITED STATES CODE, § 2778

Editor's Note

With respect to Section 38 of the Arms Export Control Act of 1976 (22 U.S.C. § 2778), only the importation provisions are administered by ATF. Export provisions are administered by the Department of State. Importation regulations issued under this law are in Part 47 of 27 CFR, and are included in this publication.

§ 2778. Control of arms exports and imports

(a) Presidential control of exports and imports of defense articles and services, guidance of policy, etc.; designation of United States Munitions List; issuance of export licenses; condition for export; negotiations information

(1) In furtherance of world peace and the security and foreign policy of the United States, the President is authorized to control the import and the export of defense articles and defense services and to provide foreign policy guidance to persons of the United States involved in the export and import of such articles and services. The President is authorized to designate those items which shall be considered as defense articles and defense services for the purposes of this section and to promulgate regulations for the import and export of such articles and services. The items so designated shall constitute the United States Munitions List.

(2) Decisions on issuing export licenses under this section shall be made in coordination with the Director of the United States Arms Control and Disarmament Agency, taking into account the Director's assessment as to whether the export of an article would contribute to an arms race, aid in the development of weapons of mass destruction, support international terrorism, increase the possibility of outbreak or escalation of conflict, or prejudice the development of bilateral or multilateral arms control or nonproliferation agreements or other arrangements. The Director of the Arms Control and Disarmament Agency is authorized, whenever the Director determines that the issuance of an export license under this section would be detrimental to the national security of the United States, to recommend to the President that such export license be disapproved.

(3) In exercising the authorities conferred by this section, the President may require that any defense article or defense service be sold under this chapter as a condition of its eligibility for export, and may require that persons engaged in the negotiation for the export of defense articles and services keep the President fully and currently informed of the progress and future prospects of such negotiations.

(b) Registration and licensing requirements for manufacturers, exporters, or importers of designated defense articles and defense services

(1)(A)(i) As prescribed in regulations issued under this section, every person (other than an officer or employee of the United States Government acting in an official capacity) who engages in the business of manufacturing, exporting, or importing any defense articles or defense services designated by the President under subsection (a)(1) of this section shall register with the United States Government agency charged with the administration of this section, and shall pay a registration fee which shall be prescribed by such regulations. Such regulations shall prohibit the return to the United States for sale in the United States (other than for the Armed Forces of the United States and its allies or for any State or local law enforcement agency) of any military firearms or ammunition of United States manufacture furnished to foreign governments by the United States under this chapter or any other foreign assistance or sales program of the United States, whether or not enhanced in value or improved in condition in a foreign country. This prohibition shall not extend to similar firearms that have been so substantially transformed as to become, in effect, articles of foreign manufacture.

(ii)(I) As prescribed in regulations issued under this section, every person (other than an officer or employee of the United States Government acting in official capacity) who engages in the business of brokering activities with respect to the manufacture, export, import, or transfer of any defense article or defense service designated by the President under subsection (a)(1) of this section, or in the business of brokering activities with respect to the manufacture, export, import, or transfer of any foreign defense article or defense service (as defined in subclause (IV)), shall register with the United States Government agency charged with the administration of this section, and shall pay a registration fee which shall be prescribed by such regulations.

(II) Such brokering activities shall include the financing, transportation, freight forwarding, or taking of any other action that facilitates the manufacture, export, or import of a defense article or defense service.

(III) No person may engage in the business of brokering activities described in subclause (I) without a license, issued in accordance with this chapter, except that no license shall be required for such activities undertaken by or for an agency of the United States Government —

(aa) for use by an agency of the United States Government; or

(bb) for carrying out any foreign assistance or sales program authorized by law and subject to the control of the President by other means.

(IV) For purposes of this clause, the term "foreign defense article or defense service" includes any non-United States defense article or defense service of a nature described on the United States Munitions List regardless of whether such article or service is of United States origin or whether such article or service contains United States origin components.

(B)[1] The prohibition under such regulations required by the second sentence of subparagraph (A) shall not extend to any military firearms (or ammunition, components, parts, accessories, and attachments for such firearms) of United States manufacture furnished to any foreign government by the United States under this chapter or any other foreign assistance or sales program of the United States if –

(i) such firearms are among those firearms that the Secretary of the Treasury is, or was at any time, required to authorize the importation of by reason of the provisions of section 925(e) of Title 18, (including the requirement for the listing of such firearms as curios or relics under section 921(a)(13) of that title); and

(ii) such foreign government certifies to the United States Government that such firearms are owned by such foreign government.

(B)[1] A copy of each registration made under this paragraph shall be transmitted to the Secretary of the Treasury for review regarding law enforcement concerns. The Secretary shall report to the President regarding such concerns as necessary.

(2) Except as otherwise specifically provided in regulations issued under subsection (a)(1) of this section, no defense articles or defense services designated by the President under subsection (a)(1) of this section may be exported or imported without a license for such export or import, issued in accordance with this chapter and regulations issued under this chapter, except that no license shall be required for exports or imports made by or for an agency of the United States Government **(A)** for official use by a department or agency of the United States Government, or **(B)** for carrying out any foreign assistance or sales program authorized by law and subject to the control of the President by other means.

(3) (A) For each of the fiscal years 1988 and 1989, $250,000 of registration fees collected pursuant to paragraph (1) shall be credited to a Department of State account, to be available without fiscal year limitation. Fees credited to that account shall be available only for the payment of expenses incurred for -

(i) contract personnel to assist in the evaluation of munitions control license applications, reduce processing time for license applications, and improve monitoring of compliance with the terms of licenses; and

(ii) the automation of munitions control functions and the processing of munitions control license applications, including the development, procurement, and utilization of computer equipment and related software.

(B) The authority of this paragraph may be exercised only to such extent or in such amounts as are provided in advance in appropriation Acts.

(c) Criminal violations; punishment

Any person who willfully violates any provision of this section or section 2779 of this title, or any rule or regulation issued under either section, or who willfully, in a registration or license application or required report, makes any untrue statement of a material fact or omits to state a material fact required to be stated therein or necessary to make the statements therein not misleading, shall upon conviction be fined for each violation not more than $1,000,000 or imprisoned not more than ten years, or both.

(d) Repealed

(e) Enforcement powers of President

In carrying out functions under this section with respect to the export of defense articles and defense services, the President is authorized to exercise the same powers concerning violations and enforcement which are conferred upon departments, agencies and officials by subsections (c), (d), (e), and (g) of section 11 of the Export Administration Act of 1979 [50 App. U.S.C.A. § 2410(c), (d), (e), and (g)], and by subsections (a) and (c) of section 12 of such Act [50 App. U.S.C.A. § 2411(a) and (c)], subject to the same terms and conditions as are applicable to such powers under such Act. [50 App. U.S.C.A. § 2401 et seq.], except that the names of the countries and the types and quantities of defense articles for which licenses are issued under this section shall not be withheld from public disclosure unless the President determines that the release of such information would be contrary to the national interest. Nothing in this subsection shall be construed as authorizing the withholding of information from the Congress. Notwithstanding section 11(c) of the Export Administration Act of 1979 [50 App. U.S.C.A. § 2410(c)], the civil penalty for each violation involving controls imposed on the export of defense articles and defense services under this section may not exceed $500,000.

(f) Periodic review of items on Munitions List

The President shall periodically review the items on the United States Munitions List to determine what items, if any, no longer warrant export controls under this section. The results of such reviews shall be reported to the Speaker of the House of Representatives and to the Committee on Foreign Relations and the Committee on Banking, Housing, and Urban Affairs of the Senate. Such a report shall be submitted at least 30 days before any item is removed from the Munitions List and shall describe the nature of any controls to be imposed on that item under the Export Administration Act of 1979 [50 App. U.S.C.A. § 2401 et seq.]

* * * *

(Pub.L. 90-629, c. 3, § 38, as added Pub.L. 94-329, Title II, § 212(a)(1), June 30, 1976, 90 Stat. 744, and amended Pub.L. 95-92, § 20, Aug. 4, 1977, 91 Stat. 623;

Pub.L. 96-70, Title III, § 3303(a)(4), Sept. 27, 1979, 93 Stat. 499; Pub.L. 96-72, § 22(a), Sept. 29, 1979, 93 Stat. 535; Pub.L. 96-92, § 21, Oct. 29, 1979, 93 Stat. 710; Pub.L. 96-533, Title I, § 107(a), (c), Dec. 16, 1980, 94 Stat. 3136; Pub.L. 97-113, Title I, §§ 106, 107, Dec. 29, 1981, 95 Stat. 1522; Pub.L. 99-64, Title I, § 123(a), July 12, 1985, 99 Stat. 156; Pub.L. 99-83, Title I, § 119(a), (b), Aug. 8, 1985, 99 Stat. 203; Pub.L. 100-202, § 101(b), [Title VII, § 8142(a)], Dec. 22, 1987, 101 Stat. 1329-88; Pub.L. 100-204, Title XII, § 1255, Dec. 22, 1987, 101 Stat. 1429; Pub.L. 101-222, §§ 3(a), 6, Dec. 12, 1989, 103 Stat. 1896, 1899.)

---

[1] So in original. There are two subparagraphs designated "(B)".

CERTAIN FIREARMS LAWS ADMINISTERED BY OTHER FEDERAL AGENCIES

POSTAL SERVICE

TITLE 18, UNITED STATES CODE, CHAPTER 83

§ 1715. FIREARMS AS NONMAILABLE; REGULATIONS

Pistols, revolvers, and other firearms capable of being concealed on the person are nonmailable and shall not be deposited in or carried by the mails or delivered by any officer or employee of the Postal Service. Such articles may be conveyed in the mails, under such regulations as the Postal Service shall prescribe, for use in connection with their official duty, to officers of the Army, Navy, Air Force, Coast Guard, Marine Corps, or Organized Reserve Corps; to officers of the National Guard or Militia of a State, Territory, Commonwealth, Possession, or District; to officers of the United States or of a State, Territory, Commonwealth, Possession, or District whose official duty is to serve warrants of arrest or commitments; to employees of the Postal Service; to officers and employees of enforcement agencies of the United States; and to watchmen engaged in guarding the property of the United States, a State, Territory, Commonwealth, Possession, or District. Such articles also may be conveyed in the mails to manufacturers of firearms or bona fide dealers therein in customary trade shipments, including such articles for repairs or replacement of parts, from one to the other, under such regulations as the Postal Service shall prescribe.

Whoever knowingly deposits for mailing or delivery, or knowingly causes to be delivered by mail according to the direction thereon, or at any place to which it is directed to be delivered by the person to whom it is addressed, any pistol, revolver, or firearm declared nonmailable by this section, shall be fined under this title or imprisoned not more than two years, or both.

(June 25, 1948, c. 645, 62 Stat. 781; May 24, 1949, c. 139, § 40, 63 Stat. 95; Aug. 12, 1970, Pub.L. 91-375, § 6(j)(24), 84 Stat. 779; Sept. 13, 1994, Pub.L. 103-322, Title XXXIII, § 330016(1)(H), 108 Stat. 2147; Oct. 11, 1996, Pub.L. 104-294, Title VI, § 607(f), 110 Stat. 3511.)

PART 47—IMPORTATION OF ARMS, AMMUNITION AND IMPLEMENTS OF WAR

Subpart A—Scope

Subpart B—Definitions

Subpart C—The U.S. Munitions Import List

Subpart D—Registration

Subpart E—Permits

Subpart F—Miscellaneous Provisions

Subpart G—Penalties, Seizures and Forfeitures

47.63   Seizure and forfeiture.

Authority: 22 U.S.C. 2778.

Source: T.D. ATF-8, 39 FR 3251, Jan. 25, 1974, unless otherwise noted.

Subpart A-Scope

§ 47.1   General.

The regulations in this part relate to that portion of Section 38, Arms Export Control Act of 1976, as amended, which is concerned with the importation of arms, ammunition and implements of war. This part contains the U.S. Munitions Import List and includes procedural and administrative requirements and provisions relating to registration of importers, permits, articles in transit, import certification, delivery verification, import restrictions applicable to certain countries, exemptions, U.S. military firearms or ammunition, penalties, seizures, and forfeitures. All designations and changes in designation of articles subject to import control under Section 414 of the Mutual Security Act of 1954, as amended, have the concurrence of the Secretary of State and the Secretary of Defense.

[T.D. ATF-8, 39 FR 3251, Jan. 25, 1974, as amended by T.D. ATF-215, 50 FR 42158, Oct. 18, 1985]

§ 47.2   Relation to other laws and regulations.

(a) All of those items on the U.S. Munitions Import List (see § 47.21) which are **"firearms"** or **"ammunition"** as defined in 18 U.S.C. 921(a) are subject to the interstate and foreign commerce controls contained in Chapter 44 of Title 18 U.S.C. and 27 CFR Part 178 and if they are **"firearms"** within the definition set out in 26 U.S.C. 5845(a) are also subject to the provisions of 27 CFR Part 179. Any person engaged in the business of importing firearms or ammunition as defined in 18 U.S.C. 921(a) must obtain a license under the provisions of 27 CFR Part 178, and if he imports firearms which fall within the definition of 26 U.S.C. 5845(a) must also register and pay special tax pursuant to the provisions of 27 CFR Part 179. Such licensing, registration and special tax requirements are in addition to registration under subpart D of this part.

(b) The permit procedures of subpart E of this part are applicable to all importations of articles on the U.S. Munitions Import List not subject to controls under 27 CFR Part 178 or 179. U.S. Munitions Import List articles subject to controls under 27 CFR Part 178 or 27 CFR Part 179 are subject to the import permit procedures of those regulations if imported into the United States (within the meaning of 27 CFR Parts 178 and

179).

(c) Articles on the U.S. Munitions Import List imported for the United States or any State or political subdivision thereof are exempt from the import controls of 27 CFR Part 178 but are not exempt from control under Section 38, Arms Export Control Act of 1976, unless imported by the United States or any agency thereof. All such importations not imported by the United States or any agency thereof shall be subject to the import permit procedures of subpart E of this part.

[T.D. ATF-215, 50 FR 42158, Oct. 18, 1985]

Subpart B-Definitions

§ 47.11 Meaning of terms.

When used in this part and in forms prescribed under this part, where not otherwise distinctly expressed or manifestly incompatible with the intent thereof, terms shall have the meanings ascribed in this section. Words in the plural form shall include the singular, and vice versa, and words imparting the masculine gender shall include the feminine. The terms **"includes"** and **"including"** do not exclude other things not enumerated which are in the same general class or are otherwise within the scope thereof.

ATF officer. An officer or employee of the Bureau of Alcohol, Tobacco and Firearms (ATF) authorized to perform any functions relating to the administration or enforcement of this part.

Article. Any of the arms, ammunition, and implements of war enumerated in the U.S. Munitions Import List.

Bureau. Bureau of Alcohol, Tobacco and Firearms, the Department of the Treasury.

Carbine. A short-barrelled rifle whose barrel is generally not longer than 22 inches and is characterized by light weight.

CFR. The Code of Federal Regulations.

Chemical agent. A substance useful in war which, by its ordinary and direct chemical action, produces a powerful physiological effect.

Defense articles. Any item designated in § 47.21 or § 47.22. This term includes models, mockups, and other such items which reveal technical data directly relating to § 47.21 or § 47.22. For purposes of Category XXII, any item enumerated on the U.S. Munitions List (22 CFR Part 121).

Defense services. **(a)** The furnishing of assistance, including training, to foreign persons in the design, engineering, development, production, processing, manufacture, use, operation, overhaul, repair, maintenance, modification, or reconstruction of defense articles, whether in the United States or abroad; or

(b) The furnishing to foreign persons of any technical data, whether in the United States or abroad.

Director. The Director, Bureau of Alcohol, Tobacco and Firearms, the Department of the Treasury, Washington, DC 20226.

Executed under the penalties of perjury. Signed with the prescribed declaration under the penalties of perjury as provided on or with respect to the application, form, or other document or, where no form of declaration is prescribed, with the declaration: **"I declare under the penalties of perjury that this _____ (insert type of document such as statement, certificate, application, or other document), including the documents submitted in support thereof, has been examined by me and, to best of my knowledge and belief, is true, correct, and complete."**

Firearms. A weapon, and all components and parts therefor, not over .50 caliber which will or is designed to or may be readily converted to expel a projectile by the action of an explosive, but shall not include BB and pellet guns, and muzzle loading (black powder) firearms (including any firearm with a matchlock, flintlock, percussion cap, or similar type of ignition system) or firearms covered by Category I(a) established to have been manufactured in or before 1898.

Import or importation. Bringing into the United States from a foreign country any of the articles on the Import List, but shall not include intransit, temporary import or temporary export transactions subject to Department of State controls under Title 22, Code of Federal Regulations.

Import List. The list of articles contained in § 47.21 and identified therein **as "The U.S. Munitions Import List".**

Machine gun. A **"machine gun"**, **"machine pistol"**, **"submachine gun"**, or **"automatic rifle"** is a firearm originally designed to fire, or capable of being fired fully automatically by a single pull of the trigger.

Permit. The same as **"license"** for purposes of 22 U.S.C. 1934(c).

Person. A partnership, company, association, or corporation, as well as a natural person.

Pistol. A hand-operated firearm having a chamber integral with, or permanently aligned with, the bore.

Regional director (compliance). The principal ATF regional official responsible for administering regulations in this part.

Revolver. A hand-operated firearm with a revolving cylinder containing chambers for individual cartridges.

Rifle. A shoulder firearm discharging bullets through a rifled barrel at least 16 inches in length, including combination and drilling guns.

Sporting type sight including optical. A telescopic sight suitable for daylight use on a rifle, shotgun, pistol, or revolver for hunting or target shooting.

This chapter. Title 27, Code of Federal Regulations, Chapter I (27 CFR Chapter I).

United States. When used in the geographical sense, includes the several States, the Commonwealth of Puerto Rico, the insular possessions of the United States, the District of Columbia, and any territory over which the United States exercises any powers of administration, legislation, and jurisdiction.

(26 U.S.C. 7805 (68A Stat. 917), 27 U.S.C. 205 (49 Stat. 981 as amended), 18 U.S.C. 926 (82 Stat. 959), and sec. 38, Arms Export Control Act (22 U.S.C. 2778, 90 Stat. 744))

[T.D. ATF-48, 43 FR 13535, Mar. 31, 1978; 44 FR 55840, Sept. 28, 1979, as amended by T.D. ATF-202, 50 FR 14382, Apr. 12, 1985; T.D. ATF-215, 50 FR 42158, Oct. 18, 1985]

Subpart C-The U.S. Munitions Import List

§ 47.21   The U.S. Munitions Import List.

The U.S. Munitions List compiled by the Department of State, Office of Munitions Control, and published at 22 CFR 121.1, with the deletions indicated, has been adopted as an enumeration of the defense articles subject to controls under this part.   The expurgated list, set out below, shall, for the purposes of this part, be known as the U.S. Munitions Import List:

THE U.S. MUNITIONS IMPORT LIST

CATEGORY I-FIREARMS

(a) Nonautomatic and semiautomatic firearms, to caliber .50 inclusive, combat shotguns, and shotguns with barrels less than 18 inches in length, and all components and parts for such firearms.

(b) Automatic firearms and all components and parts for such firearms to caliber .50 inclusive.

(c) Insurgency-counterinsurgency type firearms of other weapons having a special military application (e.g. close assault weapons systems) regardless of caliber and all components and parts for such firearms.

(d) Firearms silencers and suppressors, including flash suppressors.

(e) Riflescopes manufactured to military specifications and specifically designed or modified components therefor.

NOTE: Rifles, carbines, revolvers, and pistols, to caliber .50 inclusive, combat shotguns, and shotguns with barrels less than 18 inches in length are included under Category I(a). Machine guns, submachine guns, machine pistols and fully automatic rifles to caliber .50 inclusive are included under Category I(b).

CATEGORY II-ARTILLERY PROJECTORS

(a) Guns over caliber .50, howitzers, mortars, and recoilless rifles.

(b) Military flamethrowers and projectors.

(c) Components, parts, accessories, and attachments for the articles in paragraphs (a) and (b) of this category, including but not limited to mounts and carriages for these articles.

CATEGORY III-AMMUNITION

(a) Ammunition for the arms in Categories I and II of this section.

(b) Components, parts, accessories, and attachments for articles in paragraph (a) of this category, including but not limited to cartridge cases, powder bags, bullets, jackets, cores, shells (excluding shotgun shells), projectiles, boosters, fuzes and components therefor, primers, and other detonating devices for such ammunition.

(c) Ammunition belting and linking machines.

(d) Ammunition manufacturing machines and ammunition loading machines (except handloading ones).

NOTE: Cartridge and shell casings are included under Category III unless, prior to their importation, they have been rendered useless beyond the possibility of restoration for use as a cartridge or shell casing by means of heating, flame treatment, mangling, crushing, cutting, or popping.

CATEGORY IV—LAUNCH VEHICLES, GUIDED MISSILES, BALLISTIC MISSILES, ROCKETS, TORPEDOES, BOMBS AND MINES

(a) Rockets (including but not limited to meteorological and other sounding rockets), bombs, grenades, torpedoes, depth charges, land and naval mines, as well as launchers for such defense articles, and demolition blocks and blasting caps.

(b) Launch vehicles and missile and anti-missile systems including but not limited to guided, tactical and strategic missiles, launchers, and systems.

(c) Apparatus, devices, and materials for the handling, control, activation, monitoring, detection, protection, discharge, or detonation of the articles in paragraphs (a) and (b) of this category. Articles in this category include, but are not limited to, the following: Fuses and components for the items in this category, bomb racks and shackles, bomb shackle release units, bomb ejectors, torpedo tubes, torpedo and guided missile boosters, guidance system equipment and parts, launching racks and projectors, pistols (exploders), igniters, fuze arming devices, intervalometers, guided missile launchers and specialized handling equipment, and hardened missile launching facilities.

(d) Missile and space vehicle powerplants.

(e) Military explosive excavating devices.

(f) Ablative materials fabricated or semifabricated from advanced composites (e.g., silica, graphite, carbon, carbon/carbon, and boron filaments) for the articles in this category that are derived directly from or specifically developed or modified for defense articles.

(g) Non/nuclear warheads for rockets and guided missiles.

(h) All specifically designed components or modified components, parts, accessories, attachments, and associated equipment for the articles in this category.

NOTE: Military demolition blocks and blasting caps referred to in Category IV(a) do not include the following articles:

(a) Electric squibs.

(b) No. 6 and No. 8 blasting caps, including electric ones.

(c) Delay electric blasting caps (including No. 6 and No.

8 millisecond ones).

   (d) Seismograph electric blasting caps (including SSS, Static-Master, Vibrocap SR, and SEISMO SR).

   (e) Oil well perforating devices.

NOTE: Category V of **"Munitions List"** deleted as inapplicable to imports.

CATEGORY VI-VESSELS OF WAR AND SPECIAL NAVAL EQUIPMENT

   (a) Warships, amphibious warfare vessels, landing craft, mine warfare vessels, patrol vessels, auxiliary vessels and service craft, experimental types of naval ships and any vessels specifically designed or modified for military purposes.

   (b) Turrets and gun mounts, arresting gear, special weapons systems, protective systems, submarine storage batteries, catapults and other components, parts, attachments, and accessories specifically designed or modified for combatant vessels.

   (c) Mine sweeping equipment, components, parts, attachments and accessories specifically designed or modified therefor.

   (d) Harbor entrance detection devices (magnetic, pressure, and acoustic ones) and controls and components therefor.

   (e) Naval nuclear propulsion plants, their land prototypes and special facilities for their construction, support and maintenance. This includes any machinery, device, component, or equipment specifically developed or designed or modified for use in such plants or facilities.

NOTE: The term **"vessels of war"** includes, but is not limited to the following:

   (a) Combatant vessels:

   (1) Warships (including nuclear-powered versions):

   (i) Aircraft carriers (CV, CVN)

   (ii) Battleships (BB)

   (iii) Cruisers (CA, CG, CGN)

   (iv) Destroyers (DD, DDG)

   (v) Frigates (FF, FFG)

(vi) Submarines (SS, SSN, SSBN, SSG, SSAG).

(2) Other Combatant Classifications:

(i) Patrol Combatants (PC, PHM)

(ii) Amphibious Helicopter/Landing Craft Carriers (LHA, LPD, LPH)

(iii) Amphibious Landing Craft Carriers (LKA, LPA, LSD, LST)

(iv) Amphibious Command Ships (LCC)

(v) Mine Warfare Ships (MSO).

(b) Auxiliaries:

(1) Mobile Logistics Support:

(i) Under way Replenishment (AD, AF, AFS, AO, AOE, AOR)

(ii) Material Support (AD, AR, AS).

(2) Support Ships:

(i) Fleet Support Ships (ARS, ASR, ATA, ATF, ATS)

(ii) Other Auxiliaries (AG, AGDS, AGF, AGM, AGOR, AGOS, AGS, AH, AK, AKR, AOG, AOT, AP, APB, ARC, ARL, AVM, AVT).

(c) Combatant Craft:

(1) Patrol Craft:

(i) Coastal Patrol Combatants (PB, PCF, PCH, PTF)

(ii) River, Roadstead Craft (ATC, PBR).

(2) Amphibious Warfare Craft:

(i) Landing Craft (AALC, LCAC, LCM, LCPL, LCPR, LCU, LWT, SLWT)

(ii) Special Warfare Craft (LSSC, MSSC, SDV, SWCL, SWCM).

(3) Mine Warfare Craft:

(i) Mine Countermeasures Craft (MSB, MSD, MSI, MSM, MSR).

(d) Support and Service Craft:

(1) Tugs (YTB, YTL, YTM)

(2) Tankers (YO, YOG, YW)

(3) Lighters (YC, YCF, YCV, YF, YFN, YFNB, YFNX, YFR, YFRN, YFU, YG, YGN, YOGN, YON, YOS, YSR, YWN)

(4) Floating Dry Docks (AFDB, AFDL, AFDM, ARD, ARDM, YFD)

(5) Miscellaneous (APL, DSRV, DSV, IX, NR, YAG, YD, YDT, YFB, YFND, YEP, YFRT, YHLC, YM, YNG, YP, YPD, YR, YRB, YRBN, YRDH, YRDM, YRR, YRST, YSD).

(e) Coast Guard Patrol and Service Vessels and Craft:

(1) Coast Guard Cutters (CGC, WHEC, WMEC)

(2) Patrol Craft (WPB)

(3) Icebreakers (WAGB)

(4) Oceanography Vessels (WAGO)

(5) Special Vessels (WIX)

(6) Buoy Tenders (WLB, WLM, WLI, WLR, WLIC)

(7) Tugs (WYTM, WYTL)

(8) Light Ships (WLV).

CATEGORY VII-TANKS AND MILITARY VEHICLES

(a) Military type armed or armored vehicles, military railway trains, and vehicles specifically designed or modified to accommodate mountings for arms or other specialized military equipment or fitted with such items.

(b) Military tanks, combat engineer vehicles, bridge launching vehicles, halftracks and gun carriers.

(c) Self-propelled guns and howitzers.

NOTE: Category VII (d) and (e) of **"Munitions List"** deleted as inapplicable to imports.

(f) Amphibious vehicles.

(g) Engines specifically designed or modified for the vehicles in paragraphs (a), (b), (c), and (f) of this category.

(h) All specifically designed or modified components and parts, accessories, attachments, and associated equipment for the articles in this category, including but not limited to military bridging and deep water fording kits.

NOTE: An **"amphibious vehicle"** in Category VII(f) is an automotive vehicle or chassis which embodies all-wheel drive, which is equipped to meet special military requirements, and which has sealed electrical systems and adaptation features for deep water fording.

CATEGORY VIII—AIRCRAFT, SPACECRAFT, AND ASSOCIATED EQUIPMENT

(a) Aircraft, including but not limited to helicopters, non-expansive balloons, drones and lighter-than-air aircraft, which are specifically designed, modified, or equipped for military purposes. This includes but is not limited to the following military purposes: gunnery, bombing, rocket or missile launching, electronic and other surveillance, reconnaissance, refueling, aerial mapping, military liaison, cargo carrying or dropping, personnel dropping, airborne warning and control, and military training.

NOTE: Category VIII (b) through (j) and Categories IX, X, XI, XII and XIII of **"Munitions List"** deleted as inapplicable to imports.

Note: In Category VIII, **"aircraft"** means aircraft designed, modified, or equipped for a military purpose, including aircraft described as **"demilitarized."** All aircraft bearing an original military designation are included in Category VIII. However, the following aircraft are not so included so long as they have not been specifically equipped, reequipped, or modified for military operations:

(a) Cargo aircraft bearing "C" designations and numbered C-45 through C-118 inclusive, and C-121 through C-125 inclusive, and C-131, using reciprocating engines only.

(b) Trainer aircraft bearing "T" designations and using reciprocating engines or turboprop engines with less than 600 horsepower (s.h.p.).

(c) Utility aircraft bearing "U" designations and using reciprocating engines only.

(d) All liaison aircraft bearing an "L" designation.

(e) All observation aircraft bearing "O" designations and using reciprocating engines.

CATEGORY XIV—TOXICOLOGICAL AGENTS AND EQUIPMENT AND RADIOLOGICAL EQUIPMENT

(a) Chemical agents, including but not limited to lung irritants, vesicants, lachrymators, and tear gases (except tear gas formulations containing 1% or less CN or CS), sternutators and irritant smoke, and nerve gases and incapacitating agents.

(b) Biological agents.

(c) Equipment for dissemination, detection, and identification of, and defense against, the articles in paragraphs (a) and (b) of this category.

(d) Nuclear radiation detection and measuring devices manufactured to military specification.

(e) Components, parts, accessories, attachments, and associated equipment specifically designed or modified for the articles in paragraphs (c) and (d) of this category.

NOTE: A chemical agent in Category XIV(a) is a substance having military application which by its ordinary and direct chemical action produces a powerful physiological effect. The term **"chemical agent"** includes, but is not limited to, the following chemical compounds:

(a) Lung irritants:

(1) Diphenylcyanoarsine (DC).

(2) Fluorine (but not fluorene).

(3) Trichloronitro methane (chloropicrin PS).

(b) Vesicants:

(1) B-Chlorovinyldichloroarsine (Lewisite, L).

(2) Bis(dichlorethyl) sulphide (Mustard Gas, HD or H).

(3) Ethyldichloroarsine (ED).

(4) Methyldichloroarsine (MD).

(c) Lachrymators and tear gases:

(1) A-Brombenzyl cyanide (BBC).

(2) Chloroacetophenone (CN).

(3) Dibromodimethyl ether.

(4) Dichlorodimethyl ether (ClCi).

(5) Ethyldibromoarsine.

(6) Phenylcarbylamine chloride.

(7) Tear gas solutions (CNB and CNS).

(8) Tear gas orthochlorobenzalmalononitrile (CS).

(d) Sternutators and irritant smokes:

(1) Diphenylamine chloroarsine (Adamsite, DM).

(2) Diphenylchloroarsine (BA).

(3) Liquid pepper.

(e) Nerve agents, gases, and aerosols. These are toxic compounds which affect the nervous system, such as:

(1) Dimethylaminoethoxycyanophosphine oxide (GA).

(2) Methylisopropoxyfluorophosphine oxide (GB).

(3) Methylpinacolyloxyfluoriphosphine oxide (GD).

(f) Antiplant chemicals, such as: Butyl 2-chloro-4-fluorophenoxyacetate (LNF).

CATEGORY XV- (RESERVED)

CATEGORY XVI-NUCLEAR WEAPONS DESIGN AND TEST EQUIPMENT

(a) Any article, material, equipment, or device, which is specifically designed or modified for use in the design, development, or fabrication of nuclear weapons or nuclear explosive devices.

(b) Any article, material, equipment, or device, which is specifically designed or modified for use in the devising, carrying out, or evaluating of nuclear weapons tests or any other nuclear explosions, except such items as are in normal commercial use for other purposes.

NOTE: Categories XVII, XVIII, and XIX of "Munitions List" deleted as inapplicable to imports.

CATEGORY XX-SUBMERSIBLE VESSELS, OCEANOGRAPHIC AND ASSOCIATED EQUIPMENT

(a) Submersible vessels, manned and unmanned, designed or modified for military purposes or having independent capability to maneuver vertically or horizontally at depths below 1,000 feet, or powered by nuclear propulsion plants.

(b) Submersible vessels, manned or unmanned, designed or modified in whole or in part from technology developed by or for the U.S. Armed Forces.

(c) Any of the articles in Category VI and elsewhere in this part specifically designed or modified for use with submersible vessels, and oceanographic or associated equipment assigned a military designation.

(d) Equipment, components, parts, accessories, and attachments specifically designed for any of the articles in paragraphs (a) and (b) of this category.

CATEGORY XXI—MISCELLANEOUS ARTICLES

Any article not specifically enumerated in the other categories of the U.S. Munitions List which has substantial military applicability and which has been specifically designed or modified for military purposes. The decision on whether any article may be included in this category shall be made by the Director, Office of Munitions Control, Department of State, with the concurrence of the Department of Defense.

CATEGORY XXII—SOUTH AFRICA

(a) Defense articles enumerated on the U.S. Munitions List (22 CFR Part 121).

(b) Technical data relating to defense articles enumerated on the U.S. Munitions List.

NOTE: This category is applicable only to South Africa.

NOTE: "**Technical data**" means, for purposes of this category:

(1) Classified information relating to defense articles and defense services;

(2) Information covered by an invention secrecy order;

(3) Information which is directly related to the design, engineering, development, production, processing, manufacture, use, operation, overhaul, repair, maintenance, modification, or reconstruction of defense articles. This includes, for example, information in the form of blueprints, drawings, photographs, plans, instructions, computer software and documentation. This also includes information which advances the state of the art of articles on the U.S. Munitions List. This does not include information concerning general scientific, mathematical or engineering principles.

[T.D. ATF-215, 50 FR 42158, Oct. 18, 1985; 50 FR 46647, Nov.

12, 1985]

§ 47.22 Forgings, castings, and machined bodies.

Articles on the U.S. Munitions Import List include articles in a partially completed state (such as forgings, castings, extrusions, and machined bodies) which have reached a stage in manufacture where they are clearly identifiable as defense articles. If the end-item is an article on the U.S. Munitions Import List, (including components, accessories, attachments and parts) then the particular forging, casting, extrusion, machined body, etc., is considered a defense article subject to the controls of this part, except for such items as are in normal commercial use.

[T.D. ATF-215, 50 FR 42160, Oct. 18, 1985]

Subpart D-Registration

§ 47.31 Registration requirement.

Persons engaged in the business, in the United States, of importing articles enumerated on the U.S. Munitions Import List must register with the Director.

[T.D. ATF-8, 39 FR 3251, Jan. 25, 1974, as amended by T.D. ATF-215, 50 FR 42161, Oct. 18, 1985]

§ 47.32 Application for registration and refund of fee.

(a) Application for registration shall be filed on Form 4587, in duplicate, with the Director, and shall be accompanied by the registration fee at the rate prescribed in this section. On approval of the application by the Director, he will return the original to the applicant.

(b) Registration may be effected for periods of from 1 to 5 years at the option of the registrant by identifying on Form 4587 the period of registration desired. The registration fees are as follows:

1 year $250

2 years.......................$500

3 years.......................$700

4 years.......................$850

5 years....................$1,000

(c) Fees paid in advance for whole future years of a multiple year registration will be refunded upon request if the registrant ceases to engage in importing articles on the U.S. Munitions Import List. A request for a refund must be submitted to the Director, Bureau of Alcohol, Tobacco and Firearms, Washington, DC 20226, Attention: Firearms and Explosives Imports Branch, prior to the beginning of any year for which a refund is claimed.

(Approved by the Office of Management and Budget under control number 1512-0021)

[T.D. ATF-8, 39 FR 3251, Jan. 25, 1974, as amended by T.D. ATF-215, 50 FR 42161, Oct. 18, 1985]

§ 47.33 Notification of changes in information furnished by registrants.

Registered persons shall notify the Director in writing, in duplicate, of significant changes in the information set forth in their registration.

(Approved by the Office of Management and Budget under control number 1512-0021)

[T.D. ATF-8, 39 FR 3251, Jan. 25, 1974, as amended by T.D. ATF-215, 50 FR 42161, Oct. 18, 1985]

§ 47.34 Maintenance of records by persons required to register as importers of Import List articles.

(a) Registrants under this part engaged in the business of importing articles subject to controls under 27 CFR Parts 178 and 179 shall maintain records in accordance with the applicable provisions of those parts.

(b) Registrants under this part engaged in importing articles on the U.S. Munitions Import List subject to the permit procedures of subpart E of this part shall maintain for a period of 6 years, subject to inspection by any ATF officer, records bearing on such articles imported, including records concerning their acquisition and disposition by the registrant. The Director may prescribe a longer or shorter period in individual cases as he deems necessary.

(Approved by the Office of Management and Budget under control number 1512-0387)

[T.D. ATF-8, 39 FR 3251, Jan. 25, 1974, as amended by T.D. ATF-172, 49 FR 14941, Apr. 16, 1984; T.D. ATF-215, 50 FR 42161, Oct. 18, 1985]

§ 47.35 Forms prescribed.

(a) The Director is authorized to prescribe all forms required by this part. All of the information called for in each form shall be furnished as indicated by the headings on the form and the instructions on or pertaining to the form. In addition, information called for in each form shall be furnished as required by this part.

(b) "Public Use Forms" (ATF Publication 1322.1) is a numerical listing of forms issued or used by the Bureau of Alcohol, Tobacco and Firearms. This publication is available from the Superintendent of Documents, U.S. Government Printing Office, Washington, DC 20402.

(c) Requests for forms should be mailed to the ATF

Distribution Center, P. O. Box 5950, Springfield, Virginia 22153.

[T.D. ATF-92, 46 FR 46914, Sept. 23, 1981, as amended by T.D. ATF-249, 52 FR 5961, Fcb. 27, 1987]

Subpart E—Permits

§ 47.41 Permit requirement.

(a) Articles on the U.S. Munitions Import List not subject to import control under 27 CFR Parts 178 and 179 shall not be imported into the United States except pursuant to a permit under this subpart issued by the Director.

(b) Articles on the U.S. Munitions Import List intended for the United States or any State or political subdivision thereof, or the District of Columbia, which are exempt from import controls of 27 CFR 178.115 shall not be imported into the United States, except by the United States or agency thereof, without first obtaining a permit issued by the Director under this subpart.

(c) A permit is not required for the importation of (1) the U.S. Munitions Import List articles from Canada not subject to the import controls of 27 CFR Part 178 or 179, except articles enumerated in Categories I, II, III, IV, VI(e), VIII(a), XVI, and XX; and nuclear weapons strategic delivery systems and all specifically designed components, parts, accessories, attachments, and associated equipment thereof (see Category XXI); or (2) minor components and parts for Category I(a) firearms, except barrels, cylinders, receivers (frames) or complete breech mechanisms, when the total value does not exceed $500 wholesale in any single transaction.

[T.D. ATF-215, 50 FR 42161, Oct. 18, 1985]

§ 47.42 Application for permit.

(a) Persons required to obtain a permit as provided in §47.41 shall file a Form 6 Part I, in triplicate, with the Director. On approval of the application by the Director, he will return the original to the applicant. Such approved application will serve as the permit.

(b) For additional requirements relating to the importation of plastic explosives into the United States on or after April 24, 1997, see § 55.183 of this title.

(Approved by the Office of Management and Budget under control number 1512-0017)

[T.D. ATF-215, 50 FR 42161, Oct. 18, 1985; as amended by T.D.

ATF-387, 62 FR 8376, Feb. 25, 1997]

§ 47.43 Terms of permit.

(a) Import permits issued under this subpart are valid for one year from their issuance date unless a different period of validity is stated thereon. They are not transferable.

(b) If shipment cannot be completed during the period of validity of the permit, another application must be submitted for permit to cover the unshipped balance. Such an application shall make reference to the previous permit and may include materials in addition to the unshipped balance.

(c) No amendments or alteration of a permit may be made, except by the Director. No photographic or other copy of an original permit, unless certified by the Director, may be used to effect Customs release.

[T.D. ATF-8, 39 FR 3251, Jan. 25, 1974, as amended by T.D. ATF-325, 57 FR 29787, July 7, 1992]

§ 47.44 Permit denial, revocation or suspension.

(a) Import permits under this subpart may be denied, revoked, suspended or revised without prior notice whenever the Director finds the proposed importation to be inconsistent with the purpose or in violation of section 38, Arms Export Control Act of 1976 or the regulations in this part.

(b) Whenever, after appropriate consideration, a permit application is denied or an outstanding permit is revoked, suspended, or revised, the applicant or permittee shall be promptly advised in writing of the Director's decision and the reasons therefor.

(c) Upon written request made within 30 days after receipt of an adverse decision, the applicant or permittee shall be accorded an opportunity to present additional information and to have a full review of his case by the Director.

(d) Unused, expired, suspended, or revoked permits must be returned immediately to the Director.

[T.D. ATF-8, 39 FR 3251, Jan. 25, 1974, as amended by T.D. ATF-215, 50 FR 42161, Oct. 18, 1985]

§ 47.45 Importation.

(a) Articles subject to the import permit procedures of this subpart imported into the United States may be released from Customs custody to the person authorized to import same upon his showing that he has a permit from the Director for the importation of the article or articles to be released. In obtaining the release from Customs custody of an article imported pursuant to permit, the permit holder shall prepare Form 6A, in duplicate, and furnish the original and copy to the Customs officer releasing the article. The Customs officer shall, after certification, forward the original Form 6A to the address specified on the form.

(b) Within 15 days of the date of their release from Customs custody, the importer of the articles released shall forward to the address specified on the form a copy of Form 6A on which shall be reported any error or discrepancy appearing on the Form 6A certified by Customs.

(Approved by the Office of Management and Budget under control number 1512-0019)

[T.D. ATF-215, 50 FR 42161, Oct. 18, 1985]

§ 47.46 Articles in transit.

Articles subject to the import permit procedures of this

subpart which enter the United States for temporary deposit pending removal therefrom and such articles which are temporarily taken out of the United States for return thereto shall be regarded as in transit and will be considered neither imported nor exported under this part. Such transactions are subject to the Intransit or Temporary Export License procedures of the Department of State (see 22 CFR Part 123).

[T.D. ATF-8, 39 FR 3251, Jan. 25, 1974, as amended by T.D. ATF-215, 50 FR 42161, Oct. 18, 1985]

Subpart F-Miscellaneous Provisions

§ 47.51 Import certification and delivery verification.

Pursuant to agreement with the United States, certain foreign countries are entitled to request certification of legality of importation of articles on the U.S. Munitions Import List. Upon request of a foreign government, the Director will certify the importation, on Form ITA-645P/ATF-4522/DSP53, for the U.S. importer. Normally, the U.S. importer will submit this form to the Director at the time he applies for an import permit. This document will serve as evidence to the government of the exporting company that the U.S. importer has complied with import regulations of the U.S. Government and is prohibited from diverting, transshipping, or reexporting the material described therein without the approval of the U.S. Government. Foreign governments may also require documentation attesting to the delivery of the material into the United States. When such delivery certification is requested by a foreign government, the U.S. importer may obtain directly from the U.S. District Director of Customs the authenticated Delivery Verification Certificate (U.S. Department of Commerce Form ITA-647P) for this purpose.

(Approved by the Office of Management and Budget under control number 0625-0064)

[T.D. ATF-215, 50 FR 42162, Oct. 18, 1985]

§ 47.52 Import restrictions applicable to certain countries.

(a) It is the policy of the United States to deny licenses and other approvals with respect to defense articles and defense services originating in certain countries or areas. This policy applies to Cuba, Iran, Iraq, Libya, Mongolia, North Korea, Sudan, Syria, Vietnam, and some of the States that comprised the former Soviet Union (Armenia, Azerbaijan, Belarus, and Tajikistan). This policy applies to countries or areas with respect to which the United States maintains an arms embargo (e.g., Burma, China, the Federal Republic of Yugoslavia (Serbia and Montenegro), Haiti, Liberia, Rwanda, Somalia, Sudan, UNITA (Angola), and Zaire). It also applies when an import would not be in furtherance of world peace and the security and foreign policy of the United States.

NOTE: Changes in foreign policy may result in additions to and deletions from the above list of countries. The ATF will publish changes to this list in the FEDERAL REGISTER. Contact the Firearms and Explosives Imports Branch at (202) 927-8320 for current information.

(b) Notwithstanding paragraph (a) of this section, the Director shall deny applications to import into the United States the following firearms and ammunition:

(1) Any firearm located or manufactured in Georgia, Kazakstan, Kyrgyzstan, Moldova, Russian Federation, Turkmenistan, Ukraine, or Uzbekistan, and any firearm previously manufactured in the Soviet Union, that is not one of the models listed below:

(i) Pistols/Revolvers:

(A) German Model P08 Pistol.
(B) IZH 34M, .22 caliber Target Pistol.
(C) IZH 35M, .22 caliber Target Pistol.
(D) Mauser Model 1896 Pistol.
(E) MC-57-1 Pistol.
(F) MC-1-5 Pistol.
(G) Polish Vis Model 35 Pistol.
(H) Soviet Nagant Revolver.
(I) TOZ 35, .22 caliber Target Pistol.

(ii) Rifles:

(A) BARS-4 Bolt Action Carbine.
(B) Biathlon Target Rifle, .22LR caliber.
(C) British Enfield Rifle.
(D) CM2, .22 caliber Target Rifle (also known as SM2, 22 caliber).
(E) German Model 98K Rifle.
(F) German Model G41 Rifle.
(G) German Model G43 Rifle.
(H) IZH-94.
(I) LOS-7 Bolt Action Rifle.
(J) MC-7-07.
(K) MC-18-3.
(L) MC-19-07.
(M) MC-105-01.
(N) MC-112-02.
(O) MC-113-02.
(P) MC-115-1.
(Q) MC-125/127.
(R) MC-126.
(S) MC-128.

(T) Saiga Rifle.
(U) Soviet Model 38 Carbine.
(V) Soviet Model 44 Carbine.
(W) Soviet Model 91/30 Rifle.
(X) TOZ 18, .22 caliber Bolt Action Rifle.
(Y) TOZ 55.
(Z) TOZ 78.
(AA) Ural Target Rifle, .22LR caliber.
(BB) VEPR Rifle.
(CC) Winchester Model 1895, Russian Model Rifle;

(2) Ammunition located or manufactured in Georgia, Kazakstan, Kyrgyzstan, Moldova, Russian Federation, Turkmenistan, Ukraine, or Uzbekistan, and ammunition previously manufactured in the Soviet Union, that is 7.62X25mm caliber (also known as 7.63X25mm caliber or .30 Mauser); or

(3) A type of firearm the manufacture of which began after February 9, 1996.

(c) The provisions of paragraph (b) of this section shall not affect the fulfillment of contracts with respect to firearms or ammunition entered or withdrawn from warehouse for consumption in the United States on or before February 9, 1996.

(d) A defense article authorized for importation under this part may not be shipped on a vessel, aircraft or other means or conveyance which is owned or operated by, or leased to or from, any of the countries or areas covered by paragraph (a) of this section.

(e) Applications for permits to import articles that were manufactured in, or have been in, a country or area proscribed under this section may be approved where the articles are covered by Category I(a) of the Import List (other than those subject to the provisions of 27 CFR Part 179), are importable as curios or relics under the provisions of 27 CFR 178.118, and meet the following criteria:

(1) The articles were manufactured in a proscribed country or area prior to the date, as established by the Department of State, the country or area became proscribed, or, were manufactured in a non-proscribed country or area; and

(2) The articles have been stored for the five year period immediately prior to importation in a non-proscribed country or area.

(f) Applicants desiring to import articles claimed to meet the criteria specified in paragraph (e) of this section shall explain, and certify to, how the firearms meet the criteria. The certification statement will be prepared in letter form, executed under the penalties of perjury, and submitted to the Director at the time application is made for an import permit. The certification statement must be accompanied by documentary information on the country or area of original manufacture and on the country or area of storage for the five year period immediately prior to importation. Such information may, for example, include a verifiable statement in the English language of a government official or any other person having knowledge of the date and place of manufacture and/or the place of storage; a warehouse receipt or other document which provides the required history of storage; and any other document that the applicant believes substantiates the place and date of manufacture and the place of storage. The Director, however, reserves the right to determine whether documentation is acceptable. Applicants shall, when required by the Director, furnish additional documentation as may be necessary to determine whether an import permit application should be approved.

[T.D. ATF-202, 50 FR 14382, Apr. 12, 1985, as amended by T.D. ATF-215, 50 FR 42162, Oct. 18, 1985; T.D. ATF-287, 54 FR 13681, Apr. 5, 1989; T.D. ATF-323, 57 FR 24189, June 8, 1992; T.D. ATF-349, 58 FR 47831, Sept. 13, 1993; T.D. ATF-367, 60 FR 47866, Sept. 15, 1995; T.D. ATF-393, 62 FR 61234, Nov. 17, 1997]

§ 47.53 Exemptions.

(a) The provisions of this part are not applicable to:

(1) Importations by the United States or any agency thereof;

(2) Importation of components for items being manufactured under contract for the Department of Defense; or

(3) Importation of articles (other than those which would be **"firearms"** as defined in 18 U.S.C. 921(a)(3) manufactured in foreign countries for persons in the United States pursuant to Department of State approval.

(b) Any person seeking to import articles on the U.S. Munitions Import List as exempt under paragraph (a)(2) or (3) of this section may obtain release of such articles from Customs custody by submitting, to the Customs officer with authority to release, a statement claiming the exemption accompanied by satisfactory proof of eligibility. Such proof may be in the form of a letter from the Department of Defense or State, as the case may be, confirming that the conditions of the exemption are met.

[T.D. ATF-8, 39 FR 3251, Jan. 25, 1974, as amended by T.D. ATF-215, 50 FR 42162, Oct. 18, 1985]

§ 47.54 Administrative procedures inapplicable.

The functions conferred under section 38, Arms Export Control Act of 1976, as amended, are excluded from the operation of Chapter 5, Title 5, United States Code, with respect to Rule Making and Adjudication, 5 U.S.C. 553 and 554.

[T.D. ATF-8, 39 FR 3251, Jan. 25, 1974, as amended by T.D. ATF-215, 50 FR 42162, Oct. 18, 1985]

§ 47.55 Departments of State and Defense consulted.

The administration of the provisions of this part will be subject to the guidance of the Secretaries of State and Defense on matters affecting world peace and the external security and foreign policy of the United States.

§ 47.56 Authority of Customs officers.

(a) Officers of the U.S. Customs Service are authorized to take appropriate action to assure compliance with this part and with 27 CFR Parts 178 and 179 as to the importation or attempted importation of articles on the U.S. Munitions Import List, whether or not authorized by permit.

(b) Upon the presentation to him of a permit or written approval authorizing importation of articles on the U.S. Munitions Import List, the Customs officer who has authority to release same may require, in addition to such documents as may be required by Customs regulations, the production of other relevant documents relating to the proposed importation, including, but not limited to, invoices, orders, packing lists, shipping documents, correspondence, and instructions.

[T.D. ATF-8, 39 FR 3251, Jan. 25, 1974, as amended by T.D. ATF-215, 50 FR 42162, Oct. 18, 1985]

§ 47.57 U.S. military firearms.

(a) Notwithstanding any other provision of this part or of 27 CFR Part 178, no military firearms or ammunition of United States manufacture may be imported for sale in the United States (other than for the Armed Forces of the United States and its allies or for any State or local law enforcement agency) if such articles were furnished to a foreign government under a foreign assistance or sales program of the United States.

(1) The restrictions in paragraph (a) of this section covers firearms which are advanced in value or improved in condition in a foreign country, but it does not include those which have been substantially transformed as to become, in effect, articles of foreign manufacture.

(2) A person desiring to import into the United States military firearms and ammunition which were manufactured in the United States must submit a statement with the application for a permit certifying that the importation of such firearms or ammunition is not prohibited by the provisions of paragraph (a) of this section. The certification statement must be accompanied by documentary information on the original foreign source of the firearms or ammunition.

(b) Paragraph (a) of this section shall not apply if such firearms are curios or relics under 18 U.S.C. 925(e) and the person seeking to import such firearms provides a certification of a foreign government that the firearms were furnished to such government under a foreign assistance or sales program of the United States and that the firearms are owned by such foreign government. (See 27 CFR 178.118 providing for the importation of certain curio or relic handguns, rifles and shotguns.)

(c) For the purpose of this section, the term "military firearms and ammunition" includes all firearms and ammunition furnished to foreign governments under a foreign assistance or sales program of the United States as set forth in paragraph (a) of this section.

(Approved by the Office of Management and Budget under OMB Control No. 1512-0017)

[T.D. ATF-287, 54 FR 13681, Apr. 5, 1989, as amended by T.D. ATF-393, 62 FR 61235, Nov. 17, 1997]

Subpart G-Penalties, Seizures and Forfeitures

§ 47.61 Unlawful importation.

Any person who willfully:

(a) Imports articles on the U.S. Munitions Import List without a permit;

(b) Engages in the business of importing articles on the U.S. Munitions Import List without registering under this part; or

(c) Otherwise violates any provisions of this part;

Shall upon conviction be fined not more than $1,000,000 or imprisoned not more than 10 years, or both.

[T.D. ATF-8, 39 FR 3251, Jan. 25, 1974, as amended at 39 FR 4760, Feb. 7, 1974; T.D. ATF-215, 50 FR 42162, Oct. 18, 1985;

T.D. ATF-287, 54 FR 13681, Apr. 5, 1989]

§ 47.62 False statements or concealment of facts.

Any person who willfully, in a registration or permit application, makes any untrue statement of a material fact or fails to state a material fact required to be stated therein or necessary to make the statements therein not misleading, shall upon conviction be fined not more than $1,000,000, or imprisoned not more than 10 years, or both.

[T.D. ATF-8, 39 FR 3251, Jan. 25, 1974, as amended by T.D. ATF-215, 50 FR 42162, Oct. 18, 1985; T.D. ATF-287, 54 FR 13681, Apr. 5, 1989]

§ 47.63 Seizure and forfeiture.

Whoever knowingly imports into the United States contrary to law any article on the U.S. Munitions Import List; or receives, conceals, buys, sells, or in any manner facilitates its transportation, concealment, or sale after importation, knowing the same to have been imported contrary to law, shall be fined not more than $10,000 or imprisoned not more than 5 years, or both; and the merchandise so imported, or the value thereof shall be forfeited to the United States.

(18 U.S.C. 545)

[T.D. ATF-8, 39 FR 3251, Jan. 25, 1974, as amended by T.D. ATF-215, 50 FR 42162, Oct. 18, 1985]

PART 178-COMMERCE IN FIREARMS AND AMMUNITION

Subpart A-Introduction

Sec.

Subpart B-Definitions

Subpart C-Administrative and Miscellaneous Provisions

Subpart D-Licenses

Subpart I-Exemptions, Seizures, and Forfeitures

Subpart J-[Reserved]

Subpart K-Exportation

AUTHORITY: 5 U.S.C. 552(a); 18 U.S.C. 847, 921-930; 44 U.S.C. 3504(h).

SOURCE: 33 FR 18555, Dec. 14, 1968, unless otherwise noted. Redesignated at 40 FR 16835, Apr. 15, 1975.

Subpart A-Introduction

§ 178.1 Scope of regulations.

(a) **General.** The regulations contained in this part relate to commerce in firearms and ammunition and are promulgated to implement Title I, State Firearms Control Assistance (18 U.S.C. Chapter 44), of the Gun Control Act of 1968 (8 Stat. 1213) as amended by Pub. L. 99-308 (100 Stat. 449), Pub. L. 99-360 (10 Stat. 766), Pub. L. 99-408 (10 Stat. 920), Pub. L. 103-159 (10 Stat. 1536), Pub. L. 103-322 (10 Stat. 1796), and Pub.L. 104-208 0(11 Stat. 3009).

(b) **Procedural and substantive requirements.** This part contains the procedural and substantive requirements relative to:

(1) The interstate or foreign commerce in firearms and ammunition;

(2) The licensing of manufacturers and importers of firearms and ammunition, collectors of firearms, and dealers in firearms;

(3) The conduct of business or activity by licensees;

(4) The importation of firearms and ammunition;

(5) The records and reports required of licensees;

(6) Relief from disabilities under this part;

(7) Exempt interstate and foreign commerce in firearms and ammunition; and

(8) Restrictions on armor piercing ammunition.

[T.D. ATF-270, 53 FR 10490, Mar. 31, 1988, as amended by T.D. ATF-354, 59 FR 7112, Feb. 14, 1994; T.D. ATF-363, 60 FR 17450, Apr. 6, 1995; T.D. ATF-401, 63 FR 35520, June 30, 1998]

§ 178.2 Relation to other provisions of law.

The provisions in this part are in addition to, and are not in lieu of, any other provision of law, or regulations, respecting commerce in firearms or ammunition. For regulations applicable to traffic in machine guns, destructive devices, and certain other firearms, see Part 179 of this chapter. For statutes applicable to the registration and licensing of persons engaged in the business of manufacturing, importing or exporting arms, ammunition, or implements of war, see section 38 of the Arms Export Control Act (22 U.S.C. 2778) and regulations thereunder and Part 47 of this chapter. For statutes applicable t nonmailable firearms, see 18 U.S.C. 1715 and regulations thereunder.

[T.D. ATF-270, 53 FR 10490, Mar. 31, 1988]

Subpart B-Definitions

§ 178.11 Meaning of terms.

When used in this part and in forms prescribed under this part, where not otherwise distinctly expressed or manifestly incompatible with the intent thereof, terms shall have the meanings ascribed in this section. Words in the plural form shall include the singular, and vice versa, and words importing the masculine gender shall include the feminine. The terms " **includes**" and " **including**" do not exclude other things not enumerated which are in the same general class or are otherwise within the scope thereof.

Act. 18 U.S.C. Chapter 44.

Adjudicated as a mental defective. (a) A determination by a court, board, commission, or other lawful authority that a person, as a result of marked subnormal intelligence, or mental illness incompetency, condition, or disease:

(1) Is a danger to himself or to others; or

(2) Lacks the mental capacity to contracteor manag his own affairs.

(b) The term shall include—

(1) A finding of insanity by a court in a criminal case; and

(2) Those persons found incompetent to stand trial or found not guilty by reason of lack of mental responsibility pursuant to articles 50a and 72b of the Uniform Code of Military Justice, 10 U.S.C. 850a, 876b.

Alien illegally or unlawfully in the United States. Aliens who are unlawfully in the United States are not in valid immigrant, nonimmigrant or parole status. The term includes any alien—

(a) Who unlawfully entered the United States without inspection and authorization by an immigration officer and who has not been paroled into the United States under section 212(d)(5) of the Immigration and Nationality Act (INA);

(b) Who is a nonimmigrant and whose authorized period of stay has expired or who has violated the terms of the nonimmigrant category in which he or she was admitted;

(c) Paroled under INA section 212(d)(5) whose authorized period of parole has expired or whose parole status has been terminated; or

(d) Under an order of deportation, exclusion, or removal, or under an order to depart the United States voluntarily, whether or not he or she has left the United States.

Ammunition. Ammunition or cartridge cases, primers, bullets, or propellent powder designed for use in any firearm other than an antique firearm. The term shall not includ **(a)** any shotgun shot or pellet not designed for use as the single, complete projectile load for one shotgun hull or casing; no **(b)** any unloaded, non-metallic shotgun hull or casing not having a primer.

Antique firearm. **(a)** Any firearm (including any firearm with a matchlock, flintlock, percussion cap, or similar type of ignition system) manufactured in or before 1898; an **(b)** any replica of any firearm described in paragraph (a) of this definition if such replic **(1)** is not designed or redesigned for usin rimfire or conventidna centerfire fixed ammunition, ro **(2)** uses rimfire or conventidna centerfire fixed ammunition which is no longer manufactured in the United States and which is not readily available in the ordinary channels of commercial trade.

Armor piercing ammunition. Projectiles or projectile cores which may be used in a handgun and which are constructed entirely (excluding the presence of traces of other substances) from one or a combination of tungsten alloys, steel, iron, brass, bronze, beryllium copper, or depleted uranium; or full jacketed projectiles larger than .22 caliber designed and intended for use in a handgun and whose jacket has a weight of more than 25 percent of the total weight of the projectile The term does not include shotgun shot required by Federal or State environmental or game regulations for hunting purposes, frangible projectiles designed for target shooting, projectiles which the Director finds are primarily intended to be used for sporting purposes, or any other projectiles or projectile cores which the Director finds are intended to be used for industrial purposes, including charges used in oil and gas well perforating devices.

ATF officer. An officer or employee of the Bureau of Alcohol, Tobacco and Firearms (ATF) authorized to perform any function relating to the administration or enforcement of this part.

Business premises. The property on which the manufacturing or importing of firearms or ammunition or the dealing in firearms is or will be conducted. A private dwelling, no part of which is open to the public, shall not be recognized as coming within the meaning of the term.

Chief, National Licensing Center. The ATF official responsible for the issuance and renewal of licenses under this part.

Collector. Any person who acquires, holds, or disposes of firearms as curios or relics.

Collection premises. The premises described on the license of a collector as the location at which he maintains his collection of curios and relics.

Commerce. Travel, trade, traffic, commerce, transportation, or communication among the several States, or between the District of Columbia and any State, or between any foreign country or any territory or possession and any State or the District of Columbia, or between points in the same State but through any other State or the District of Columbia or a foreign country.

Committed to a mental institution. A formal commitment of a person to a mental institution by a court, board, commission, or other lawful authority. The term includes a commitment to a mental institution involuntarily. The term includes a commitment for mental defectiveness or mental illness. It also includes commitments for othe reasons, such as for drug use. The term does not include a person in a mental institution for observation or a voluntary admission to a mental institution.

Controlled substance. A drug or other substance, or immediate precursor, as defined in section 102 of the Controlled Substances Act, 21 U.S.C. 802. The term includes, but is not limited to marijuana, depressants, stimulants, and narcotic drugs. The term does not include distilled spirits, wine, malt beverages, or tobacco, as those terms are defined or used in Subtitle E of the Internal Revenue Code of 1986, as amended.

Crime punishable by imprisonment for a term exceeding 1 year. Any Federal, State or foreign offense for which the maximum penalty, whether or not imposed, is capital punishment or imprisonment in excess of 1 year. The term shall not includ **(a)** any Federal or State offenses pertaining to antitrust violations, unfair trade practices, restraints of trade, or other similar offenses relating to the regulation of business practices or **(b)** any State offense classified by the laws of the State as a misdemeanor and punishable by a term of imprisonment of 2 years or less. What constitutes a conviction of such a crime shall be determined in accordance with the law of the jurisdiction in which the proceedings were held Any conviction which has been expunged or set aside or for which a person has been pardoned or has had civil rights restored shall not be considered a conviction for the purposes of the Act or this part, unless such pardon, expunction, or restoration of civil rights expressly provides that the person may not ship, transport, possess, or receive firearms, or unless the person is prohibited by the law of the jurisdiction in which the proceedings were held from receiving or possessing any firearms.

Curios or relics. Firearms which are of special interest to collectors by reason of some quality other than is associated with firearms intended for sporting use or as offensive or defensive weapons. To be recognized as curios or relics, firearms must fall within one of the following categories:

(a) Firearms which were manufactured at least 50 years prior to the current date, but not including replicas thereof;

(b) Firearms which are certified by the curator of a municipal, State, or Federal museum which exhibits firearms to be curios or relics of museum interest; and

(c) Any other firearms which derive a substantial part of their monetary value from the fact that they are novel, rare, bizarre, or because of their association with some historical figure, period, or event. Proof of qualification of a particular firearm under this category may be established by evidence of present value and evidence that like firearms are not available except as collector's items, or that the value of like firearms available in ordinary commercial channels is substantially less.

Editor's Note

ATF Publication 5300.11, **Firearms Curios and Relics List**, consists of lists of those firearms determined to be curios or relics from 1972 to the present.

Customs officer. Any officer of the Customs Service or any commissioned, warrant, or petty officer of the Coast Guard, or any agent or other person authorized by law or designated by the Secretary of the Treasury to perform any duties of an officer of the Customs Service.

Dealer. Any person engaged in the business of selling firearms at wholesale or retail; any person engaged in the business of repairing firearms or of making or fitting special barrels, stocks, or trigger mechanisms to firearms; or any person who is a pawnbroker. The term shall include any person who engages in such business or occupation on a part-time basis.

Destructive device. **(a)** Any explosive, incendiary, or poison gas **(1)** bomb, (**2**) grenade, **(3)** rocket having a propellant charge of more than 4 ounces, **(4)** missile having an explosive or incendiary charge of more than one-quarter ounce **(5)** mine, ro **(6)** device similar to any of the devices described in the preceding paragraphs of this definition **(b)** any type of weapon (other than a shotgun or a shotgun shell which the Director finds is generally recognized as particularly suitable for sporting purposes) by whatever name known which will, or which may be readily converted to, expel a projectile by the action of an explosive or other propellant, and which has any barrel with a bore of more than one-half inch in diameter; and **(c)** any combination of parts either designed or intended for use in converting any device into any destructive device described in paragraph (a) or (b) of this section and from which a destructive device may be readily assembled The term shall not include any device which is neither designed nor redesigned for use as a weapon; any device, although originally designed for use as a weapon, which is redesigned for use as a signalling, pyrotechnic, line throwing, safety, or similar device; surplus ordnance sold, loaned, or given by the Secretary of the Army pursuant to the provisions of section 4684(2), 4685, or 4686 of title 10, United States Code; or any other device which the Director finds is not likely to be used as a weapon, is an antique, or is a rifle which the owner intends to use solely for sporting, recreational, or cultural purposes.

Director. The Director, Bureau of Alcohol, Tobacco and Firearms, the Department of the Treasury, Washington, DC.

Discharged under dishonorable conditions. Separation from the U.S. Armed Forces resulting from a dishonorable discharge or dismissal adjudged by a general court-martial. The term does not include separation from the Armed Forces resulting from any other discharge, e.g., a bad conduct discharge.

Engaged in the business - **(a) Manufacturer of firearms.** A person who devotes time, attention, and labor to manufacturing firearms as a regular course of trade or business with the principal objective of livelihood and profit through the sale or distribution of the firearms manufactured;

(b) Manufacturer of ammunition. A person who devotes time, attention, and labor to manufacturing ammunition as a regular course of trade or business with the principal objective of livelihood and profit through the sale or distribution of the ammunition manufactured;

(c) Dealer in firearms other than a gunsmith or a pawnbroker. A person who devotes time, attention, and labor to dealing in firearms as a regular course of trade or business with the principal objective of livelihood and profit through the repetitive purchase and resale of firearms, but such a term shall not include a person who makes occasional sales, exchanges, or purchases of firearms for the enhancement of a personal collection or for a hobby, or who sells all or part of his personal collection of firearms;

(d) Gunsmith. A person who devotes time, attention, and labor to engaging in such activity as a regular course of trade or business with the principal objective of livelihood and profit, but such a term shall not include a person who makes occasional repairs of firearms or who occasionally fits special barrels, stocks, or trigger mechanisms to firearms;

(e) Importer of firearms. A person who devotes time, attention, and labor to importing firearms as a regular course of trade or business with the principal objective of livelihood and profit through the sale or distribution of the firearms imported; and,

(f) Importer of ammunition. A person who devotes time, attention, and labor to importing ammunition as a regular course of trade or business with the principal objective of livelihood and profit through the sale or distribution of the ammunition imported.

Executed under penalties of perjury. Signed with the prescribed declaration under the penalties of perjury as provided on or with respect to the retur form, or other document or, where no form of declaration is prescribed, with the declaration:

"I declare under the penalties of perjury that this - (insert type of document, such as, statement, application, request, certificate), including the documents submitted in support thereof, has been examined by me and, to the best of my knowledge and belief, is true, correct, and complete."

Federal Firearms Act. 15 U.S.C. Chapter 18.

Firearm. Any weapon, including a starter gun, which will or is designed to or may readily be converted to expel a projectile by the action of an explosive; the frame or receiver of any such weapon; any firearm muffler or firearm silencer; or any destructive device; but the term shall not include an antique firearm. In the case of a licensed collector, the term shall mean only curios and relics.

Firearm frame or receiver. That part of a firearm which provides housing for the hammer, bolt o breechblock, and firing mechanism, and which is usually threaded at its forward portion to receive the barrel.

Firearm muffler or firearm silencer. Any device for silencing, muffling, or diminishing the report of a portable firearm, including any combination of parts, designed or redesigned, and intended for use in assembling or fabricating a firearm silencer or firearm muffler, and any part intended only for use in such assembly or fabrication.

Fugitive from justice. Any person who has fled from any State to avoid prosecution for a felony or a misdemeanor; or any person who leaves the State to avoid giving testimony in any criminal proceeding. The term also includes any person who knows that misdemeanor or felony charges are pending against such person and who leaves the State of prosecution.

Handgun. **(a)** Any firearm which has a short stock and is designed to be held and fired by the use of a single hand; and

(b) Any combination of parts from which a firearm described in paragraph (a) can be assembled.

Identification document. A document containing the name, residence address, date of birth, and photograph of the holder and which was made or issued by or under the authority of the United States Government, a State, political subdivision of a State, a foreign government, political subdivision of a foreign government, an international governmental or an international quasi-governmental organization which, when completed with information concerning a particular individual, is of a type intended or commonly accepted for the purpose of identification of individuals.

Importation. The bringing of a firearm or ammunition into the United States; except that the bringing of a firearm or ammunition from outside the United States into a foreign-trade zone for storage pending shipment to a foreign country or subsequent importation into this country, pursuant to this part, shall not be deemed importation.

Importer. Any person engaged in the business of importing or bringing firearms or ammunition into the United States. The term shall include any person who engages in such business on a part-time basis.

Indictment. Includes an indictment or information in any court, under which a crime punishable by imprisonment for a term exceeding 1 year (as defined in this section) may be prosecuted, or in military cases to any offense punishable by imprisonment for a term exceeding 1 year which has been referred to a general court-martial. An information is a formal accusation of a crime, differing from an indittment in tha it is made by a prosecuting attorney and not a grand jury.

Interstate or foreign commerce. Includes commerce between any place in a State and any place outside of that State, or within any possession of the United States (not including the Canal Zone) or the District of Columbia. The term shall not include commerce between places within the same State but through any place outside of that State.

Intimate partner. With respect to a person, the spouse of the person, a former spouse of the person, an individual who is a parent of a child of the person, and an individual wh cohabitates orsha cohabitated with the person.

Large capacity ammunition feeding device. A magazine, belt, drum, feed strip, or similar device for a firearm manufactured after September 13, 1994, that has a capacity of, or that can be readily restored or converted to accept, more than 10 rounds of ammunition. The term does not include an attached tubular device designed to accept, and capable of operating only with, .22r calibe rimfire ammunition, or a fixed device for a manually operated firearm, or a fixed device for a firearm listed in 18 U.S.C. 922, Appendix A.

Licensed collector. A collector of curios and relics only and licensed under the provisions of this part.

Licensed dealer. A dealer licensed under the provisions of this part.

Licensed importer. An importer licensed under the provisions of this part.

Licensed manufacturer. A manufacturer licensed under the provisions of this part.

Machine gun. Any weapon which shoots, is designed to shoot, or can be readily restored to shoot, automatically more than one shot, without manual reloading, by a single function of the trigger. The term shall also include the frame or receiver of any such weapon, any part designed and intended solely and exclusively, or combination of parts designed and intended, for use in converting a weapon into a machine gun, and any combination of parts from which a machine gun can be assembled if such parts are in the possession or under the control of a person.

Manufacturer. Any person engaged in the business of manufacturing firearms or ammunition. The term shall include any person who engages in such business on a part-time basis.

Mental institution. Includes mental health facilities, mental hospitals, sanitariums, psychiatric facilities, and other facilities that provide diagnoses by licensed professionals of mental retardation or mental illness, including a psychiatric ward in a general hospital.

Misdemeanor crime of domestic violence. (a) Is a Federal, State or local offense that:

(1) Is a misdemeanor under Federal or State law or, in States which do not classify offenses as misdemeanors, is an offense punishable by imprisonment for a term of one year or less, and includes offenses that are punishable only by a fine. (This is true whether or not the State statute specifically defines the offense as a "misdemeanor" or as a "misdemeanor crime of domestic violence." The term includes all such misdemeanor convictions in Indian Courts established pursuant to 25 CFR part 11.);

(2) Has, as an element, the use or attempted use of physical force (e.g., assault and battery), or the threatened use of a deadly weapon; and

(3) Was committed by a current or former spouse, parent, or guardian of the victim, by a person with whom the victim shares a child in common, by a person who is cohabiting with or has cohabited with the victim as a spouse, parent, or guardian, (e.g., the equivalent of a "common law" marriage even if such relationship is not recognized under the law), or a person similarly situated to a spouse, parent, or guardian of the victim (e.g., two persons who are residing at the same location in an intimate relationship with the intent to make that place their home would be similarly situated to a spouse).

(b) A person shall not be considered to have been convicted of such an offense for purposes of this part unless:

(1) The person is considered to have been convicted by the jurisdiction in which the proceedings were held.

(2) The person was represented by counsel in the case, or knowingly and intelligently waived the right to counsel in the case; and

(3) In the case of a prosecution for which a person was entitled to a jury trial in the jurisdiction in which the case was tried, either

(i) The case was tried by a jury, or

(ii) The person knowingly and intelligently waived the right to have the case tried by a jury, by guilty plea or otherwise.

(c) A person shall not be considered to have been convicted of such an offense for purposes of this part if the conviction has been expunged or set aside, or is an offense for which the person has been pardoned or has had civil rights restored (if the law of the jurisdiction in which the proceedings were held provides for the loss of civil rights upon conviction for such an offense) unless the pardon, expunction, or restoration of civil rights expressly provides that the person may not ship, transport, possess, or receive firearms, and the person is not otherwise prohibited by the law of the jurisdiction in which the proceedings were held from receiving or possessing any firearms.

National Firearms Act. 26 U.S.C. Chapter 53.

NICS. The National Instant Criminal Background Check System established by the Attorney General pursuant to 18 U.S.C. 922(t).

Pawnbroker. Any person whose business or occupation includes the taking or receiving, by way of pledge or pawn, of any firearm as security for the payment or repayment of money. The term shall include any person who engages in such business on a part-time basis.

Permanently inoperable. A firearm which is incapable of discharging a shot by means of an explosive and incapable of being readily restored to a firing condition. An acceptable method of rendering most firearms permanently inoperable is to fusion weld the chamber closed and fusion weld the barrel solidly to the frame. Certain unusual firearms require other methods to render the firearm permanently inoperable. Contact ATF for instructions.

Person. Any individual, corporation, company, association, firm, partnership, society, or joint stock company.

Pistol. A weapon originally designed, made, and intended to fire a projectile (bullet) from one or more barrels when held in one hand, and having **(a)** a chamber(s) as an integral part(s) of, or permanently aligned with, the bore(s); an **(b)** a short stock designed to be gripped by one hand and at an angle to and extending below the line of the bore(s).

Principal objective of livelihood and profit. The intent underlying the sale or disposition of firearms is predominantly one of obtaining livelihood and pecuniary gain, as opposed to other intents such as improving or liquidating a personal firearms collection: **Provided,** That proof of profit shall not be required as to a person who engages in the regular and repetitive purchase and disposition of firearms for criminal purposes or terrorism. For purposes of this part, the term "terrorism" means activity, directed against United States persons, which -

(a) Is committed by an individual who is not a national or permanent resident alien of the United States;

(b) Involves violent acts or acts dangerous to human life which would be a criminal violation if committed within the jurisdiction of the United States; and

(c) Is intended -

(1) To intimidate or coerce a civilian population;

(2) To influence the policy of a government by intimidation or coercion; or

(3) To affect the conduct of a government by assassination or kidnapping.

Published ordinance. A published law of any political subdivision of a State which the Director determines to be relevant to the enforcement of this part and which is contained on a list compiled by the Director, which list is incorporated by reference in the FEDERAL REGISTER, revised annually, and furnished to licensees under this part.

Division. A Bureau of Alcohol, Tobacco and Firearms Region.

Director of Industry Operations. The principal ATF regional official responsible for administering regulations in this part.

Renounced U.S. citizenship. (a) A person has renounced his U.S. citizenship if the person, having been a citizen of the United States, has renounced citizenship either —

(1) Before a diplomatic or consular officer of the United States in a foreign state pursuant to 8 U.S.C. 1481(a)(5); or

(2) Before an officer designated by the Attorney General when the United States is in a state of war pursuant to 8 U.S.C. 1481(a)(6).

(b) The term shall not include any renunciation of citizenship that has been reversed as a result of administrative or judicial appeal.

Revolver. A projectile weapon, of the pistol type, having a breechloading chambered cylinder so arranged that the cocking of the hammer or movement of the trigger rotates it and brings the next cartridge in line with the barrel for firing.

Rifle. A weapon designed or redesigned, made or remade, and intended to be fired from the shoulder, and designed or redesigned and made or remade to use the energy of the explosive in a fixed metallic cartridge to fire only a single projectile through a rifled bore for each single pull of the trigger.

Semiautomatic assault weapon.

(a) Any of the firearms, or copies or duplicates of the firearms in any caliber, known as:

(1) Norinco, Mitchell, an Poly Technologie Avtomat Kalashnikovs (all models),

(2) Action Arms Israeli Military Industries UZI an Galil,

(3) Beretta Ar70 (SC-70),

(4) Colt AR-15,

(5) Fabrique National FN/FAL, FN/LAR, and FNC,

(6) SWD M-10, M-11, M-11/9, and M-12,

(7) Steyr AUG,

(8) INTRATEC TEC-9, TEC-DC9 and TEC-22, and

(9) Revolving cylinder shotguns, such as (or similar to) the Street Sweeper and Striker 12;

Editor's Note

"Norinco"refers only to AK 47 rifles or variations thereof.

(b) A semiautomatic rifle that has an ability to accept a detachable magazine and has at least 2 of -

(1) A folding or telescopin stock,

(2) A pistol grip that protrudes conspicuously beneath the action of the weapon,

(3) A bayonet mount,

(4) A flash suppressor or threaded barrel designed to accommodate a flash suppressor, and

(5) A grenade launcher;

(c) A semiautomatic pistol that has an ability to accept a detachable magazine and has at least 2 of -

(1) An ammunition magazine that attaches to the pistol outside of the pistol grip,

(2) A threaded barrel capable of accepting a barrel extender, flash suppressor, forward handgrip, or silencer,

(3) A shroud that is attached to, or partially or completely encircles, the barrel and that permits the shooter to hold the firearm with th nontrigger hand without being burned,

(4) A manufactured weight of 50 ounces or more when the pistol is unloaded, and

(5) A semiautomatic version of an automatic firearm; and

(d) A semiautomatic shotgun that has at least 2 of -

(1) A folding or telescopin stock,

(2) A pistol grip that protrudes conspicuously beneath the action of the weapon,

(3) A fixed magazine capacity in excess of 5 rounds, and

(4) An ability to accept a detachable magazine.

Semiautomatic pistol. Any repeating pistol which utilizes a portion of the energy of a firing cartridge to extract the fired cartridge case and chamber the next round, and which requires a separate pull of the trigger to fire each cartridge.

Semiautomatic rifle. Any repeating rifle which utilizes a portion of the energy of a firing cartridge to extract the fired cartridge case and chamber the next round, and which requires a separate pull of the trigger to fire each cartridge.

Semiautomatic shotgun. Any repeating shotgun which utilizes a portion of the energy of a firing cartridge to extract the fired cartridge case and chamber the next round, and which requires a separate pull of the trigger to fire each cartridge.

Short-barreled rifle. A rifle having one or more barrels less than 16 inches in length, and any weapon made from a rifle, whether by alteration, modification, or otherwise, if such weapon, as modified, has an overall length of less than 26 inches.

Short-barreled shotgun. A shotgun having one or more barrels less than 18 inches in length, and any weapon made from a shotgun, whether by alteration, modification, or otherwise, if such weapon as modified has an overall length of less than 26 inches.

Shotgun. A weapon designed or redesigned, made or remade, and intended to be fired from the shoulder, and designed or redesigned and made or remade to use the energy of the explosive in a fixed shotgun shell to fire through a smooth bore either a number of ball shot or a single projectile for each single pull of the trigger.

State. A State of the United States. The term shall include the District of Columbia, the Commonwealth of Puerto Rico, and the possessions of the United States (not including the Canal Zone).

State of residence. The State in which an individual resides. An individual resides in a State if he or she is present in a State with the intention of making a home in that State. If an individual is on active duty as a member of the Armed Forces, the individual's State of residence is the State in which his or her permanent duty station is located An alien who is legally in the United States shall be considered to be a resident of a State only if the alien is residing in the State and has resided in the State for a period of at least 90 days prior to the date of sale or delivery of a firearm. The following are examples that illustrate this definition:

Example 1. A maintains a home in State X. A travels to State Y on a hunting, fishing, business, or other type of trip. A does not become a resident of State Y by reason of such trip.

Example 2. A is a U.S. citizen and maintains a home in State X and a home in State Y. A resides in State X except for weekends or the summer months of the year and in State Y for the weekends or the summer months of the year. During the time that A actually resides in State X, A is a resident of State X, and during the time that A actually resides in State Y, A is a resident of State Y.

Example 3. A, an alien, travels on vacation or on a business trip to State X. Regardless of the length of time A spends in State X, A does not have a State of residence in State X. This is because A does not have a home in State X at which he has resided for at least 90 days.

Unlawful user of or addicted to any controlled substance. A person who uses a controlled substance and has lost the power of self-control with reference to the use of controlled substance; and any person who is a current user of a controlled substance in a manner other than as prescribed by a licensed physician. Such use is not limited to the use of drugs on a particular day, or within a matter of days or weeks before, but rather than the unlawful use has occurred recently enough to indicate that the individual is actively engaged in such conduct. A person may be an unlawful current user of a controlled substance even though the substance is not being used at the precise time the person seeks to acquire a firearm or receives or possesses a. firearm An inference of current use may be drawn from evidence of a recent use or possession of a controlled substance or a pattern of use or possession that reasonably covers the present time, e.g., a conviction for use or possession of a controlled substance within the past year; multiple arrests for such offenses within the past 5 years if the most recent arrest occurred within the past year; or persons found through a drug test to use a controlled substance unlawfully, provided that the test was administered within the past year. For a current or former member of the Armed Forces, an inference of current use may be drawn from recent disciplinary or other administrative action based on confirmed drug use, e.g., court-martial conviction, nonjudicial punishment, or an administrative discharge based on drug use or drug rehabilitation failure.

Unserviceable firearm. A firearm which is incapable of discharging a shot by means of an explosive and is incapable of being readily restored to a firing condition.

U.S.C. The United States Code.

(5 U.S.C. 522(a), 8 Stat. 383, as amended; 18 U.S.C4 847 (8 Stat. 959); 18 U.S.C. 92ℓ (8 Stat. 1226))

[T.D. ATF-48, 43 FR 13536, Mar. 31 1978; 44FR 55842, Sept. 28, 1979, as amended by T.D. ATF-135, 48 FR 24067, May 31, 1983; T.D. ATF-160, 48 FR 52909, Nov. 23, 1983; T.D. ATF-200, 50 FR 10498, Mar. 15, 1985; T.D. ATF-241, 51 FR 39615,.Oct 29, 1986; T.D. ATF-247, 52 FR 2049, Jan. 16, 1987; T.D. ATF290, 54 FR 53054, Dec. 27, 1989; T.D. ATF-313, 56 FR 32508, July 17, 1991; T.D. ATF-346, 58 FR 40589, July 29, 1993; T.D. ATF-354, 59 FR 7112, Feb.14, 1994; T.D. 361, 60 FR 10786, February 27, 1995; T.D. ATF-363, 60 FR 17450, April 6, 1995; T.D. ATF-389, 62 FR 19442, Apr. 21, 1997; T.D. ATF-391, 62 FR 34634, June 27, 1997; T.D. ATF-401, 63 FR 35520, June 30, 1998; T.D. ATF-415, 63 FR 58272, Oct. 29, 1998]

Subpart C-Administrative and Miscellaneous Provisions

§ 178.21 Forms prescribed.

(a) The Director is authorized to prescribe all forms required by this part. All of the information called for in each form shall be furnished as indicated by the headings on the form and the instructions on or pertaining to the form. In addition, information called for in each form shall be furnished as required by this part.

(b) Requests for forms should be mailed to the ATF Distribution Center, 943 Angus Street, Springfield, Virginia 22153.

(5 U.S.C. 552(a); 8 Stat. 383, as amended)

[T.D. ATF-92, 46 FR 46915, Sept. 23, 1981, as amended by T.D. ATF-249, 52 FR 5962, Feb. 27, 1987; T.D. ATF-270, 53 FR 10492, Mar. 31, 1988; T.D. 372, 61 FR 20724, May 8, 1996]

§ 178.22 Alternate methods or procedures; emergency variations from requirements.

(a) Alternate methods or procedures. The licensee, on specific approval by the Director as provided in this paragraph, may use an alternate method or procedure in lieu of a method or procedure specifically prescribed in this part. The Director may approve an alternate method or procedure, subject to stated conditions, when it is found that:

(1) Good cause is shown for the use of the alternate method or procedure;

(2) The alternate method or procedure is within the purpose of, and consistent with the effect intended by, the specifically prescribed method or procedure and that the alternate method or procedure is substantially equivalent to that specifically prescribed method or procedure; and

(3) The alternate method or procedure will not be contrary to any provision of law and will not result in an increase in cost to the Government or hinder the effective administration of this part. Where the licensee desires to employ an alternate method or procedure, a written application shall be submitted to the appropriate Director of Industry Operations, for transmittal to the Director. The application shall specifically describe the proposed alternate method or procedure and shall set forth the reasons for it. Alternate methods or procedures may not be employed until the application is approved by the Director. The licensee shall, during the period of authorization of an alternate method or procedure, comply with the terms of the approved application. Authorization of any alternate method or procedure may be withdrawn whenever, in the judgment of the Director, the effective administration of this part is hindered by the continuation of the authorization.

(b) **Emergency variations from requirements.** The Director may approve a method of operation other than as specified in this part, where it is found that an emergency exists and the proposed variation from the specified requirements are necessary and the proposed variations (1) will not hinder the effective administration of this part, and (2) will not be contrary to any provisions of law. Variations from requirements granted under this paragraph are conditioned on compliance with the procedures, conditions, and limitations set forth in the approval of the application. Failure to comply in good faith with the procedures, conditions, and limitations shall automatically terminate the authority for the variations, and the licensee shall fully comply with the prescribed requirements of regulations from which the variations were authorized. Authority for any variation may be withdrawn whenever, in the judgment of the Director, the effective administration of this part is hindered by the continuation of the variation. Where the licensee desires to employ an emergency variation, a written application shall be submitted to the appropriate regional director (compliance) for transmittal to the Director. The application shall describe the proposed variation and set forth the reasons for it. Variations may not be employed until the application is approved.

(c) **Retention of approved variations.** The licensee shall retain, as part of the licensee's records, available for examination by ATF officers, any application approved by the Director under this section.

[T.D. ATF-270, 53 FR 10492, Mar. 31, 1988]

§ 178.23 Right of entry and examination.

(a) Except as provided in paragraph (b), any ATF officer, when there is reasonable cause to believe a violation of the Act has occurred and that evidence of the violation may be found on the premises of any licensed manufacturer, licensed importer, licensed dealer, or licensed collector, may, upon demonstrating such cause before a Federal magistrate and obtaining from the magistrate a warrant authorizing entry, enter during business hours (or, in the case of a licensed collector, the hours of operation) the premises, including places of storage, of any such licensee for the purpose of inspecting or examining:

(1) Any records or documents required to be kept by such licensee under this part and

(2) Any inventory of firearms or ammunition kept or stored by any licensed manufacturer, licensed importer, or licensed dealer at such premises or any firearms curios or relics or ammunition kept or stored by

any licensed collector at such premises.

(b) Any ATF officer, without having reasonable cause to believe a violation of the Act has occurred or that evidence of the violation may be found and without demonstrating such cause before a Federal magistrate or obtaining from the magistrate a warrant authorizing entry, may enter during business hours the premises, including places of storage, of any licensed manufacturer, licensed importer, or licensed dealer for the purpose of inspecting or examining the records, documents, ammunition and firearms referred to in paragraph (a) of this section:

(1) In the course of a reasonable inquiry during the course of a criminal investigation of a person or persons other than the licensee,

(2) For insuring compliance with the recordkeeping requirements of this part:

(i) Not more than once during any 12-month period, or

(ii) At any time with respect to records relating to a firearm involved in a criminal investigation that is traced to the licensee, or

(3) When such inspection or examination may be required for determining the disposition of one or more particular firearms in the course of a bona fide criminal investigation.

(c) Any ATF officer, without having reasonable cause to believe a violation of the Act has occurred or that evidence of the violation may be found and without demonstrating such cause before a Federal magistrate or obtaining from the magistrate a warrant authorizing entry, may enter during hours of operation the premises, including places of storage, of any licensed collector for the purpose of inspecting or examining the records, documents, firearms, and ammunition referred to in paragraph (a) of this sectio (1) for ensuring compliance with the recordkeeping requirements of this part not more than once during any 12-month period or (2) when such inspection or examination may be required for determining the disposition of one or more particular firearms in the course of a bona fide criminal investigation. At the election of the licensed collector, the annual inspection permitted by this paragraph shall be performed at the ATF office responsible for conducting such inspection in closest proximity to th collectors premises.

(d) The inspections and examinations provided by this section do not authorize an ATF officer to seize any records or documents other than those records or documents constituting material evidence of a violation of law. If an ATF officer seizes such records or documents, copies shall be provided the licensee within a reasonable time.

[T.D. ATF-270, 53 FR 10492, Mar. 31, 1988; as amended by T.D. ATF-363, 60 FR 17450, Apr. 6, 1995]

§ 178.24 Compilation of State laws and published ordinances.

(a) The Director shall annually revise and furhish Federa firearms licensees with a compilation of State laws and published ordinances which are relevant to the enforcement of this part. The Director annually revises the compilation and publishes it a **"State Laws and Published Ordinances - Firearms"** which is furnished free of charge to licensees under this part. Where the compilation has previously been furnished to licensees, the Director need only furnish amendments of the relevant laws and ordinances to such licensees.

(b) **"State Laws and Published Ordinances - Firearms"** is incorporated by reference in this part. It is ATF Publication 5300.5, revised yearly. The current edition is available from the Superintendent of Documents, U.S. Government Printing Office, Washington, DC 20402. It is also available for inspection at the Office of the Federal Register, 800 North Capitol Street, NW., suite 700, Washington, DC This incorporation by reference was approved by the Director of the Federal Register.

[T.D. ATF-270, 53 FR 10493, Mar. 31, 1988]

§ 178.25 Disclosure of information.

The Director of Industry Operations may make available to any Federal, State or local law enforcement agency any information which is obtained by reason of the provisions of the Act with respect to the identification of persons prohibited from purchasing or receiving firearms or ammunition who have purchased or received firearms or ammunition, together with a description of such firearms or ammunition. Upon the request of any Federal, State or local law enforcement agency, the Director of Industry Operations may provide such agency any information contained in the records required to be maintained by the Act or this part.

[T.D. ATF-270, 53 FR 10493, Mar. 31, 1988]

§ 178.25a Responses to requests for information.

Each licensee shall respond immediately to, and in no event later than 24 hours after the receipt of, a request by an ATF officer at the National Tracing Center for information contained in the records required to be kept by this part for determining the disposition of one or more firearms in the course of a bona fide criminal investigation. The requested information shall be provided orally to the ATF officer within the 24-hour period. *Verification of the identity and employment of National Tracing Center personnel requesting information may be established at the time the requested information is provided by telephoning the toll-free number 1-800-788-7132 or using the toll-free facsimile (FAX) number 1-800-578-7223.

(Approved by the Office of Management and Budget under control number 1512-0387)

[T.D. ATF-363, 60 FR 17451, Apr. 6, 1996, as amended by T.D. ATF-396, 63 FR 12646, Mar. 16, 1998]

Editor's Note

The toll-free telephone number set forth in section 178.25a has been changed. The toll-free number that should be called to verify the identity of the person seeking firearms trace information is 1-800-788-7133.

§ 178.26 Curio and relic determination.

Any person who desires to obtain a determination whether a particular firearm is a curio or relic shall submit a written request, in duplicate, for a ruling thereon to the Director. Each such request shall be executed under the penalties of perjury and shall contain a complete and accurate description of the firearm, and such photographs, diagrams, or drawings as may be necessary to enable the Director to make a determination. The Director may require the submission of the firearm for examination and evaluation. If the submission of the firearm is impractical, the person requesting the determination shall so advise the Director and designate the place where the firearm will be available for examination and evaluation.

[T.D. ATF-270, 53 FR 10493, Mar. 31, 1988]

§ 178.27 Destructive device determination.

The Director shall determin in accordance with 18 U.S.C. 921(a)(4) whether a device is excluded from the definition of a destructive device. A person who desires to obtain a determination under that provision of law for any device which he believes is not likely to be used as a weapon shall submit a written request, in triplicate, for a ruling thereon to the Director Each such request shall be executed under the penalties of perjury and contain a complete and accurate description of the device, the name and address of the manufacturer or importer thereof, the purpose of and use for which it is intended, and such photographs, diagrams, or drawings as may be necessary to enable the Director to make his determination. The Director may require the submission to him, of a sample of such device for examination and evaluation. If the submission of such device is impracticable, the person requesting the ruling shall so advise the Director and designate the place where the device will be available for examination and evaluation.

§ 178.28 Transportation of destructive devices and certain firearms.

(a) The Director may authorize a person to transport in interstate or foreign commerce any destructive device, machine gun, short-barreled shotgun, or short-barreled rifle, if he finds that such transportation is reasonably necessary and is consistent with public safety and applicable State and local law. A person who desires to transport in interstate or foreign commerce any such device or weapon shall submit a written request so to do, in duplicate, to the Director. The request shall contain:

(1) A complete description and identification of the device or weapon to be transported;

(2) A statement whether such transportation involves a transfer of title;

(3) The need for such transportation;

(4) The approximate date such transportation is to take place;

(5) The present location of such device or weapon and the place to which it is to be transported;

(6) The mode of transportation to be used (including, if by common or contract carrier, the name and address of such carrier); and

(7) Evidence that the transportation or possession of such device or weapon is not inconsistent with the laws at the place of destination.

(b) No person shall transport any destructive device, machine gun, short-barreled shotgun, or short-barreled rifle in interstate or foreign commerce under the provisions of this section until he has received specific authorization so to do from the Director. Authorization granted under this section does not carry or import relief from any other statutory or regulatory provision relating to firearms.

(c) This section shall not be construed as requiring licensees to obtain authorization to transport destructive devices, machine guns, short-barreled shotguns, and short-barreled rifles in interstate or foreign commerce : **Provided,** That in the case of a licensed importer, licensed manufacturer, or licensed dealer, such a licensee is qualified under the National Firearms Act (see also Part 179 of this chapter) and this part to engage in the business with respect to the device or weapon to be transported, and that in the case of a licensed collector, the device or weapon to be transported is a curio or relic.

[33 FR 18555, Dec. 14,. 1968 Redesignated at 40 FR 16385, Apr. 15, 1975, and amended by T.D. ATF-138, 48 FR 35399, Aug. 4, 1983]

§ 178.29 Out-of-State acquisition of firearms by nonlicensees.

No person, other than a licensed importer, licensed manufacturer, licensed dealer, or licensed collector, shall transport into or receive in the State where the person resides (or if a corporation or other business entity, where it maintains a place of business) any firearm purchased or otherwise obtained by such person outside that State: **Provided**, That the provisions of this section:

(a) Shall not preclude any person who lawfully acquires a firearm by bequest or intestate succession in a State other than his State of residence from transporting the firearm into or receiving it in that State, if it is lawful for such person to purchase or possess such firearm in that State,

(b) Shall not apply to the transportation or receipt of a rifle or shotgun obtained from a licensed manufacturer, licensed importer, licensed dealer, or licensed collector in a State other than the transferee's State of residence in an over-the-counter transaction at the licensee's premises obtained in conformity with the provisions of § 178.96(c) and

(c) Shall not apply to the transportation or receipt of a firearm obtained in conformity with the provisions of §§ 178.30 and 178.97.

[T.D. ATF-270, 53 FR 10493, Mar. 31, 1988]

§ 178.29a Acquisition of firearms by nonresidents.

No person, other than a licensed importer, licensed manufacturer, licensed dealer, or licensed collector, who does not reside in any State shall receive any firearms unless such receipt is for lawful sporting proposes.

[T.D. ATF-363, 60 FR 17451, Apr. 6, 1995]

§ 178.30 Out-of-State disposition of firearms by nonlicensees.

No nonlicensee shall transfer, sell, trade, give, transport, or deliver any firearm to any other nonlicensee, who the transferor knows or has reasonable cause to believe does not reside in (or if the person is a corporation or other business entity, does not maintain a place of business in) the State in which the transferor resides: **Provided**, That the provisions of this section:

(a) shall not apply to the transfer, transportation, or delivery of a firearm made to carry out a bequest of a firearm to, or any acquisition by intestate succession of a firearm by, a person who is permitted to acquire or possess a firearm under the laws of the State of his residence; and

(b) shall not apply to the loan or rental of a firearm to any person for temporary use for lawful sporting purposes.

[T.D. ATF-313, 56 FR 32508, July 17, 1991; 57 FR 1205, Jan. 10, 1992]

§ 178.31 Delivery by common or contract carrier.

(a) No person shall knowingly deliver or cause to be delivered to any common or contract carrier for transportation or shipment in interstate or foreign commerce to any person other than a licensed importer, licensed manufacturer, licensed dealer, or licensed collector, any package or other container in which there is any firearm or ammunition without written notice to the carrier that such firearm or ammunition is being transported or shipped:

Provided, That any passenger who owns or legally possesses a firearm or ammunition being transported aboard any common or contract carrier for movement with the passenger in interstate or foreign commerce may deliver said firearm or ammunition into the custody of the pilot, captain, conductor or operator of such common or contract carrier for the duration of that trip without violating any provision of this part.

(b) No common or contract carrier shall require or cause any label, tag, or other written notice to be placed on the outside of any package, luggage, or other container indicating that such package, luggage, or other container contains a firearm.

(c) No common or contract carrier shall transport or deliver in interstate or foreign commerce any firearm or ammunition with knowledge or reasonable cause to believe that the shipment, transportation, or receipt thereof would be in violation of any provision of this part:

Provided, however, That the provisions of this paragraph shall not apply in respect to the transportation of firearms or ammunition in in-bond shipment under Customs laws and regulations.

(d) No common or contract carrier shall knowingly deliver in interstate or foreign commerce any firearm without obtaining written acknowledgement of receipt from the recipient of the package or other container in which there is a firearm: **Provided,** That this paragraph shall not apply with respect to the return of a firearm to a passenger who places firearms in the carrier's custody for the duration of the trip.

[33 FR 18555, Dec. 14,. 1968 Redesignated at 40 FR 16385, Apr. 15, 1975, and amended by T.D. ATF-354, 59 FR 7112, Feb. 14, 1994; T.D. ATF-361, 60 FR 10786, Feb. 27, 1995]

§ 178.32 Prohibited shipment, transportation, possession, or receipt of firearms and ammunition by certain persons.

(a) No person may ship or transport any firearm or ammunition in interstate or foreign commerce, or receive any firearm or ammunition which has been shipped or transported in interstate or foreign commerce, or possess any firearm or ammunition in or affecting commerce, who:

(1) Has been convicted in any court of a crime punishable by imprisonment for a term exceeding 1 year,

(2) Is a fugitive from justice,

(3) Is an unlawful user of or addicted to any controlled substance (as defined in section 102 of the Controlled Substances Act, 21 U.S.C. 802),

(4) Has been adjudicated as a mental defective or has been committed to a mental institution,

(5) Is an alien illegally or unlawfully in the United States,

(6) Has been discharged from the Armed Forces under dishonorable conditions,

(7) Having been a citizen of the United States, has renounced citizenship,

(8) Is subject to a court order that -

(i) Was issued after a hearing of which such person received actual notice, and at which such person had an opportunity to participate;

(ii) Restrains such person from harassing, stalking, or threatening an intimate partner of such person or child of such intimate partner or person, or engaging in other conduct that would place an intimate partner in reasonable fear of bodily injury to the partner or child; and

(iii)(A) Includes a finding that such person represents a credible threat to the physical safety of such intimate partner or child; or

(B) By its terms explicitly prohibits the use, attempted use, or threatened use of physical force against such intimate partner or child that would reasonably be expected to cause bodily injury, or

(9) Has been convicted of a misdemeanor crime of domestic violence.

(b) No person who is under indictment for a crime punishable by imprisonment for a term exceeding one year may ship or transport any firearm or ammunition in interstate or foreign commerce or receive any firearm or ammunition which has been shipped or transported in interstate or foreign commerce.

(c) Any individual, who to that individual's knowledge and while being employed by any person described in paragraph (a) of this section, may not in the course of such employment receive, possess, or transport any firearm or ammunition in commerce or affecting commerce or receive any firearm or ammunition which has been shipped or transported in interstate or foreign commerce.

(d) No person may sell or otherwise dispose of any firearm or ammunition to any person knowing or having reasonable cause to believe that such person:

(1) Is under indictment for, or has been convicted in any court of, a crime punishable by imprisonment for a term exceeding 1 year,

(2) Is a fugitive from justice,

(3) Is an unlawful user of or addicted to any controlled substance (as defined in section 102 of the Controlled Substances Act, 21 U.S.C. 802),

(4) Has been adjudicated as a mental defective or has been committed to a mental institution,

(5) Is an alien illegally or unlawfully in the United States,

(6) Has been discharged from the Armed Forces under dishonorable conditions,

(7) Having been a citizen of the United States, has renounced citizenship,

(8) Is subject to a court order that restrains such person from harassing, stalking, or threatening an intimate partner of such person or child of such intimate partner or person, or engaging in other conduct that would place an intimate partner in reasonable fear of bodily injury to the partner or child: **Provided,** That the provisions of this paragraph shall only apply to a court order that -

    (i) Was issued after a hearing of which such person received actual notice, and at which such person had the opportunity to participate; and

    (ii)(A) Includes a finding that such person represents a credible threat to the physical safety of such intimate partner or child; or

    (B) By its terms explicitly prohibits the use, attempted use, or threatened use of physical force against such intimate partner or child that would reasonably be expected to cause bodily injury, or

    (9) Has been convicted of a misdemeanor crime of domestic violence.

    (e) The actual notice required by paragraphs (a)(8)(i) and (d)(8)(i) of this section is notice expressly and actually given, and brought home to the party directly, including service of process personally served on the party and service by mail. Actual notice also includes proof of facts and circumstances that raise the inference that the party received notice including, but not limited to, proof that notice was left at the party's dwelling house or usual place of abode with some person of suitable age and discretion residing therein; or proof that the party signed a return receipt for a hearing notice which had been mailed to the party. It does not include notice published in a newspaper.

[T.D. ATF-270, 53 FR 10493, Mar. 31, 1988, T.D. as amended by ATF-363, 60 FR 17451, Apr. 6, 1995; T.D. ATF-391, 62 FR 34639, June 27, 1997; T.D. ATF-401, 63 FR 35520, June 30, 1998]

§ 178.33   Stolen firearms and ammunition.

    No person shall transport or ship in interstate or foreign commerce any stolen firearm or stolen ammunition knowing or having reasonable cause to believe that the firearm or ammunition was stolen, and no person shall receive, possess, conceal, store, barter, sell, or dispose of any stolen firearm or stolen ammunition, or pledge or accept as security for a loan any stolen firearm or stolen ammunition, which is moving as, which is a part of, which constitutes, or which has been shipped or transported in, interstate or foreign commerce, either before or after it was stolen, knowing or having reasonable cause to believe that the firearm or ammunition was stolen.

§ 178.33a   Theft of firearms.

    No person shall steal or unlawfully take or carry away from the person or the premises of a person who is licensed to engage in the business of importing, manufacturing, or dealing in firearms, any firearm in the licensee's business inventory that has been shipped or transported in interstate or foreign commerce.

[T.D. ATF-354, 59 FR 7112, Feb. 14, 1994]

§ 178.34   Removed, obliterated, or altered serial number.

    No person shall knowingly transport, ship, or receive in interstate or foreign commerce any firearm which has had the importer's or manufacturer's serial number removed, obliterated, or altered, or possess or receive any firearm which has had the importer's or manufacturer's serial number removed, obliterated, or altered and has, at any time, been shipped or transported in interstate or foreign commerce.

[T.D. ATF-313, 56 FR 32508, July 17, 1991]

§ 178.35   Skeet, trap, target, and similar shooting activities.

Licensing and recordkeeping requirements, including permissible alternate records, for skeet, trap, target, and similar organized activities shall be determined by Director of Industry Operations on a case by case basis.

§ 178.36 Transfer or possession of machineguns.

No person shall transfer or possess a machine gun except:

(a) A transfer to or by, or possession by or under the authority of, the United States, or any department or agency thereof, or a State, or a department, agency, or political subdivision thereof. (See Part 179 of this chapter); or

(b) Any lawful transfer or lawful possession of a machinegun that was lawfully possessed before May 19, 1986 (See Part 179 of this chapter).

[T.D. ATF-270, 53 FR 10494, Mar. 31, 1988]

§ 178.37 Manufacture, importation and sale of armor piercing ammunition.

No person shall manufacture or import, and no manufacturer or importer shall sell or deliver, armor piercing ammunition, except:

(a) The manufacture or importation, or the sale or delivery by any manufacturer or importer, of armor piercing ammunition for the use of the United States or any department or agency thereof or any State or any department, agency or political subdivision thereof;

(b) The manufacture, or the sale or delivery by a manufacturer or importer, of armor piercing ammunition for the purpose of exportation; or

(c) The sale or delivery by a manufacturer or importer of armor piercing ammunition for the purposes of testing or experimentation as authorized by the Director under the provisions of § 178.149.

[T.D. ATF-270, 53 FR 10494, Mar. 31, 1988]

§ 178.38 Transportation of firearms.

Notwithstanding any other provision of any law or any rule or regulation of a State or any political subdivision thereof, any person who is not otherwise prohibited by this chapter from transporting, shipping, or receiving a firearm shall be entitled to transport a firearm for any lawful purpose from any place where such person may lawfully possess and carry such firearm to any other place where such person may lawfully possess and carry such firearm if, during such transportation the firearm is unloaded, and neither the firearm nor any ammunition being transported is readily accessible or is directly accessible from the passenger compartment of such transporting vehicle **Provided,** That in the case of a vehicle without a compartment separate from the driver's compartment the firearm or ammunition shall be contained in a locked container other than the glove compartment or console.

[T.D. ATF-270, 53 FR 10494, Mar. 31, 1988]

§ 178.39 Assembly of semiautomatic rifles or shotguns.

(a) No person shall assemble a semiautomatic rifle or any shotgun using more than 10 of the imported parts listed in paragraph (c) of this section if the assembled firearm is prohibited from importation under section 925(d)(3) as not being particularly suitable for or readily adaptable to sporting purposes.

(b) The provisions of this section shall not apply to:

(1) The assembly of such rifle or shotgun for sale or distribution by a licensed manufacturer to the United States or any department or agency thereof or to any State or any department, agency, or political subdivision thereof; or

(2) The assembly of such rifle or shotgun for the purposes of testing or experimentation authorized by the Director under the provisions of §178.151; or

(3) The repair of any rifle or shotgun which had been imported into or assembled in the United States prior to November 30, 1990, or the replacement of any part of such firearm.

(c) For purposes of this section, the ter **imported parts** are:

(1) Frames, receivers, receiver castings forgings or stampings

(2) Barrels

(3) Barrel extensions

(4) Mounting blocks (trunions)

(5) Muzzle attachments

(6) Bolts

(7) Bolt carriers

(8) Operating rods

(9) Gas pistons

(10) Trigger housings

(11) Triggers

(12) Hammers

(13) Sears

(14) Disconnectors

(15) Buttstocks

(16) Pistol grips

(17) Forearms handguards

(18) Magazine bodies

(19) Followers

(20) Floorplates

[T.D. ATF-346, 58 FR 40589, July 29, 1993]

§ 178.39a Reporting theft or loss of firearms.

Each licensee shall report the theft or loss of a firearm from the licensee's inventory (including any firearm which has been transferred from the licensee's inventory to a personal collection and held as a personal firearm for at least 1 year), or from the collection of a licensed collector, within 48 hours after the theft or loss is discovered. Licensees shall report thefts or losses by telephoning 1-800-800-3855 (nationwide toll free number) and by preparing ATF Form 3310.11, Federal Firearms Licensee Theft/Loss Report, in accordance with the instructions on the form. The original of the report shall be forwarded to the office specified thereon, and Copy 1 shall be retained by the licensee as part of the licensee's permanent records. Theft or loss of any firearm shall also be reported to the appropriate local authorities.

(Approved by the Office of Management and Budget under control number 1512-0524)

[T.D. ATF-363, 60 FR 17451, Apr. 6, 1995]

§ 178.40 Manufacture, transfer, and possession of semiautomatic assault weapons.

(a) Prohibition. No person shall manufacture, transfer, or possess a semiautomatic assault weapon.

(b) Exceptions. The provisions of paragraph (a) of this section shall not apply to:

(1) The possession or transfer of any semiautomatic assault weapon otherwise lawfully possessed in the United States under Federal law on September 13, 1994;

(2) Any of the firearms, or replicas or duplicates of the firearms, specified in 18 U.S.C. 922, Appendix A, as such firearms existed on October 1, 1993;

(3) Any firearm that -

(i) Is manually operated by bolt, pump, lever, or slide action;

(ii) Has been rendered permanently inoperable; or

(iii) Is an antique firearm;

(4) Any semiautomatic rifle that cannot accept a detachable magazine that holds more than 5 rounds of ammunition;

(5) Any semiautomatic shotgun that cannot hold more than 5 rounds of ammunition in a fixed or detachable magazine;

(6) The manufacture for, transfer to, or possession by the United States or a department or agency of the United States or a State or a department, agency, or political subdivision of a State, or a transfer to or possession by a law enforcement officer employed by such an entity for purposes of law enforcement;

(7) The transfer to a licensee under title I of the Atomic Energy Act of 1954 (42 U.S.C. 2011 et seq.) for purposes of establishing and maintaining an on-site physical protection system and security organization required by Federal law, or possession by an employee or contractor of such licensee on-site for such purposes or off-site for purposes of licensee-authorized training or transportation of nuclear materials;

(8) The possession, by an individual who is retired from service with a law enforcement agency and is not otherwise prohibited from receiving a firearm, of a semiautomatic assault weapon transferred to the individual by the agency upon such retirement;

(9) The manufacture, transfer, or possession of a semiautomatic assault weapon by a licensed manufacturer or licensed importer for the purposes of testing or experimentation as authorized by the Director under the provisions of § 178.153; or

(10) The manufacture, transfer, or possession of a semiautomatic assault weapon by a licensed manufacturer, licensed importer, or licensed dealer for the purpose of exportation in compliance with the Arms Export Control Act (22 U.S.C. 2778).

(c) **Manufacture and dealing in semiautomatic assault weapons.** Subject to compliance with the provisions of this part, licensed manufacturers and licensed dealers in semiautomatic assault weapons may manufacture and deal in such weapons manufactured after September 13, 1994: **Provided,** The licensee obtains evidence that the weapons will be disposed of in accordance with paragraph (b) of this section. Examples of acceptable evidence include the following:

(1) Contracts between the manufacturer and dealers stating that the weapons may only be sold to law enforcement agencies, law enforcement officers, or other purchasers specified in paragraph (b) of this section;

(2) Copies of purchase orders submitted to the manufacturer or dealer by law enforcement agencies or other purchasers specified in paragraph (b) of this section;

(3) Copies of letters submitted to the manufacturer or dealer by government agencies, law enforcement officers, or other purchasers specified in paragraph (b) of this section expressing an interest in purchasing the semiautomatic assault weapons;

(4) Letters from dealers to the manufacturer stating that sales will only be made to law enforcement agencies, law enforcement officers, or other purchasers specified in paragraph (b) of this section; and

(5) Letters from law enforcement officers purchasing in accordance with paragraph (b)(6) of this section and § 178.132.

(Paragraph (c) approved by the Office of Management and Budget under control number 1512-0526)

[T.D. ATF-363, 60 FR 17451, Apr. 6, 1995]

§ 178.40a **Transfer and possession of large capacity ammunition feeding devices.**

(a) **Prohibition.** No person shall transfer or possess a large capacity ammunition feeding device.

(b) **Exceptions.** The provisions of paragraph (a) of this section shall not apply to:

(1) The possession or transfer of any large capacity ammunition feeding device otherwise lawfully possessed on September 13, 1994;

(2) The manufacture for, transfer to, or possession by the United States or a department or agency of the United States or a State or a department, agency, or political subdivision of a State, or a transfer to or possession by a law enforcement officer employed by such an entity for purposes of law enforcement;

(3) The transfer to a licensee under title I of the Atomic Energy Act of 1954 for purposes of establishing and maintaining an on-site physical protection system and security organization required by Federal law, or possession by an employee or contractor of such licensee on-site for such purposes or off-site for purposes of licensee-authorized training or transportation of nuclear materials;

(4) The possession, by an individual who is retired from service with a law enforcement agency and is not otherwise prohibited from receiving ammunition, of a large capacity ammunition feeding device transferred to the individual by the agency upon such retirement;

(5) The manufacture, transfer, or possession of any large capacity ammunition feeding device by a manufacturer or importer for the purpose of testing or experimentation in accordance with § 178.153; or

(6) The manufacture, transfer, or possession of any large capacity ammunition feeding device by a manufacturer or importer for the purpose of exportation in accordance with the Arms Export Control Act (22 U.S.C. 2778).

(c) **Importation, manufacture, and dealing in large capacity ammunition feeding devices.** Possession and transfer of large capacity ammunition feeding devices by persons who manufacture, import, or deal in such devices will be presumed to be lawful if such persons maintain evidence establishing that the devices are possessed and transferred for sale to purchasers specified in paragraph (b) of this section. Examples of acceptable evidence include the following:

(1) Contracts between persons who import or manufacture such devices and persons who deal in such devices stating that the devices may only be sold to law enforcement agencies or other purchasers specified in paragraph (b) of this section;

(2) Copies of purchase orders submitted to persons who manufacture, import, or deal in such devices by law enforcement agencies or other purchasers specified in paragraph (b) of this section;

(3) Copies of letters submitted to persons who manufacture, import, or deal in such devices by government agencies or other purchasers specified in paragraph (b) of this section expressing an interest in purchasing the devices;

(4) Letters from persons who deal in such devices to persons who import or manufacture such devices stating that sales will only be made to law enforcement agencies or other purchasers specified in paragraph (b) of this section; and

(5) Letters from law enforcement officers purchasing in accordance with paragraph (b)(2) of this section and § 178.132.

(Paragraph (c) approved by the Office of Management and Budget under control number 1512-0526)

[T.D. ATF-363, 60 FR 17452, Apr. 6, 1995, as amended by T.D. ATF-383, 61 FR 39321, July 29, 1996]

Subpart D-Licenses

§ 178.41   General.

(a) Each person intending to engage in business as an importer or manufacturer of firearms or ammunition, or a dealer in firearms shall, before commencing such business, obtain the license required by this subpart for the business to be operated. Each person who desires to obtain a license as a collector of curios or relics may obtain such a license under the provisions of this subpart.

(b) Each person intending to engage in business as a firearms or ammunition importer or manufacturer, or dealer in firearms shall file an application, with the required fee (see § 178.42), with ATF in accordance with the instructions on the form (see § 178.44), and, pursuant to § 178.47, receive the license required for such business from the Chief, National Licensing Center. Except as provided in § 178.50, a license must be obtained for each business and each place at which the applicant is to do business A license as an importer or manufacturer of firearms or ammunition, or a dealer in firearms shall, subject to the provisions of the Act and other applicable provisions of law, entitle the licensee to transport, ship, and receive firearms and ammunition covered by such license in interstate or foreign commerce and to engage in the business specified by the license, at the location described on the license, and for the period stated on the license. However, it shall not be necessary for a licensed importer or a licensed manufacturer to also obtain a dealer's license in order to engage in business on the licensed premises as a dealer in the same type of firearms authorized by the license to be imported or manufactured Payment of the license fee as an importer or manufacturer of destructive devices, ammunition for destructive devices or armor piercing ammunition or as a dealer in destructive devices includes the privilege of importing or manufacturing firearms other than destructive devices and ammunition for other than destructive devices or ammunition other than armor piercing ammunition, or dealing in firearms other than destructive devices, as the case may be, by such a licensee at the licensed premises.

(c) Each person seeking the privileges of a collector licensed under this part shall file an application, with the required fee (see § 178.42), with ATF in accordance with the instructions on the form (see § 178.44), and pursuant to § 178.47, receive from the Chief, National Licensing Center, the license covering the collection of curios and relics A separate license may be obtained for each collection premises, and such license shall, subject to the provisions of the Act and other applicable provisions of law, entitle the licensee to transport, ship, receive, and acquire curios and relics in interstate or foreign commerce, and to make disposition of curios and relics in interstate or foreign commerce, to any other person licensed under the provisions of this part, for the period stated on the license.

(d) The collector license provided by this part shall apply only to transactions related to a collector's activity in acquiring, holding or disposing of curios and relics. A collector's license does not authorize the collector to engage in a business required to be licensed under the Act or this part. Therefore, if the acquisitions and dispositions of curios and relics by a collector bring the collector within the definition of a manufacturer, importer, or dealer under this part, he shall qualify as such. (See also § 178.93 of this part.)

(18 U.S.C. 847 (8 Stat. 959); 18 U.S.C.2926 (8 Stat. 1226))

[33 FR 18555, Dec. 14,.1968 Redesignated at 40 FR 16835, Apr. 15, 1975, and amended by T.D. ATF-270, 53 FR 10494, Mar. 31, 1988; T.D. ATF-290, 54 FR 53054, Dec. 27, 1989]

§ 178.42 License fees.

Each applicant shall pay a fee for obtaining a firearms license or ammunition license, a separate fee being required for each business or collecting activity at each place of such business or activity, as follows:

(a) For a manufacturer:

(1) Of destructive devices, ammunition for destructive devices or armor piercing ammunition-$1,000 per year.

(2) Of firearms other than destructive devices-$50 per year.

(3) Of ammunition for firearms other than ammunition for destructive devices or armor piercing ammunition-$10 per year.

(b) For an importer:

(1) Of destructive devices, ammunition for destructive devices or armor piercing ammunition-$1,000 per year.

(2) Of firearms other than destructive devices or ammunition for firearms other than destructive devices or ammunition other than armor piercing ammunition-$50 per year.

(c) For a dealer:

(1) In destructive devices-$1,000 per year.

(2) Who is not a dealer in destructive devices-$200 for 3 years, except that the fee for renewal of a valid license shall be $90 for 3 years.

(d) For a collector of curios and relics-$10 per year.

[T.D. ATF-270, 53 FR 10494, Mar. 31, 1988, as amended by T.D. ATF-354, 59 FR 7112, Feb. 14, 1994]

§ 178.43 License fee not refundable.

No refund of any part of the amount paid as a license fee shall be made where the operations of the licensee are, for any reason, discontinued during the period of an issued license. However, the license fee submitted with an application for a license shall be refunded if that application is denied or withdrawn by the applicant prior to being acted upon.

[T.D. ATF-270, 53 FR 10494, Mar. 31, 1988]

§ 178.44 Original license.

(a) Any person who intends to engage in business as a firearms or ammunition importer or manufacturer, or firearms dealer, or who has not previously been licensed under the provisions of this part to so engage in business, or who has not timely submitted an application for renewal of the previous license issued under this part, shall file an application for license, ATF Form 7 (Firearms), in duplicate, with ATF in accordance with the instructions on the form. The application must be executed under the penalties of perjury and the penalties imposed by 18 U.S.C. 924. The application shall include a photograph and fingerprints as required in the instruction on the .form The application shall be accompanied by a completed ATF Form 5300.37 (Certification of Compliance with State and Local Law) and ATF Form 5300.36 (Notification of Intent to Apply for a Federal Firearms License), and shall include the appropriate fee in the form of money order or check made payable to the Bureau of Alcohol, Tobacco and Firearms. ATF Forms 7 (Firearms), ATF Forms 5300.37, and ATF

Forms 5300.36 may be obtained by contacting any ATF office.

(b) Any person who desires to obtain a license as a collector under the Act and this part, or who has not timely submitted an application for renewal of the previous license issued under this part, shall file an application, ATF Form 7CR (Curios and Relics), with ATF in accordance with the instructions on the form. The application must be executed under the penalties of perjury and the penalties imposed by 18 U.S.C. 924. The application shall be accompanied by a completed ATF Form 5300.37 and ATF Form 5300.36 and shall include the appropriate fee in the form of a money order or check made payable to the Bureau of Alcohol, Tobacco and Firearms. ATF Forms 7CR (Curios and Relics), ATF Forms 5300.37, and ATF Forms 5300.36 may be obtained by contacting any ATF office.

[T.D. ATF-363, 60 FR 17453, Apr. 6, 1995]

§ 178.45   Renewal of license.

If a licensee intends to continue the business or activity described on a license issued under this part during any portion of the ensuing year, the licensee shall, unless otherwise notified in writing by the Chief, National Licensing Center, execute and file with ATF prior to the expiration of the license an application for a license renewal, ATF Form 8 Part II, accompanied by a completed ATF Form 5300.37 and ATF Form 5300.36, in accordance with the instructions on the forms, and the required fee. The Chief, National Licensing Center, may, in writing, require the applicant for license renewal to also file completed ATF Form 7 or ATF Form 7CR in the manner required by § 178.44. In the event the licensee does not timely file an ATF Form 8 Part II, the licensee must file an ATF Form 7 or an ATF Form 7CR as required by § 178.44, and obtain the required license before continuing business or collecting activity   If an ATF Form 8 Part II is not timely received through the mails, the licensee should so notify the Chief, Firearms and Explosives Licensing Center.

(18 U.S.C. 923 (8 Stat. 1226))

[T.D. ATF-270, 53 FR 10495, Mar. 31, 1988, as amended by T.D. ATF-290, 54 FR 53054, Dec. 27, 1989; T.D. ATF-363, 60 FR 17453, April 6, 1995]

§ 178.46   Insufficient fee.

If an application is filed with an insufficient fee, the application and any fee submitted will be returned to the applicant.

(18 U.S.C. 847 (8 Stat. 959); 18 U.S.C.2926 (8 Stat. 1226))

[T.D. ATF-200, 50 FR 10498, Mar. 15, 1985]

§ 178.47   Issuance of license.

(a) Upon receipt of a properly executed application for a license on ATF Form 7, ATF Form 7CR, or ATF Form 8 Part II, the Chief, National Licensing Center, shall, upon finding through further inquiry or investigation, or otherwise, that the applicant is qualified, issue the appropriate license. Each license shall bear a serial number and such number may be assigned to the licensee to whom issued for so long as the licensee maintains continuity of renewal in the same location (State).

(b) The Chief, National Licensing Center, shall approve a properly executed application for license on ATF Form 7, ATF Form 7CR, or ATF Form 8 Part II, if:

(1) The applicant is 21 years of age or over;

(2) The applicant (including, in the case of a corporation, partnership, or association, any individual possessing, directly or indirectly, the power to direct or cause the direction of the management and policies of the corporation, partnership, or association) is not prohibited under the provisions of the Act from shipping or transporting in interstate or foreign commerce, or possessing in or affecting commerce, any firearm or ammunition, or from receiving any firearm or ammunition which has been shipped or transported in interstate or foreign commerce;

(3) The applicant has not willfully violated any of the provisions of the Act or this part;

(4) The applicant has not willfully failed to disclose any material information required, or has not made any false statement as to any material fact, in connection with his application;

(5) The applicant has in a State

(i) premises from which he conducts business subject to license under the Act or from which he intends to conduct such business within a reasonable period of time, or in the case of a collector, premises from which he conducts his collecting subject to license under the Act or from which he intends to conduct such collecting within a reasonable period of time; and

(6) The applicant has filed an ATF Form 5300.37 (Certification of Compliance with State and Local Law) with ATF in accordance with the instructions on the form certifying under the penalties of perjury that -

(i) The business to be conducted under the license is not prohibited by State or local law in the place where the licensed premises are located;

(ii) Within 30 days after the application is approved the business will comply with the requirements of State and local law applicable to the conduct of business;

(iii) The business will not be conducted under the license until the requirements of State and local law applicable to the business have been met; and

(iv) The applicant has completed and sent or delivered ATF F 5300.36 (Notification of Intent to Apply for a Federal Firearms License) to the chief law enforcement officer of the locality in which the premises are located, which indicates that the applicant intends to apply for a Federal firearms license. For purposes of this paragraph, the "chief law enforcement officer" is the chief of police, the sheriff, or an equivalent officer.

(c) The Chief, National Licensing Center, shall approve or the Director of Industry Operations shall deny an application for a license within the 60-day period beginning on the date the properly executed application was received: **Provided**, That when an applicant for license renewal is a person who is, pursuant to the provisions of § 178.78, § 178.143, or § 178.144, conducting business or collecting activity under a previously issued license, action regarding the application will be held in abeyance pending the completion of the proceedings against the applicant's existing license or license application, final determination of the applicant's criminal case, or final action by the Director on an application for relief submitted pursuant to §178.144, as the case may be.

(d) When the Director of Industry Operations or the Chief, National Licensing Center fails to act on an application for a license within the

60-day period prescribed by paragraph (c) of this section, the applicant may file an action under section 1361 of title 28, United States Code, to compel ATF to act upon the application.

(Paragraph (b)(6) approved by the Office of Management and Budget under control numbers 1512-0522 and 1512-0523)

(18 U.S.C. 847 (8 Stat. 959); 18 U.S.C.2926 (8 Stat. 1226))

[33 FR 18555, Dec. 14,.1968 Redesignated at 40 FR 16835, Apr. 15, 1975, and amended by T.D. ATF-135, 48 FR 24068, May 31, 1983; T.D. ATF-241, 51 FR 39619, Oct. 29, 1986; T.D. ATF-270, 53 FR 10495, Mar. 31, 1988; T.D. ATF-290, 54 FR 53054, Dec. 27, 1989; T.D. ATF-363, 60 FR 17453, Apr. 6, 1995]

§ 178.48 Correction of error on license.

(a) Upon receipt of a license issued under the provisions of this part, each licensee shall eexamin same to ensure that the information contained thereon is accurate. If the information is incorrect, the licensee shall return the license to the Chief, National Licensing Center, with a statement showing the nature of the error. The Chief, National Licensing Center, shall correct the error, if the error was made in his office, and return the license. However, if the error resulted from information contained in the licensee's application for the license, the Chief, National Licensing Center, shall require the licensee to file an amended application setting forth the correct information and a statement explaining the error contained in the application. Upon receipt of the amended application and a satisfactory explanation of the error, the Chief, National Licensin Center, shall make the correction on the license and return same to the licensee.

(b) When the Chief, National Licensing Center, finds through any means other than notice from the licensee that an incorrect license has been issued, the Chief, National Licensing Center, may require the holder of the incorrect license to:

(1) return the license for correction, and

(2) if the error resulted from information contained in the licensee's application for the license, the Chief, National Licensing Center, shall require the licensee to file an amended application setting forth the correct information, and a statement explaining the error contained in the application. The Chief, National Licensing Center, then shall make the correction on the license and return same to the licensee.

[33 FR 18555, Dec. 14, 1968, unless otherwise noted Redesignated at 40 FR 16835, Apr. 15, 1975, and amended by T.D. ATF-290, 54 FR 53054, Dec. 27, 1989]

§ 178.49   Duration of license.

The license entitles the person to whom issued to engage in the business or activity specified on the license, within the limitations of the Act and the regulations contained in this part, for a three year period, unless terminated sooner.

[T.D. ATF-270, 53 FR 10495, Mar. 31, 1988]

§ 178.50   Locations covered by license.

The license covers the class of business or the activity specified in the license at the address specified therein. A separate license must be obtained for each location at which a firearms or ammunition business or activity requiring a license under this part is conducted except:

(a) No license is required to cover a separate warehouse used by the licensee solely for storage of firearms or ammunition if the records required by this part are maintained at the licensed premises served by such warehouse;

(b) A licensed collector may acquire curios and relics at any location, and dispose of curios or relics to any licensee or to other persons who are residents of the State where the collector's license is held and the disposition is made;

(c) A licensee may conduct business at a gun show pursuant to the provision of § 178.100; or

(d) A licensed importer, manufacturer, or dealer may engage in the business of dealing in curio or relic firearms with another licensee at any location pursuant to the provisions of § 178.100.

[T.D. ATF-191, 49 FR 46890, Nov. 29, 1984; T.D. ATF-401, 63 FR 35520, June 30, 1998]

§ 178.51 License not transferable.

Licenses issued under this part are not transferable. In the event of the lease, sale, or other transfer of the operations authorized by the license, the successor must obtain the license required by this part prior to commencing such operations. However, for rules on right of succession, see § 178.56.

§ 178.52 Change of address.

(a) Licensees may during the term of their current license remove their business or activity to a new location at which they intend regularly to carry on such business or activity by filing an Application for an Amended Federal Firearms License, ATF Form 5300.38, in duplicate, not less than 30 days prior to such removal with the Chief, National Licensing Center. The ATF Form 5300.38 shall be completed in accordance with the instructions on the form. The application must be executed under the penalties of perjury and penalties imposed by 18 U.S.C. 924. The application shall be accompanied by the licensee's original license. The Chief, National Licensing Center, may, in writing, require the applicant for an amended license to also file completed ATF Form 7 or ATF Form 7CR, or portions thereof, in the manner required by § 178.44.

Editor's Note

Tear-out Forms 5300.38 for changing the location of a licensed business are in the back of this publication.

(b) Upon receipt of a properly executed application for an amended license, the Chief, National Licensing Center, shall, upon finding through further inquiry or investigation, or otherwise, that the applicant is qualified at the new location, issue the amended license, and return it to the applicant. The license shall be valid for the remainder of the term of the original license. The Chief, National Licensing Center, shall, if the applicant is not qualified refer the application for amended license to the Director of Industry Operations for denial in accordance with § 178.71.

(Approved by the Office of Management and Budget under control number 1512-0525)

[T.D. ATF-363, 60 FR 17453, Apr. 6, 1995]

§ 178.53 Change in trade name.

A licensee continuing to conduct business at the location shown on his license is not required to obtain a new license by reason of a mere change in trade name under which he conducts his business: **Provided**, That such licensee furnishes his license for endorsement of such change to the Chief, National Licensing Center within 30 days from the date the licensee begins his business under the new trade name.

[33 FR 18555, Dec. 14,. 1968 Redesignated at 40 FR 16835, Apr. 15, 1975, and amended by T.D. ATF-48, 44 FR 55842, Sept. 28, 1979; T.D. ATF-290, 54 FR 53055, Dec. 27, 1989]

§ 178.54 Change of control.

In the case of a corporation or association holding a license under this part, if actual or legal control of the corporation or association changes, directly or indirectly, whether by reason of change in stock ownership or control (in the licensed corporation or in any other corporation), by operations of law, or in any other manner, the licensee shall, within 30 days of such change, give written notification thereof, executed under the penalties of perjury, to the Chief, National Licensing Center. Upon expiration of the license, the corporation or association must file a Form 7 (Firearms) as required by § 178.44.

[33 FR 18555, Dec. 14, 1968, unless otherwise noted Redesignated at 40 FR 16835, Apr. 15, 1975, and amended by T.D. ATF-290, 54 FR 53054, Dec. 27, 1989]

§ 178.55 Continuing partnerships.

Where, under the laws of the particular State, the partnership is not terminated on death or insolvency of a partner, but continues until the winding up of the partnership affairs is completed, and the surviving partner has the exclusive right to the control and possession of the partnership assets for the purpose of liquidation and settlement, such surviving partner may continue to operate the business under the license of the partnership. If such surviving partner acquires the business on completion of the settlement of the partnership, he shall obtain a license in his own name from the date of acquisition, as provided in § 178.44. The rule set forth in this section shall also apply where there is more than one surviving partner.

§ 178.56 Right of succession by certain persons.

(a) Certain persons other than the licensee may secure the right to carry on the same firearms or ammunition business at the same address shown on, and for the remainder of the term of, a current license. Such persons are:

(1) The surviving spouse or child, or executor, administrator, or other legal representative of a deceased licensee; and

(2) A receiver or trustee in bankruptcy, or an assignee for benefit of creditors.

(b) In order to secure the right provided by this section, the person or persons continuing the business shall furnish the license for that business for endorsement of such succession to the Chief, National Licensing Center, within 30 days from the date on which the successor begins to carry on the business.

[33 FR 18555, Dec. 14,. 1968 Redesignated at 40 FR 16835, Apr. 15, 1975, and amended by T.D. ATF-48, 44 FR 55842, Sept. 28, 1979; T.D. ATF-290, 54 FR 53055, Dec. 27, 1989]

§ 178.57 Discontinuance of business.

(a) Where a firearm or ammunition business is either discontinued or succeeded by a new owner, the owner of the business discontinued or succeeded shall within 30 days thereof furnish to the Chief, National Licensing Center notification of the discontinuance or succession. (See also § 178.127.)

(b) Since section 922(v), Title 18, U.S.C., makes it unlawful to transfer or possess a semiautomatic assault weapon, except as provided in the law, any licensed manufacturer, licensed importer, or licensed dealer intending to discontinue business shall, prior to going out of business, transfer in compliance with the provisions of this part any semiautomatic assault weapon manufactured or imported after September 13, 1994, to a person specified in § 178.40(b), or, subject to the provisions of §§ 178.40(c) and 178.132, a licensed manufacturer, a licensed importer, or a licensed dealer.

(c) Since section 922(w), Title 18, U.S.C., makes it unlawful to transfer or possess a large capacity ammunition feeding device, except as provided in the law, any person who manufactures, imports, or deals in such devices and who intends to discontinue business shall, prior to going out of business, transfer in compliance with the provisions of this part any large capacity ammunition feeding device manufactured after September 13, 1994, to a person specified in § 178.40a(b), or, subject to the provisions of §§ 178.40a(c) and 178.132, a person who manufactures, imports, or deals in such devices.

[33 FR 18555, Dec. 14, 1968 Redesignated at 40 FR 16835, Apr. 15, 1975, and amended by T.D. ATF-48, 44 FR 55842, Sept. 28, 1979; T.D. ATF-290, 54 FR 53055, Dec. 27, 1989; T.D. ATF-363, 60 FR 17453, April 6, 1995; T.D. ATF-383, 61 FR 39321, July 29, 1996]

§ 178.58 State or other law.

A license issued under this part confers no right or privilege to conduct business or activity contrary to State or other law. The holder of such a license is not by reason of the rights and privileges granted by that license immune from punishment for operating a firearm or ammunition business or activity in violation of the provisions of any State or other law. Similarly, compliance with the provisions of any State or other law affords no immunity under Federal law or regulations.

§ 178.59 Abandoned application.

Upon receipt of an incomplete or improperly executed application on ATF Form 7 (5310.12), or ATF Form 8 (5310.11) Part II, the applicant shall be notified of the deficiency in the application. If the application is not corrected and returned within 30 days following the date of notification, the application shall be considered as having been abandoned and the license fee returned.

[T.D. ATF-135, 48 FR 24068, May 31, 1983]

§ 178.60 Certain continuances of business.

A licensee who furnishes his license to the Chief, Firearms and Explosives Licensing Center for correction or endorsement in compliance with the provisions contained in this subpart may continue his operations while awaiting its return.

[33 FR 18555, Dec. 14, 1968, unless otherwise noted   Redesignated at 40 FR 16835, Apr. 15, 1975, and amended by T.D. ATF-290, 54 FR 53054, Dec. 27, 1989]

Subpart E-License Proceedings

§ 178.71 Denial of an application for license.

Whenever the Director of Industry Operations has reason to believe that an applicant is not qualified to receive a license under the provisions of § 178.47, he may issue a notice of denial, on Form 4498, to the applicant. The notice shall set forth the matters of fact and law relied upon in determining that the application should be denied, and shall afford the applicant 15 days from the date of receipt of the notice in which to request a hearing to review the denial. If no request for a hearing is filed within such time, the application shall be disapproved and a copy, so marked, shall be returned to the applicant.

[33 FR 18555, Dec. 14,. 1968 Redesignated at 40 FR 16835, Apr. 15, 1975, and amended by T.D. ATF-270, 53 FR 10495, Mar. 31, 1988]

§ 178.72 Hearing after application denial.

If the applicant for an original or renewal license desires a hearing to review the denial of his application, he shall file a request therefor, in duplicate, with the Director of Industry Operations within 15 days after receipt of the notice of denial. The request should include a statement of the reasons therefor. On receipt of the request, the Regional director (compliance) shall, as expeditiously as possible, make the necessary arrangements for the hearing and advise the applicant of the date, time, location, and the name of the officer before whom the hearing will be held. Such notification shall be made not less than 10 days in advance of the date set for the hearing. On conclusion of the hearing and consideration of all relevant facts and circumstances presented by the applicant or his representative, the Director of Industry Operations shall render his decision confirming or reversing the denial of the application. If the decision is that the denial should stand, a certified copy of the Director of Industry Operations findings and conclusions shall be furnished to the applicant with a final notice of denial, Form 4501. A copy of the application, marked "Disapproved," will be returned to the applicant. If the decision is that the license applied for should be issued, the applicant shall be so notified, in writing, and the license shall be issued as provided by § 178.47.

§ 178.73 Notice of revocation, suspension, or imposition of civil fine.

(a) **Basis for action.** Whenever the Director of Industry Operations has reason to believe that a licensee has willfully violated any provision of the Act or this part, a notice of revocation of the license, ATF Form 4500, may be issued. In addition, a notice of revocation, suspension, or imposition of a civil fine may be issued on ATF Form 4500 whenever the Director of Industr Operations has reason to believe that a licensee has knowingly transferred a firearm to an unlicensed person and knowingly failed to comply with the requirements of 18 U.S.C. 922(t)(1) with respect to the transfer and, at the time that the transferee most recently proposed the transfer, the national instant criminal background check system was operating and information was available to the system demonstrating that the transferee's receipt of a firearm would violate 18 U.S.C. 922(g) or 922(n) or State law.

(b) **Issuance of notice.** The notice shall set forth the matters of fact constituting the violations specified, dates, places, and the sections of law and regulations violated. The Director of Industry

Operations shall afford the licensee 15 days from the date of receipt of the notice in which to request a hearing prior to suspension or revocation of the license, or imposition of a civil fine. If the licensee does not file a timely request for a hearing, the Director of Industr Operations shall issue a final notice of suspension or revocation and/or imposition of a civil fine on ATF Form 4501, as provided in § 178.74.

[T.D. ATF-270, 53 FR 10495, Mar. 31, 1988; T.D. ATF-415, 63 FR 58272, Oct. 29, 1998]

§ 178.74 Request for hearing after notice of suspension, revocation or imposition of civil fine.

If a licensee desires a hearing after receipt of a notice of suspension or revocation of a license, or imposition of a civil fine, the licensee shall file a request, in duplicate, with the regional director (compliance) within 15 days after receipt of the notice of suspension or revocation of a license, or imposition of a civil fine. On receipt of such request, the regional director (compliance) shall, as expeditiously as possible, make necessary arrangements for the hearing and advise the licensee of the date, time, location and the name of the officer before whom the hearing will be held. Such notification shall be made no less than 10 days in advance of the date set for the hearing. On conclusion of the hearing and consideration of all the relevant presentations made by the licensee or the licensee's representative, the regional director (compliance) shall render a decision and shall prepare a brief summary of the findings and conclusions on which the decision is based If the decision is that the license should be revoked, or, in actions under 18 U.S.C. 922(t)(5), that the license should be revoked or suspended, and/or that a civil fine should be imposed, a certified copy of the summary shall be furnished to the licensee with the final notice of revocation, suspension, or imposition of a civil fine on ATF Form 4501. If the decision is that the license should not be revoked, or in actions under 18 U.S.C. 922(t)(5), that the license should not be revoked or suspended, and a civil fine should not be imposed, the licensee shall be notified in writing.

[T.D. ATF-270, 53 FR 10495, Mar. 31, 1988; T.D. ATF-415, 63 FR 58272, Oct. 29, 1998]

§ 178.75 Service on applicant or licensee.

All notices and other documents required to be served on an applicant or licensee under this subpart shall be served by certified mail or by personal delivery. Where service is by certified mail, a signed duplicate original copy of the formal document shall be mailed, with return receipt requested, to the applicant or licensee at the address stated in his application or license, or at his last known address. Where service is by personal delivery, a signed duplicate original copy of the formal document shall be delivered to the applicant or licensee, or, in the case of a corporation, partnership, or association, by delivering it to an officer, manager, or general agent thereof, or to its attorney of record.

[33 FR 18555, Dec. 14,. 1968 Redesignated at 40 FR 16835, Apr. 15, 1975, and furthe redesignated by T.D. ATF-241, 51 FR 39619, Oct. 29, 1986; T.D. ATF-270, 53 FR 10496, Mar. 31, 1988]

§ 178.76 Representation at a hearing.

An applicant or licensee may be represented by an attorney, certified public accountant, or other person recognized to practice before the Bureau of Alcohol, Tobacco and Firearms as provided in 31 CFR Part 8 (Practice Before the Bureau of Alcohol, Tobacco and Firearms), if he has otherwise complied with the applicable requirements of 26 CFR 601.521

through 601.527 (conference and practice requirements for alcohol, tobacco, and firearms activities) of this chapter The Director of Industry Operations may be represented in proceedings by an attorney in the office of the Assistant Chief Counsel or Division Counsel who is authorized to execute and file motions, briefs and other papers in the proceeding, on behalf of the Director of Industry Operations, in his own name as "Attorney for the Government."

[33 FR 18555, Dec. 14,. 1968 Redesignated at 40 FR 16835, Apr. 15, 1975, and amended by T.D. ATF-48, 44 FR 55842, Sept. 28, 1979; T.D. ATF-92, 46 FR 46916, Sept. 23, 1981. rFurthe redesignated by T.D. ATF-241, 51 FR 39619, Oct. 29, 1986; T.D. ATF-270, 53 FR 10496, Mar. 31, 1988]]

§ 178.77 Designated place of hearing.

The designated place of the hearing shall be a location convenient to the aggrieved party.

[T.D. ATF-270, 53 FR 10496, Mar. 31, 1988]

§ 178.78 Operations by licensee after notice.

In any case where denial, suspension, or revocation proceedings are pending before the Bureau of Alcohol, Tobacco and Firearms, or notice of denial, suspension, or revocation has been served on the licensee and he has filed timely request for a hearing, the license in the possession of the licensee shall remain in effect even though such license has expired, or the suspension or revocation date specified in the notice of revocation on Form 4500 served on the licensee has passed: **Provided**, That with respect to a license that has expired, the licensee has timely filed an application for the renewal of his license If a licensee is dissatisfied witha posthearing decision revoking or suspending the license or denying the application or imposing a civil fine, as the case may be, he may, pursuant to 18 U.S.C. 923(f)(3), within 60 days after receipt of the final notice denying the application or revoking or suspending the license or imposing a civil fine, file a petition for judicial review of such action. Such petition should be filed with the U.S. district court for the district in which the applicant or licensee resides or has his principal place of business. In such case, when the Director of Industry Operations finds that justice so requires, he may postpone the effective date of suspension or revocation of a license or authorize continued operations under the expired license, as applicable, pending judicial review.

[33 FR 18555, Dec. 14,. 1968 Redesignated at 40 FR 16835, Apr. 15, 1975, and amended by T.D. ATF-48, 44 FR 55842, Sept. 28, 1979. Further redesignated by T.D. ATF-241, 51 FR 39619, Oct. 29, 1986; T.D. ATF-415, 63 FR 58272, Oct. 29, 1998]

Subpart F-Conduct of Business

§ 178.91 Posting of license.

Any license issued under this part shall be kept posted and kept available for inspection on the premises covered by the license.

§ 178.92 Identification of firearms, armor piercing ammunition, and large capacity ammunition feeding devices.

(a)(1) **Firearms**. Each licensed manufacturer or licensed importer of any firearm manufactured or imported shall legibly identify each such firearm by engraving, casting, stamping (impressing), or otherwise conspicuously placing or causing to be engraved, cast, stamped (impressed) or placed on the frame or receiver thereof in a manner not susceptible of being readily obliterated, altered, or removed, an individual serial number not duplicating any serial number placed by the manufacturer or importer on any other firearm, and by engraving, casting, stamping (impressing), or otherwise conspicuously placing or causing to be engraved, cast, stamped (impressed) or placed on the frame, receiver, or barrel thereof in a manner not susceptible of being readily obliterated, altered or removed, the model, if such designation has been made; the caliber or gauge; the name (or recognized abbreviation of same) of the manufacturer and also, when applicable, of the importer; in the case of a domestically made firearm, the city and State (or recognized abbreviation thereof) wherein the licensed manufacturer maintains its place of business; and in the case of an imported firearm, the name of the country in which manufactured and the city and State (or recognized abbreviation thereof) of the importer.

(2) Special markings for semiautomatic assault weapons, effective July 5, 1995. In the case of any semiautomatic assault weapon manufactured after September 13, 1994, the frame or receiver shall be marked **"RESTRICTED LAW ENFORCEMENT/GOVERNMENT USE ONLY"** or, in the case of weapons manufactured for export **"FOR EXPORT ONLY,"** in the manner prescribed in paragraph (a)(1) of this section.

(3) Exceptions.

(i) Alternate means of identification. The Director may authorize other means of identification of the licensed manufacturer or licensed importer upon receipt of a letter application, in duplicate, showing that such other identification is reasonable and will not hinder the effective administration of this part.

(ii) Destructive devices. In the case of a destructive device, the Director may authorize other means of identifying that weapon upon receipt of a letter application, in duplicate, from the licensed manufacturer or licensed importer showing that engraving, casting, or stamping (impressing) such a weapon would be dangerous or impracticable.

(iii) Machine guns, silencers, and parts. A firearm frame or receiver, or any part defined as a machine gun, firearm muffler, or firearm silencer in § 178.11, which is not a component part of a complete weapon at the time it is sold, shipped, or otherwise disposed of by a licensed manufacturer or licensed importer, shall be identified as required by this section The Director may authorize other means of identification of parts defined as machine guns other than frames or receivers and parts defined as mufflers or silencers upon receipt of a letter application, in duplicate, showing that such other identification is reasonable and will not hinder the effective administration of this part.

(b) Armor piercing ammunition. -

(1) Marking of ammunition. Each licensed manufacturer or licensed importer of armor piercing ammunition shall identify such ammunition by means of painting, staining or dying the exterior of the projectile with an opaque black coloring. This coloring must completely cover the point of the projectile and at least 50 percent of that portion of the projectile which is visible when the projectile is loaded into a cartridge case.

(2) Labeling of packages. Each licensed manufacturer or licensed importer of armor piercing ammunition shall clearly and conspicuously label each package in which armor piercing ammunition is contained, e.g., each box, carton, case, or other container. The label shall include the words **" ARMOR PIERCING"** in block letters at least 1/4 inch in height. The lettering shall be located on the exterior surface of th package which contains information concerning the caliber or gauge of the ammunition. There shall also be placed on the same surface of the package in block lettering at least 1/8 inch in height the words **"FOR GOVERNMENTAL ENTITIES OR EXPORTATION ONLY."** The statements required by this subparagraph shall be on a contrasting background.

(c) **Large capacity ammunition feeding devices manufactured after September 13, 1994.** **(1)** Each person who manufactures or imports any large capacity ammunition feeding device manufactured after September 13, 1994, shall legibly identify each such device with a serial number. Such person may use the same serial number for all large capacity ammunition feeding devices produced.

(i) Additionally, in the case of a domestically made large capacity ammunition feeding device, such device shall be marked with the name, city and State (or recognized abbreviation thereof) of the manufacturer;

(ii) And in the case of an imported large capacit ammunition feeding device, such device shall be marked:

(A) With the name of the manufacturer, country of origin, and,

(B) Effective July 5 1995, the name, city and State (or recognized abbreviation thereof) of the importer.

(iii) Further, large capacity ammunition feeding devices manufactured after September 13, 1994, shall be d marke **"RESTRICTED LAW ENFORCEMENT/GOVERNMENT USE ONLY"** or, in the case of devices manufactured or imported for export, effective July 5, 1995 **"FOR EXPORT ONLY."**

(2) All markings required by this paragraph (c shall be cast, stamped, or engraved on the exterior of the device. In the case of a magazine, the markings shall be placed on the magazine body.

(3) Exceptions.

(i) Metallic links. Persons who manufacture or import metallic links for use in the assembly of belted ammunition are only required to place the identification marks prescribed in paragraph (c)(1) of this section on the containers used for the packaging of the links.

(ii) Alternate means of identification. The Director may authorize other means of identifying large capacity ammunition feeding devices upon receipt of a letter application, in duplicate, from the manufacturer or importer showing that such other identification is reasonable and will not hinder the effective administration of this part.

[T.D. ATF-270, 53 FR 10496, Mar. 31, 1988, as amended by T.D. ATF-363, 60 FR 17454, April 6, 1995; T.D. ATF-383, 61 FR 39321, July 29, 1996; T.D. ATF-396, 63 FR 12646, March 16, 1998]

§ 178.93   Authorized operations by a licensed collector.

The license issued to a collector of curios or relics under the provisions of this part shall cover only transactions by the licensed collector in curios and relics The collector's license is of no force or effect and a licensed collector is of the same status under the Act and this part as a nonlicensee with respect t **(a)** any acquisition or disposition of firearms other than curios or relics, or any transportation, shipment, or receipt of firearms other than curios or relics in interstate or foreign commerce, an **(b)** any transaction with a nonlicensee involving any firearm other than a curio or relic. (See also § 178.50.) A collector's license is not necessary to receive or dispose of ammunition, an a licensed collector is not precluded by law from receiving or disposing of armor piercing ammunition. However, a licensed collector may not dispose of any ammunition to a person prohibited from receiving or possessing ammunition (See § 178.99(c)). Any licensed collector who disposes of armor piercing ammunition must record the disposition as required by § 178.125 (a) and (b).

[T.D. ATF-270, 53 FR 10496, Mar. 31, 1988]

§ 178.94 Sales or deliveries between licensees.

A licensed importer, licensed manufacturer, or licensed dealer selling or otherwise disposing of firearms, and a licensed collector selling or otherwise disposing of curios or relics, to another licensee shall verify the identity and licensed status of the transferee prior to making the transaction Verification shall be established by the transferee furnishing to the transferor a certified copy of the transferee's license and by such other means as the transferor deems necessary Provided, That it shall not be require **(a)** for a transferee who has furnished a certified copy of its license to a transferor to again furnish such certified copy to that transferor during the term of the transferee's current license **(b)** for a licensee to furnish a certified copy of its license to another licensee if a firearm is being returned either directly or through another licensee to such licensee an **(c)** for licensees fo multilicensed business organizations to furnish certified copies of their licenses to other licensed locations operated by such organization Provided further, That a multilicensed business organization may furnish to a transferor, in lieu of a certified copy of each license, a list, certified to be true, correct and complete, containing the name, address, license number, and the date of license expiration of each licensed location operated by such organization, and the transferor may sell or otherwise dispose of firearms as provided by this section to any licensee appearing on such list without requiring a certified copy of a license therefrom. A transferor licensee who has the certified information required by this section may sell or dispose of firearms to a licensee for not more than 45 days following the expiration date of the transferee's license.

Editor's Note

Also see ATF Rul. 75-27, Operations under pending renewal application.

(Approved by the Office of Management and Budget under control number 1512-0387)

[T.D. ATF-270, 53 FR 10496, Mar. 31, 1988]

§ 178.95 Certified copy of license.

The license furnished to each person licensed under the provisions of this part contains a purchasing certification statement. This original license may be reproduced and the reproduction then certified by the licensee for use pursuant to § 178.94. If the licensee desires an additional copy of the license for certification (instead of making a reproduction of the original license), the licensee may submit a request, in writing, for a certified copy or copies of the license to the Chief, National Licensing Center. The request must set forth the name, trade name (if any) and address of the licensee, and the number of license copies desired. There is a charge of $1 for each copy. The fee paid for copies of the license must accompany the request .for copies The fee may be paid yb **(a)** cash, ro **(b)** money order or check made payable to the Bureau of Alcohol, Tobacco and Firearms.

(Approved by the Office of Management and Budget under control number 1512-0387)

[T.D. ATF-270, 53 FR 10497, Mar. 31, 1988, as amended by T.D. ATF-290, 54 FR 53055, Dec. 27, 1989]

§ 178.96 Out-of-State and mail order sales.

(a) The provisions of this section shall apply when a firearm is purchased by or delivered to a person not otherwise prohibited by the Act from purchasing or receiving it.

(b) A licensed importer, licensed manufacturer, or licensed dealer may sell a firearm that is not subject to the provisions of § 178.102(a) to a nonlicensee who does not appear in person at the licensee's business premises if the nonlicensee is a resident of the same State in which the licensee's business premises are located, and the nonlicensee furnishes to the licensee the firearms transaction record, Form 4473, required by § 178.124. The nonlicensee shall attach to such record a true copy of any permit or other information required pursuant to any statute of the State and published ordinance applicable to the locality in which he resides. The licensee shall prior to shipment or delivery of the firearm, forward by registered or certified mail (return receipt requested) a copy of the record, Form 4473, to the chief law enforcement officer named on such record, and delay shipment or delivery of the firearm for a period of at least 7 days following receipt by the licensee of the return receipt evidencing delivery of the copy of the record to such chief law enforcement officer, or the return of the copy of the record to him due to the refusal of such chief law enforcement officer to accept same in accordance with U.S. Postal Service regulations The original Form 4473, and evidence of receipt or rejection of delivery of the copy of the Form 4473 sent to the chief law enforcement officer shall be retained by the licensee as a part of the records required of him to be kept under the provisions of subpart H of this part.

(c) (1) A licensed importer, licensed manufacturer, or licensed dealer may sell or deliver a rifle or shotgun, and a licensed collector may sell or deliver a rifle or shotgun that is a curio or relic to nonlicensed resident of a State other than the State in which the licensee's place of business is located if -

(i) The purchaser meets with the licensee in person at the licensee's premises to accomplish the transfer, sale, and delivery of the rifle or shotgun;

(ii) The licensed importer, licensed manufacturer, or licensed dealer complies with the provisions of § 178.102;

(iii) The purchaser furnishes to the licensed importer, licensed manufacturer, or licensed dealer the firearms transaction record, Form 4473, required by § 178.124; and

(iv) The sale, delivery, and receipt of the rifle or shotgun fully comply with the legal conditions of sale in both such States.

(2) For purposes of paragraph (c) of this section, any licensed manufacturer, licensed importer, or licensed dealer is presumed, in the absence of evidence to the contrary, to have had actual knowledge of the State laws and published ordinances of both such States.

(Approved by the Office of Management and Budget under control number 1512-0130)

[33 FR 18555, Dec. 14,. 1968 Redesignated at 40 FR 16835, Apr. 15, 1975, and amended by T.D. ATF-48, 44 FR 55842, Sept. 28, 1979; T.D. ATF-241, 51 FR 39620, Oct. 29, 1986; T.D. ATF-270, 53 FR 10497, Mar. 31, 1988; T.D. ATF-354, 59 FR 7112, Feb. 14, 1994; T.D. ATF-415, 63 FR 58272, Oct. 29, 1998]

§ 178.97 Loan or rental of firearms.

(a) A licensee may lend or rent a firearm to any person for temporary use off the premises of the licensee for lawful sporting purposes: **Provided,** That the delivery of the firearm to such person is not prohibited by § 178.99(b) or § 178.99(c), the licensee complies with the requirements of § 178.102, and the licensee records such loan or rental in the records required to be kept by him under Subpart H of this part.

(b) A club, association, or similar organization temporarily furnishing firearms (whether by loan, rental, or otherwise) to participants in a skeet, trap, target, or similar shooting activity for use at the time and place such activity is held does not, unattended by other circumstances, cause such club, association, or similar organization to be engaged in the business of a dealer in firearms or as engaging in firearms transactions. Therefore, licensing and recordkeeping requirements contained in this part pertaining to firearms transactions would not apply to this temporary furnishing of firearms for use on premises on which such an activity is conducted.

§ 178.98 Sales or deliveries of destructive devices and certain firearms.

The sale or delivery by a licensee of any destructive device, machine gun, short-barreled shotgun, or short-barreled rifle, to any person other than another licensee who is licensed under this part to deal in such device or firearm, is prohibited unless the person to receive such device or firearm furnishes to the licensee a sworn statement setting forth

(a) The reasons why there is a reasonable necessity for such person to purchase or otherwise acquire the device or weapon; and

(b) That such person's receipt or possession of the device or weapon would be consistent with public safety. Such sworn statement shall be made on the application to transfer and register the firearm required by Part 179 of this chapter. The sale or delivery of the device or weapon shall not be made until the application for transfer is approved by the Director and returned to the licensee (transferor) as provided in Part 179 of this chapter.

[T.D. ATF-270, 53 FR 10497, Mar. 31, 1988]

§ 178.99 Certain prohibited sales or deliveries.

(a) **Interstate sales or deliveries.** A licensed importer, licensed manufacturer, licensed dealer, or licensed collector shall not sell or deliver any firearm to any person not licensed under this part and who the licensee knows or has reasonable cause to believe does not reside in (or if a corporation or other business entity, does not maintain a place of business in) the State in which the licensee's place of business or activity is located: **Provided,** That the foregoing provisions of this paragraph **(1)** shall not apply to the sale or delivery of a rifle or shotgun (curio or relic, in the case of a licensed collector) to a resident of a State other than the State in which the licensee's place of business or collection premises is located if the requirements of § 178.96(c) are fully met, an **(2)** shall not apply to the loan or rental of a firearm to any person for temporary use for lawful sporting purposes (see § 178.97).

(b) **Sales or deliveries to underage persons.** A licensed importer, licensed manufacturer, licensed dealer, or licensed collector shall not sell or delive **(1)** any firearm or ammunition to any individual who the importer, manufacturer, dealer, or collector knows or has reasonable cause to believe is less than 18 years of age, and, if the firearm, or ammunition, is other than a shotgun or rifle, or ammunition for a shotgun or rifle, to any individual who the importer, manufacturer, dealer, or collector knows or has reasonable cause to believe is less than 21 years of age,ro **(2)** any firearm to any person in any State where the purchase or possession by such person of such firearm would be in violation of any State law or any published ordinance applicable at the place of sale, delivery, or other disposition, unless the importer, manufacturer, dealer, or collector knows or has reasonable cause to believe that the purchase or possession would not be in violation of such State law or such published ordinance.

(c) Sales or deliveries to prohibited categories of persons. A licensed manufacturer, licensed importer, licensed dealer, or licensed collector shall not sell or otherwise dispose of any firearm or ammunition to any person knowing or having reasonable cause to believe that such person:

(1) Is, except as provided by § 178.143, under indictment for, or, except as provided by § 178.144, has been convicted in any court of a crime punishable by imprisonment for a term exceeding 1 year;

(2) Is a fugitive from justice;

(3) Is an unlawful user of or addicted to any controlled substance (as defined in section 102 of the Controlled Substance Act, 21 U.S.C. 802);

(4) Has been adjudicated as a mental defective or has been committed to any mental institution;

(5) Is an alien illegally or unlawfully in the United States;

(6) Has been discharged from the Armed Forces under dishonorable condition ;

(7) Who, having been a citizen of the United States, has renounced citizenship;

(8) Is subject to a court order that restrains such person from harassing, stalking, or threatening an intimate partner of such person or child of such intimate partner or person, or engaging in other conduct that would place an intimate partner in reasonable fear of bodily injury to the partner or child, except that this paragraph shall only apply to a court order that -

(i) Was issued after a hearing of which such person received actual notice, and at which such person had the opportunity to participate; and

(ii)(A) Includes a finding that such person represents a credible threat to the physical safety of such intimate partner or child; or

(B) By its terms explicitly prohibits the use, attempted use, or threatened use of physical force against such intimate partner or child that would reasonably be expected to cause bodily injury, or

(9) Has been convicted of a misdemeanor crime of domestic violence.

(d) Manufacture, importation, and sale of armor piercing ammunition by licensed importers and licensed manufacturers. A licensed importer or licensed manufacturer shall not import or manufacture armor piercing ammunition or sell or deliver such ammunition, except:

(1) For use of the United States or any department or agency thereof or any State or any department, agency, or political subdivision thereof;

(2) For the purpose of exportation; or

(3) For the purpose of testing or experimentation authorized by the Director under the provisions of § 178.149.

(e) Transfer of armor piercing ammunition by licensed dealers. A licensed dealer shall not willfully transfer armor piercing ammunition: **Provided,** That armor piercing ammunition received and maintained by the licensed dealer as business inventory prior to August 28, 1986, may be transferred to any department or agency of the United States or any State or political subdivision thereof if a record of such ammunition is maintained in the form and manner prescribed by § 178.125(c). Any licensed dealer who violates this paragraph is subject to license revocation. See subpart E of this part. For purposes of this paragraph, the Director shall furnish each licensed dealer information defining which projectiles are considered armor piercing. Such information may not be all-inclusive for purposes of the prohibition on manufacture, importation, or sale or delivery by a manufacturer or importer of such ammunition or 18 U.S.C. 929 relating to criminal misuse of such ammunition.

[T.D. ATF-270, 53 FR 10497, Mar. 31, 1988, as amended by T.D. ATF-363, 60 FR 17454, April 6, 1995; T.D. ATF-401, 63 FR 35520, June 30, 1998]

§ 178.100 Conduct of business away from licensed premises.

(a)(1) A licensee may conduct business temporarily at a gun show or event as defined in paragraph (b) if the gun show or event is located in the same State specified on the license: **Provided,** That such business shall not be conducted from any motorized or towed vehicle. The premises of the gun show or event at which the licensee conducts business shall be considered part of the licensed premises. Accordingly, no separate fee or license is required for the gun show or event locations. However, licensees shall comply with the provisions of § 178.91 relating to posting of licenses (or a copy thereof) while conducting business at the gun show or event.

(2) A licensed importer, manufacturer, or dealer may engage in the business of dealing in curio or relic firearms with another licensee at any location.

(b) A gun show or an event is a function sponsored by any national, State, or local organization, devoted to the collection, competitive use, or other sporting use of firearms, or an organization or association that sponsors functions devoted to the collection, competitive use, or other sporting use of firearms in the community.

(c) Licensees conducting business at locations other than the premises specified on their license under the provisions of paragraph (a) of this section shall maintai firearms records in the form and manner prescribed by Subpart H of this part. In addition, records of firearms transactions conducted at such locations shall include the location of the sale or other disposition, be entered in the acquisition and disposition records of the licensee, and retained on the premises specified on the license.

[T.D. ATF-270, 53 FR 10498, Mar. 31, 1988; T.D. ATF-401, 63 FR 35520, June 30, 1998]

§ 178.101 Record of transactions.

Every licensee shall maintain firearms and armor piercing ammunition records in such form and manner as is prescribed by subpart H of this part.

[T.D. ATF-270, 53 FR 10498, Mar. 31, 1988]

§ 178.102 Sales or deliveries of firearms on and after November 30, 1998.

(a) **Background check.** Except as provided in paragraph (d) of this section, a licensed importer, licensed manufacturer, or licensed dealer (the licensee) shall not sell, deliver, or transfer a firearm to any other person who is not licensed under this part unless the licensee meets the following requirements:

(1) Before the completion of the transfer, the licensee has contacted NICS;

(2)(i) NICS informs the licensee that it has no information that receipt of the firearm by the transferee would be in violation of Federal or State law and provides the licensee with a unique identification number; or

(ii) Three business days (meaning days on which State offices are open) have elapsed from the date the licensee contacted NICS and NICS has not notified the licensee that receipt of the firearm by the transferee would be in violation of law; and

(3) The licensee verifies the identity of the transferee by examining the identification document presented in accordance with the provisions of § 178.124(c).

Example for paragraph (a). A licensee contacts NICS on Thursday, and gets a "delayed" response. The licensee does not get a further response from NICS. If State offices are not open on Saturday and Sunday, 3 business days would have elapsed on the following Tuesday. The licensee may transfer the firearm on the next day, Wednesday.

(b) **Transaction number.** In any transaction for which a licensee receives a transaction number from NICS (which shall include either a NICS transaction number or, in States where the State is recognized as a point of contact for NICS checks, a State transaction number), such number shall be recorded on a firearms transaction record, Form 4473, which shall be retained in the records of the licensee in accordance with the provisions of § 178.129. This applies regardless of whether the transaction is approved or denied b NICS, and regardless of whether the firearm is actually transferred.

(c) **Time limitation on NICS checks.** A NICS check conducted in accordance with paragraph (a) of this section may be relied upon by the licensee only for use in a single transaction, and for a period not to exceed 30 calendar days from the date that NICS was initially contacted. If the transaction is not completed within the 30-day period, the licensee shall initiate a new NICS check prior to completion of the transfer.

Example 1 for paragraph (c). A purchaser completes the Form 4473 on December 15, 1998, dan a NICS check is initiated by the licensee on that date. The licensee is informed by the NICS that the information available to the system does not indicate that receipt of the firearm by the transferee would be in violation of law, and a unique identification number is provided. However, the State imposes a 7-day waiting period on all firearms transactions, and the purchaser does not return to pick up the firearm until January 22, 1999. The licensee must conduct another

NICS check before transferring the firearm to the purchaser.

Example 2 for paragraph (c). A purchaser completes the Form 4473 on January 25, 1999, and arranges for the purchase of a single firearm A NICS check is initiated by the licensee on that date. The licensee is informed by NICS that the information available to the system does not indicate that receipt of the firearm by the transferee would be in violation of law, and a unique identification number is provided. The State imposes a 7-day waiting peribd on al firearms transactions, and the purchaser returns to pick up the firearm on February 15, 1999. Before the licensee executes the Form 4473, and the firearm is transferred, the purchaser decides to purchase an additional firearm. The transfer of these two firearms is considered a single transaction; accordingly, the licensee may add the second firearm to the Form 4473, and transfer that firearm without conducting another NICS check.

Example 3 for paragraph (c). A purchaser completes a Form 4473 on February 15, 1999. The licensee receives a unique identification number from NICS on that date, the Form 4473 is executed by the licensee, and the firearm is transferred. On February 20, 1999, the purchaser returns to the licensee's premises and wishes to purchase a second firearm. The purchase of the second firearm is a separate tranşaction; thus a new NICS check must be initiated by the licensee.

(d) Exceptions to NICS check. The provisions of paragraph (a) of this section shall not apply if -

(1) The transferee has presented to the licedsee a vali permit or license that -

(i) Allows the transferee to possess, acquire, or carry a firearm;

(ii) Was issued not more than 5 years earlier by the State in which the transfer is to take place; and

(iii) The law of the State provides that such a permit or license is to be issued only after an authorized government official has verified that the information available to such official does not indicate that possession of a firearm by the transferee would be in violation of Federal, State, or local law **Provided**, That on and after November 30, 1998, the information available to such official includes the NICS;

(2) The firearm is subject to the provisions of the National Firearms Act and has been approved for transfer under 27 CFR Part 179; or

(3) On application of the licensee, in accordance with the provisions of § 178.150, the Director has certified that compliance with paragraph (a)(1) of this section is impracticable.

(e) The document referred to in paragraph (d)(1) of this section (or a copy thereof) shall be retained or the required information from the document shall be recorded on the firearms transaction record in accordance with the provisions of § 178.131.

(Approved by the Office of Management and Budget under control number 1512 -0544)

[T.D. ATF-361, 60 FR 10786, Feb. 27, 1995; T.D. ATF-415, 63 FR 58272, Oct. 29, 1998]

§ 178.103 Posting of signs and written notification to purchasers of handguns.

(a) Each licensed importer, manufacturer, dealer, or collector who delivers a handgun to a nonlicensee shall provide such nonlicensee with written notification as described in paragraph (b) of this section.

(b) The written notification (ATF I 5300.2) required by paragraph (a) of this section shall state as follows:

(1) The misuse of handguns is a leading contributor to juvenile violence and fatalities.

(2) Safely storing and securing firearms away from children will help prevent the unlawful possession of handguns by juveniles, stop accidents, and save lives.

(3) Federal law prohibits, except in certain limited circumstances, anyone under 18 years of age from knowingly possessing a handgun, or any person from transferring a handgun to a person under 18.

(4) A knowing violation of the prohibition against selling, delivering, or otherwise transferring a handgun to a person under the age of 18 is, under certain circumstances, punishable by up to 10 years in prison.

FEDERAL LAW

The Gun Control Act of 1968, 18 U.S.C. Chapter 44, provides in pertinent part as follows:

18 U.S.C. 922(x)

(x)(1) It shall be unlawful for a person to sell, deliver, or otherwise transfer to a person who the transferor knows or has reasonable cause to believe is a juvenile—

(A) a handgun; or

(B) ammunition that is suitable for use only in a handgun.

(2) It shall be unlawful for any person who is a juvenile to knowingly possess—

(A) a handgun; or

(B) ammunition that is suitable for use only in a handgun.

(3) This subsection does not apply to—

(A) a temporary transfer of a handgun or ammunition to a juvenile or to the possession or use of a handgun or ammunition by a juvenile if the handgun and ammunition are possessed and used by the juvenile—

(i) in the course of employment, in the course of ranching or farming related to activities at the residence of the juvenile (or on property used for ranching or farming at which the juvenile, with the permission of the property owner or lessee, is performing activities related to the operation of the farm or ranch), target practice, hunting, or a course of instruction in the safe and lawful use of a handgun;

(ii) with the prior written consent of the juvenile's parent or guardian who is not prohibited by Federal, State, or local law from possessing a firearm, except—

(I during transportation by the juvenile of an unloaded handgun in a locked container directly from the place of transfer to a place at which an activity described in clause (i) is to take place and transportation by the juvenile of that handgun, unloaded and in a locked container, directly from the place at which such an activity took place to the transferor; or

(II with respect to ranching or farming activities as described in clause (i) a juvenile may possess and use a handgun or ammunition with the prior written approval of the juvenile's parent or legal guardian and at the direction of an adult who is not prohibited by Federal, State, or local law from possessing a firearm;

(iii the juvenile has the prior written consent in the juvenile's possession at all times when a handgun is in the possession of the juvenile; and

(iv in accordance with State and local law;

(B) a juvenile who is a member of the Armed Forces of the United States or the National Guard who possesses or is armed with a handgun in the line of duty;

(C) a transfer by inheritance of title (but not possession) of a handgun or ammunition to a juvenile; or

(D) the possession of a handgun or ammunition by a juvenile taken in defense of the juvenile or other persons against an intruder into the residence of the juvenile or a residence in which the juvenile is an invited guest.

(4) A handgun or ammunition, the possession of which is transferred to a juvenile in circumstances in which the transferor is not in violation of this subsection shall not be subject to permanent confiscation by the Government if its possession by the juvenile subsequently becomes unlawful because of the conduct of the juvenile, but shall be returned to the lawful owner when such handgun or ammunition is no longer required by the Government for the purposes of investigation or prosecution.

(5) For purposes of this subsection, the term "juvenile" means a person who is less than 18 years of age.

(6)(A) In a prosecution of a violation of this subsection, the court shall require the presence of a juvenile defendant's parent or legal guardian at all proceedings.

(B) The court may use the contempt power to enforce subparagraph (A).

(C) The court may excuse attendance of a parent or legal guardia of a juvenile defendant at a proceeding in a prosecution of a violation of this subsection for good cause shown.

18 U.S.C. 924(a)(6)

(6)(A (i) A juvenile who violates section 922(x) shall be fined under this title, imprisoned not more than 1 year, or both, except that a juvenile described in clause (ii) shall be sentenced to probation on appropriate conditions and shall not be incarcerated unless the juvenile fails to comply with a condition of probation.

(ii) A juvenile is described in this clause if—

(I the offense of which the juvenile is charged is possession of a handgun or ammunition in violation of section 922(x)(2); and

(II the juvenile has not been convicted in any court of an offense (including an offense under section 922(x) or a similar State law, but not including any other offense consisting of conduct that if engaged in by an adult would not constitute an offense) or adjudicated as a juvenile delinquent for conduct that if engaged in by an adult would constitute an offense.

(B) A person other than a juvenile who knowingly violates section 922(x)—

(i) shall be fined under this title, imprisoned not more than 1 year, or both; and

(ii) if the person sold, delivered, or otherwise transferred a handgun or ammunition to a juvenile knowing or having reasonable cause to know that the juvenile intended to carry or otherwise possess or discharge or otherwise use the handgun or ammunition in the commission of a crime of violence, shall be fined under this title, imprisoned not more than 10 years, or both.

(c) This written notification shall be delivered to the nonlicensee on ATF I 5300.2, or in the alternative, the same written notification may be delivered to the nonlicensee on another type of written notification, such as a manufacturer's or importer's brochure accompanying the handgun; a manufacturer's or importer's operational manual accompanying the handgun; or a sales receipt or invoice applied to the handgun package or container delivered to a nonlicensee. Any written notification delivered to a nonlicensee other than on ATF I 5300.2 shall include the language set forth in paragraph (b) of this section in its entirety. Any written notification other than ATF I 5300.2 shall be legible, clear, and conspicuous, and the required language shall appear in type size no smaller than 10-point type.

(d) Except as provided in paragraph (f) of this section, each licensed importer, manufacturer, or dealer who delivers a handgun to a nonlicensee shall display at its licensed premises (including temporary business locations at gun shows) a sign as described in paragraph (e) of this section. The sign shall be displayed where customers can readily see it. Licensed importers, manufacturers, and dealers will be provided with such signs by ATF. Replacement signs may be requested from the ATF Distribution Center.

(e) The sign (ATF I 5300.1) required by paragraph (d) of this section shall state as follows:

(1) The misuse of handguns is a leading contributor to juvenile violence and fatalities.

(2) Safely storing and securing firearms away from children will help prevent the unlawful possession of handguns by juveniles, stop accidents, and save lives.

(3) Federal law prohibits, except in certain limited circumstances, anyone under 18 years of age from knowingly possessing a handgun, or any person from transferring a handgun to a person under 18.

(4) A knowing violation of the prohibition against selling, delivering, or otherwise transferring a handgun to a person under the age of 18 is, under certain circumstances, punishable by up to 10 years in prison.

Note: ATF I 5300.2 provides the complete language of the statutory prohibitions and exceptions provided in 18 U.S.C. 922(x) and the penalty provisions of 18 U.S.C. 924(a)(6). The Federal firearms licensee posting this sign will provide you with a copy of this publication upon request. Requests for additional copies of ATF I 5300.2 should be mailed to the ATF Distribution Center, P.O. Box 5950 Springfield, Virginia 22150-5950.

(f) The sign required by paragraph (d) of this section need not be posted on the premises of any licensed importer, manufacturer, or dealer whose only dispositions of handguns t nonlicensees are t nonlicensees who do not appear at the licensed premises and the dispositions otherwise comply with the provisions of this part.

[T.D. ATF-402, 63 F 37740, July 13, 1998]

Subpart G-Importation

§ 178.111   General.

(a) Section 922(a)(3) of the Act makes it unlawful, with certain exceptions not pertinent here, for any person other than a licensee to transport into or receive in the State where the person resides any firearm purchased or otherwise obtained by the person outside of that State. However, section 925(a)(4) provides a limited exception for the transportation, shipment, receipt or importation of certain firearms and ammunition by certain members of the United States Armed Forces Section 922(1) of the Act makes it unlawful for any person knowingly to import or bring into the United States or any possession thereof any firearm or ammunition except as provided by section 925(d) of the Act, which section provides standards for importing or bringing firearms or ammunition into the United States. Section 925(d) also provides standards for importing or bringing firearm barrels into the United States. Accordingly, no firearm, firearm barrel, or ammunition may be imported or brought into the United States except as provided by this part.

(b) Where a firearm, firearm barrel, or ammunition is imported and the authorization for importation required by this subpart has not been obtained by the person importin same, such person shall:

(1) Store, at the person's expense, such firearm, firearm barrel, or ammunition at a facility designated by U.S Customs or the Director of Industry Operations to await the issuance of the required authorization or other disposition; or

(2) Abandon such firearm, firearm barrel, or ammunition to the U.S. Government; or

(3) Export such firearm, firearm barrel, or ammunition.

(c) Any inquiry relative to the provisions or procedures under this subpart, other than that pertaining to the payment of customs duties or the release from Customs custody of firearms, firearm barrels, or ammunition authorized by the Director to be imported, shall be directed to the Director of Industry Operations for reply.

[T.D. ATF-270, 53 FR 10498, Mar. 31, 1988]

§ 178.112 Importation by a licensed importer.

(a) No firearm, firearm barrel, or ammunition shall be imported or brought into the United States by a licensed importer (as defined in § 178.11) unless the Director has authorized the importation of the firearm, firearm barrel, or ammunition.

(b) An application for a permit, ATF Form 6, to import or bring a firearm, firearm barrel, or ammunition into the United States or a possession thereof under this section shall be filed, in triplicate, with the Director The application shall contain (1) the name, address, and license number of the importer, (2) a description of the firearm, firearm barrel, or ammunition to be imported, including type (e.g., rifle, shotgun, pistol, revolver; and in the case of ammunition only, ball, wadcutter), model, caliber, size or gauge, barrel length (if a firearm or firearm barrel), country of manufacture, and name of the manufacturer, (3) the unit cost of the firearm, firearm barrel, or ammunition to be imported, (4) the country from which to be imported, (5) the name and address of the foreign seller and the foreign shipper, (6) verification that if a firearm, it will be identified as required by this part, and (7)(i) if a firearm or ammunition imported or brought in for scientific or research purposes, a statement describing such purposes, or (ii) if a firearm or ammunition for use in connection with competition or training pursuant to Chapter 401 of Title 10, U.S.C., a statement describing such intended use, or (iii) if an unserviceable firearm (other than a machine gun) being imported as a curio or museum piece, a description of how it was rendered unserviceable and an explanation of why it is a curio or museum piece, or (iv) if a firearm other than a surplus military firearm, of a type that does not fall within the definition of a firearm under section 5845(a) of the Internal Revenue Code of 1986, and is for sporting purposes, an explanation of why the firearm is generally recognized as particularly suitable for or readily adaptable to sporting purposes, or (v) if ammunition being imported for sporting purposes, a statement why the ammunition is particularly suitable for or readily adaptable to sporting purposes, or (vi) if a firearm barrel, and is for a handgun, an explanation why the handgun is generally recognized as particularly suitable for or readily adaptable to sporting purposes If the Director approves the application, such approved application shall serve as the permit to import the firearm, firearm barrel, or ammunition described therein, and importation of such firearms, firearm barrels, or ammunition may continue to be made by the licensed importer under the approved application (permit) during the period specified thereon. The Director shall furnish the approved application (permit) to the applicant and retain two copies thereof for administrative use. If the Director disapproves the application, the licensed importer shall be notified of the basis for the disapproval.

(c) A firearm, firearm barrel, or ammunition imported or brought into the United States by a licensed importer may be released from Customs custody to the licensed importer upon showing that the importer has obtained a permit from the Director for the importation of the firearm, firearm barrel, or ammunition to be.released In obtaining the release from Customs custody of a firearm, firearm barrel, or ammunition authorized by this section to be imported through use of a permit, the licensed importer shall prepare ATF Form 6A, in duplicate, and furnish the original ATF Form 6A to the Customs officer releasing the firearm, firearm barrel, or ammunition. The Customs officer shall, after certification, forward the ATF Form 6A to the address specified on the form The ATF Form 6A shall show the name, address, and license number of the importer, the name of the manufacturer of the firearm, firearm barrel, or ammunition, the country of manufacture, the type, model, and caliber, size or gauge, and the number of firearms, firearm barrels, or rounds of ammunition released.

(d) Within 15 days of the date of release from Customs custody, the licensed importer shal (1) forward to the address specified on the form a copy of ATF Form 6A on which shall be reported any error or discrepancy appearing on the ATF Form 6A certified by ,Customs (2) pursuant to § 178.92, place all required identification data on each imported firearm if same did not bear such identification data at the time of its release from Customs custody, an (3) post in the records required to be maintained by the importer under subpart H of this part all required information regarding the importation.

[T.D. ATF-270, 53 FR 10498, Mar. 31, 1988]

§ 178.113 Importation by other licensees.

(a) No person other than a licensed importer (as defined in § 178.11) shall engage in the business of importing firearms or ammunition. Therefore, no firearm or ammunition shall be imported or brought into the United States or a possession thereof by any licensee other than a licensed importer unless the Director issues a permit authorizing the importation of the firearm or ammunition. No barrel nfor a handgu not generally recognized as particularly suitable for or readily adaptable to sporting purposes shall be imported or brought into the United States or a possession thereof by any person. Therefore, no firearm barrel shall be imported or brought into the United States or possession thereof by any licensee other than a licensed importer unless the Director issues a permit authorizing the importation of the firearm barrel.

(b) An application for a permit, ATF Form 6, to import or bring a firearm, firearm barrel or ammunition into the United States or a possession thereof by a licensee, other than a licensed importer, shall be filed, in triplicate, with the Director. The application shall contain (1) the name, address, and license number of the applicant, (2) a description of the firearm, firearm barrel or ammunition to be imported, including type (e.g., rifle, shotgun, pistol, revolver; and in the case of ammunition only, ,ball wadcutter), model, caliber, size or gauge, barrel length (if a firearm or firearm barrel), country of manufacture, and name of the manufacturer, (3) the unit cost of the firearm, firearm barrel or ammunition to be imported, (4) the name and address of the foreign seller and the foreign shipper, (5) the country from which the firearm, firearm barrel, or ammunition is to be imported, and (6)(i) if a firearm or ammunition imported or brought in for scientific or research purposes, a statement describing such purposes, or (ii) if a firearm or ammunition for use in connection with competition or training pursuant to Chapter 401 of Title 10, U.S.C., a statement describing such intended use, or (iii) if an unserviceable firearm (other than a machine gun) being imported as a curio or museum piece, a description of how it was rendered unserviceable and an explanation of why it is a curio or museum piece, or (iv) if a firearm other than a surplus military firearm, of a type that does not fall within the definition of a firearm under section 5845(a) of the Internal Revenue Code of 1986 and is for sporting purposes, an explanation of why the firearm is generally recognized as particularly suitable for or readily adaptable to sporting purposes; or (v) if ammunition being imported for sporting purposes, a statement why the ammunition is generally recognized as particularly suitable for or readily adaptable to sporting purposes, or (vi) if a firearm barrel, and is for a handgun, an explanation why the handgun is generally recognized as particularly suitable for or readily adaptable to sporting purposes. If the Director approves the application, such approved application shall serve as the permit to import the firearm, firearm barrel or ammunition described therein. The Director shall furnish the approved application (permit) to the applicant and retain two copies thereof for administrative use. If the Directo disapproves the application, the applicant shall be notified of the basis for the disapproval.

(c) A firearm, firearm barrel, or ammunition imported or brought into the United States or a possession thereof under the provisions of this section may be released from Customs custody to the licensee importing the firoarm, firearm barrel, or ammunition upon showing that the licensee has obtained a permit from the Director for the importation. In obtaining the release of the firearm, firearm barrel, or ammunition from Customs custody, the licensee importin same shall furnish ATF Form 6A to the Customs officer releasing the firearm, firearm barrel, or ammunition. The Customs officer shall, after certification, forward the ATF Form 6A to the address specified on the.form The ATF Form 6A shall show the name, address, and the license number of the licensee, the name of the manufacturer, the country of manufacture, and the type, model, and caliber, size (if ammunition) or gauge of the firearm, firearm barrel or ammunition so released, and, if applicable, the number of firearms, firearm barrels, or rounds of ammunition released.

[T.D. ATF-270, 53 FR 10499, Mar. 31, 1988]

§ 178.113a Importation of firearm barrels by nonlicensees.

(a) A permit will not be issued for a firearm barrel for a handgun not generally recognized as particularly suitable for or readily adaptable to sporting purposes. No firearm barrel shall be imported or brought into the United States or possession thereof by any nonlicensee unless the Director issues a permit authorizing the importation of the firearm barrel.

(b) An application for a permit, ATF Form 6, to import or bring a firearm barrel into the United States or a possession thereof under this section shall be filed, in triplicate, with the Director The application shall contain (1) the name and address of the applicant, (2) a description of the firearm barrel to be imported, including type (e.g.; rifle, shotgun, pistol, revolver), model, caliber, size or gauge, barrel length, country of manufacture, and name of the manufacturer, (3) the unit cost of the firearm barrel, (4) the name and address of the foreign seller and the foreign shipper, (5) the country from which the firearm barrel is to be imported, and (6) if a handgun barrel, an explanation of why the barrel is for a handgun that is generally recognized as particularly suitable for or readily adaptable to sporting purposes. If the Director approves the application, such approved application shall serve as the permit to import the firearm barrel. The Director shall furnish the approved application (permit) to the applicant and retain two copies thereof for administrative use. If the Directo disapproves the application, the applicant shall be notified of the basis for the disapproval.

(c) A firearm barrel imported or brought into the United States or a possession thereof under the provisions of this section may be released from Customs custody to the person importing the firearm barrel upon showing that the person has obtained a permit from the Director for the importation. In obtaining the release of the firearm barrel from Customs custody, the person importin same shall furnish ATF Form 6A to the Customs officer releasing the firearm barrel. The Customs officer shall, after certification, forward the ATF Form 6A to the address specified on the form. The ATF Form 6A shall show the name and address of the person importing the firearm barrel, the name of the manufacturer, the country of manufacture, and the type, model, and caliber or gauge of the firearm barrel so released, and, if applicable, the number of firearm barrels released.

[T.D. ATF-270, 53 FR 10499, Mar. 31, 1988]

§ 178.114 Importation by members of the U.S. Armed Forces.

(a) The Director may issue a permit authorizing the importation of a firearm or ammunition into the United States to the place of residence of any military member of the U.S. Armed Forces who is on active duty outside the United States, or who has been on active duty outside the United States within the 60-day period immediately preceding the intended importation: **Provided,** That such firearm or ammunition is generally recognized as particularly suitable for or readily adaptable to sporting purposes and is intended for the personal use of such member. An application for such a permit, ATF Form 6, shall be filed, in triplicate, with the Director The application shall contain (1) the name and current address of the applicant, (2) certification that the transportation, receipt, or possession of the firearm or ammunition to be imported would not constitute a violation of any provision of the Act or of any State law or local ordinance at the place of the applicant's residence, (3) a description of the firearm or ammunition to be imported, including type (e.g.; rifle, shotgun, pistol, revolver; and in the case of ammunition only, ball wadcutte), model, caliber, size or gauge, barrel length (if a firearm), country of manufacture, and the name of the manufacturer, (4) the unit cost of the firearm or ammunition to be imported, (5) the name and address of the foreign seller (if applicable) and the foreign shipper, (6) the country from which the firearm or ammunition is to be imported, (7)(i) that the firearm or ammunition being imported is for the personal use of the applicant, and (ii) if a firearm, a statement that it is not a surplus military firearm, that it does not fall within the definition of a firearm under section 5845(a) of the Internal Revenue Code of 1986, and an explanation of why the firearm is generally recognized as particularly suitable for or readily adaptable to sporting purposes, or (iii) if ammunition, a statement why it is generally recognized as particularly suitable for or readily adaptable to sporting purposes, and (8) the applicant's date of birth, rank or grade, place of residence, present foreign duty station or last foreign duty station, as the case may be, the date of the applicant's reassignment to a duty station within the United States, if applicable, and the military branch of which the applicant is a member. If the Director approves the application, such approved application shall serve as the permit to import the firearm or ammunition described therein. The Director shall furnish the approved application (permit) to the applicant and shall retain the two copies thereof for administrative purposes. If the Directo disapproves the application, the applicant shall be notified of the basis for the disapproval.

(b) Upon receipt of an approved application (permit) to import the firearm or ammunition, the applicant may obtain the release of same from Customs custody upon showing that the applicant has obtained a permit from the Director for the importation. In obtaining the release of the firearm or ammunition from Customs custody, the military member of the U.S. Armed Forces importqin same shall furnish ATF Form 6A to the Customs officer releasing the firearm or ammunition. The Customs officer shall, after certification, forward the ATF Form 6A to the address specified on the form. The ATF Form 6A shall show the name and address of such military member, the name of the manufacturer, the country of manufacture, and the type, model, and caliber, size or gauge of the firearm or ammunition so released, and, if applicable, the number of firearms or rounds of ammunition released. However, when such military member is on active duty outside the United States, the military member may appoint, in writing, an agent to obtain the release of the firearm or ammunition from Customs custody for such member. Such agent shall present sufficient identification of the agent and the written authorization to act on behalf of such military member to the Customs officer who is to release the firearm or ammunition.

(c) Firearms determined by the Department of Defense to be war souvenirs may be imported into the United States by the military members of the U.S. Armed Forces under such provisions and procedures as the

Department of Defense may issue.

[T.D. ATF-270, 53 FR 10500, Mar. 31, 1988]

§ 178.115 Exempt importation.

(a) Firearms and ammunition may be brought into the United States or any possession thereof by any person who can establish to the satisfaction of Customs that such firearm or ammunition was previously taken out of the United States or any possession thereof by such person. Registration on Customs Form 4457 or on any other registration document available for this purpose may be completed before departure from the United States at any U.S. customhouse or any office of a Director of Industry Operations. A bill of sale or other commercial document showing transfer of the firearm or ammunition in the United States to such person also may be used to establish proof that the firearm or ammunition was taken out of the United States by such person Firearms and ammunition furnished under the provisions of section 925(a)(3) of the Act to military members of the U.S. Armed Forces on active duty outside of the United States also may be imported into the United States or any possession thereof by such military members upon establishing to the satisfaction of Customs that such firearms and ammunition were so obtained.

(b) Firearms, firearm barrels, and ammunition may be imported or brought into the United States by or for the United States or any department or agency thereof, or any State or any department, agency, or political subdivision thereof. A firearm, firearm barrel or ammunition imported or brought into the United States under this paragraph may be released from Customs custody upon a showing that the firearm, firearm barrel or ammunition is being imported or brought into the United States by or for such a governmental entity.

(c) The provisions of this subpart shall not apply with respect to the importation into the United States of any antique firearm.

(d) Firearms and ammunition are not imported into the United States, and the provisions of this subpart shall not apply, when such firearms and ammunition are brought into the United States by:

(1) A nonresident of the United States for legitimate hunting or lawful sporting purposes, and such firearms and such ammunition as remains following such shooting activity are to be taken back out of the territorial limits of the United States by such person upon conclusion of the shooting activity;

(2) Foreign military personnel on official assignment to the United States who bring such firearms or ammunition into the United States for their exclusive use while on official duty in the United States;

(3) Official representatives of foreign governments who are accredited to the U.S. Government or are en route to or from other countries to which accredited;

(4) Officials of foreign governments and distinguished foreign visitors who have been so designated by the Department of State; and

(5) Foreign law enforcement officers of friendly foreign governments entering the United States on official law enforcement business.

(e) Notwithstanding the provisions of paragraphs (d)(2), (3), (4) and (5) of this section, the Secretary of the Treasury or his delegate may in the interest of public safety and necessity require a permit for the importation or bringing into the United States of any firearms or ammunition.

[33 FR 18555, Dec. 14,. 1968 Redesignated at 40 FR 16835, Apr. 15, 1975, and amended by T.D. ATF-58, 44 FR 32367, June 6, 1979; T.D. ATF-270, 53 FR 10500, Mar. 31, 1988]

§ 178.116   Conditional importation.

The Director shall permit the conditional importation or bringing into the United States or any possession thereof of any firearm, firearm barrel, ammunition, or ammunition feeding device as defined in §178.119(b) for the purpose of examining and testing the firearm, firearm barrel, ammunition, or ammunition feeding device in connection with making a determination as to whether the importation or bringing in of such firearm, firearm barrel, ammunition, or ammunition feeding device will be authorized under this part. An application on ATF Form 6 for such conditional importation shall be filed, in duplicate, with the Director. The Director may impose conditions upon any importation under this section including a requirement that the firearm, firearm barrel, ammunition, or ammunition feeding device be shipped directly from Customs custody to the Director and that the person importing or bringing in the firearm, firearm barrel, ammunition, or ammunition feeding device must agree to either export the firearm, firearm barrel, ammunition, or ammunition feeding device or destroy same if a determination is made that the firearm, firearm barrel, ammunition, or ammunition feeding device may not be imported or brought in under this part A firearm, firearm barrel, ammunition, or ammunition feeding device imported or brought into the United States or any possession thereof under the provisions of this section shall be released from Customs custody upon the payment of customs duties, if applicable, and in the manner prescribed in the conditional authorization issued by the Director.

[T.D. ATF-383, 61 FR 39321, July 29, 1996]

§ 178.117   Function outside a customs territory.

In the insular possessions of the United States outside customs territory, the functions performed by U.S. Customs officers under this subpart within a customs territory may be performed by the appropriate authorities of a territorial government or other officers of the United States who have been designated to perform such functions. For the purpose of this subpart, the term customs territory means the United States, the District of Columbia, and the Commonwealth of Puerto Rico.

§ 178.118   Importation of certain firearms classified as curios or relics.

Notwithstanding any other provision of this part, a licensed importer may import all rifles and shotguns classified by the Director as curios or relics, and all handguns classified by the Director as curios or relics that are determined to be generally recognized as particularly suitable for or readily adaptable to sporting purposes. The importation of such curio or relic firearms must be in accordance with the applicable importation provisions of this part and the importation provisions of 27 CFR Part 47. Curios or relics which fall within the definition of "firearm" under 26 U.S.C. 5845(a) must also meet the importation provisions of 27 CFR Part 179 before they may be imported.

[T.D. ATF-202, 50 FR 14383, Apr. 12, 1985]

§ 178.119   Importation of ammunition feeding devices.

(a) No ammunition feeding device shall be imported or brought into the United States unless the Director has authorized the importation of such device.

(b) For purposes of this section, an "ammunition feeding device" is a magazine, belt, drum, feed strip, or similar device for a firearm that has a capacity of, or that can be readily restored or converted to accept, more than 10 rounds of ammunition. The term does not include an attached tubular device designed to accept, and capable of operating only with, .22 calibre rimfire ammunition, or a fixed device for a manually operated firearm, or a fixed device for a firearm listed in 18 U.S.C. 922, Appendix A.

(c) An application for a permit, ATF Form 6, to import or bring a ammunition feeding device into the United States or a possession thereof under this section shall be filed, in triplicate, with the Director. The application shall contain:

(1) The name and address of the person importing the device,

(2) A description of the device to be imported, including type and cartridge capacity, model and caliber of firearm for which the device was made, country of manufacture, and name of the manufacturer if known,

(3) The unit cost of the device to be imported,

(4) The country from which to be imported,

(5) The name and address of the foreign seller and the foreign shipper,

(6) Verification that such device will be marked as required by this part, and

(7) A statement by the importer that the device is being imported for sale to purchasers specified in § 178.40a(b) or physical or reasonable documentary evidence establishing that the magazine was manufactured on or before September 13, 1994. Any one of the following examples, which are not meant to be exhaustive, may be sufficient to establish the time of manufacture:

(i) Permanent markings or physical characteristics which establish that the magazine was manufactured on or before September 13, 1994;

(ii) A certification from the importer, under penalty of perjury, that the importer maintained continuous custody beginning on a date prior to September 14, 1994, and continuing until the date of the certification. Such certification shall also be supported by reasonable documentary evidence, such as commercial records;

(iii) A certification from the importer, under penalty of perjury, that the magazines sought to be imported were in the custody and control of a foreign Government on or before September 13, 1994, along with reasonable documentary evidence to support the certification; or

(iv) A certification from the importer, under penalty of perjury, that the magazine was in the possession of a foreign arms supplier on or before September 13, 1994, along with reasonable documentary evidence to support the certification.

(d) The Director shall act upon applications to import ammunition feeding devices as expeditiously as possible. If the Director approves the application, such approved application shall serve as the permit to import the device described therein, and importation of such devices may continue to be made by the person importing such devices under the approved application (permit) during the period specified thereon. The Director shall furnish the approved application (permit) to the applicant and retain two copies thereof for administrative use. If the Director disapproves the application, the person importing such devices shall be notified of the basis for the disapproval.

(e) An ammunition feeding device imported or brought into the United States by a person importing such a device may be released from Customs custody to the person importing such a device upon showing that such person has obtained a permit from the Director for the importation of the device to be released. In obtaining the release from Customs custody of such a device authorized by this section to be imported through use of a permit, the person importing such a device shall prepare ATF Form 6A, in duplicate, and furnish the original ATF Form 6A to the Customs officer releasing the device. The Customs officer shall, after certification, forward the ATF Form 6A to the address specified on the form. The ATF Form 6A shall show the name and address of the person importing the device, the name of the manufacturer of the device, the country of manufacture, the type, model, caliber, size, and the number of devices released.

(f) Within 15 days of the date of release from Customs custody the person importing such a device shall:

(1) Forward to the address specified on the form a copy of ATF Form 6A on which shall be reported any error or discrepancy appearing on the ATF Form 6A certified by Customs, and

(2) Pursuant to § 178.92, place all required identification data on each imported device manufactured after September 13, 1994, if same did not bear such identification data at the time of its release from Customs custody.

(g) The Director may authorize the conditional importation of an ammunition feeding device as provided in § 178.116.

(Paragraphs (a), (c), and (d) approved by the Office of Management and Budget under control numbers 1512-0017 and 1512-0018; paragraphs (e) and (f) approved by the Office of Management and Budget under control number 1512-0019)

[T.D. ATF-383, 61 FR 39320, July 29, 1996]

Subpart H-Records

§ 178.121 General.

(a) The records pertaining to firearms transactions prescribed by this part shall be retained on the licensed premises in the manner prescribed by this subpart and for the length of time prescribed by § 178.129. The records pertaining to ammunition prescribed by this part shall be retained on the licensed premises in the manner prescribed by § 178.125.

(b) ATF officers may, for the purposes and under the conditions prescribed in § 178.23, enter the premises of any licensed importer, licensed manufacturer, licensed dealer, or licensed collector for the purpose of examining or inspecting any record or document required by or obtained under this part. Section 923(g) of the Act requires licensed importers, licensed manufacturers, licensed dealers, and licensed collectors to make such records available for such examination or inspection during business hours or, in the case of licensed collectors, hours of operation, as provided in § 178.23.

(c) Each licensed importer, licensed manufacturer, licensed dealer, and licensed collector shall maintain such records of importation, production, shipment, receipt, sale, or other disposition, whether temporary or permanent, of firearms and such records of the disposition of ammunition as the regulations contained in this part prescribe. Section 922(m) of the Act makes it unlawful for any licensed importer, licensed manufacturer, licensed dealer, or licensed collector knowingly to make any false entry in, to fail to make appropriate entry in, or to fail to properly maintain any such record.

(d) For recordkeeping requirements for sales by licensees at gun shows see § 178.100(c).

(Information collection requirements in paragraph (a) approved by the Office of Management and Budget under control number 1512-0129; information collection requirements in paragraphs (b) and (c) approved by the Office of Management and Budget under control number 1512-0387)

[33 FR 18555, Dec. 14,. 1968 Redesignated at 40 FR 16835, Apr. 15, 1975, and amended by T.D. ATF-191, 49 FR 46891, Nov. 29, 1984; T.D. ATF-208, 50 FR 26703, June 28, 1985; T.D. ATF-270, 53 FR 10501, Mar. 31, 1988]

§ 178.122 Records maintained by importers.

(a) Each licensed importer shall, within 15 days of the date of importation or other acquisition, record the type, model, caliber or gauge, manufacturer, country of manufacture, and the serial number of each firearm imported or otherwise acquired, and the date such importation or other acquisition was made.

(b) A record of firearms disposed of by a licensed importer to another licensee and a separate record of armor piercing ammunition dispositions to governmental entities, for exportation, or for testing or experimentation authorized under the provisions of § 178.149 shall be maintained by the licensed importer on the licensed premises. For firearms, the record shall show the quantity, type, manufacturer, country of manufacture, caliber or gauge, model, serial number of the firearms so transferred, the name and license number of the licensee to whom the firearms were transferred, and the date of the transaction. For armor piercing ammunition, the record shall show the date of the transaction, manufacturer, caliber or gauge, quantity of projectiles, and the name and address of the purchaser. The information required by this paragraph shall be entered in the proper record book not later than the seventh day following the date of the transaction, and such information shall be recorded under the following formats:

Importer's Firearms Disposition Record

Quantity	Type	Manufacturer	Country of manufacture	Caliber or gauge	Model	Serial No.	Name and license No. of licensee to whom transferred	Date of the transaction

Importer's Armor Piercing Ammunition Disposition Record

Date	Manufacturer	Caliber or Gauge	Quantity of projectiles	Purchaser - Name and address

(c) Notwithstanding the provisions of paragraph (b) of this section, the Director of Industry Operations may authorize alternate records to be maintained by a licensed importer to record the disposal of firearms and armor piercing ammunition when it is shown by the licensed importer that such alternate records will accurately and readily disclose the information required by paragraph (b) of this section. A licensed importer who proposes to use alternate records shall submit a letter application, in duplicate, to the Director of Industry Operations and shall describe the proposed alternate records and the need therefor Such alternate records shall not be employed by the licensed importer until approval in such regard is received from the Director of Industry Operations.

   (d) Each licensed importer shall maintain separate records of the sales or other dispositions made of firearms t nonlicensees. Such records shall be maintained in the form and manner as prescribed by §§ 178.124 and 178.125 in regard to firearms transaction records and records of acquisition and disposition of firearms.

(Approved by the Office of Management and Budget under control number 1512-0387)

[T.D. ATF-270, 53 FR 10501, Mar. 31, 1988]

§ 178.123 Records maintained by manufacturers.

   (a) Each licensed manufacturer shall record the type, model, caliber or gauge, and serial number of each complete firearm manufactured or otherwise acquired, and the date such manufacture or other acquisition was made. The information required by this paragraph shall be recorded not later than the seventh day following the date such manufacture or other acquisition was made.

   (b) A record of firearms disposed of by a manufacturer to another licensee and a separate record of armor piercing ammunition dispositions to governmental entities, for exportation, or for testing or experimentation authorized under the provision of § 178.149 shall be maintained by the licensed manufacturer on the licensed premises. For firearms, the record shall show the quantity, type, model, manufacturer, caliber, size or gauge, serial number of the firearms s transferred, the name and license number of the licensee to whom the firearms were transferred, and the date of the transaction. For armor piercing ammunition, the record shall show the manufacturer, caliber or gauge, quantity, the name and address of the transferee to whom the armor piercing ammunition was transferred, and the date of the transaction The information required by this paragraph shall be entered in the proper record book not later than the seventh day following the date of the transaction, and such information shall be recorded under the format prescribed by § 178.122, except that the name of the manufacturer of a firearm or armor piercing ammunition need not be recorded if the firearm or armor piercing ammunition is of the manufacturer's own manufacture.

   (c) Notwithstanding the provisions of paragraph (b) of this section, the Director of Industry Operations may authorize alternate records to be maintained by a licensed manufacturer to record the disposal of firearms and armor piercing ammunition when it is shown by the licensed manufacturer that such alternate records will accurately and readily disclose the information required by paragraph (b) of this section. A licensed manufacturer who proposes to use alternate records shall submit a letter application, in duplicate, to the Director of Industry Operations and shall describe the proposed alternate record and the need therefor. Such alternate records shall not be employed by the licensed manufacturer until approval in such regard is received from the Director of Industry Operations.

   (d) Each licensed manufacturer shall maintain separate records of the sales or other dispositions made of firearms t nonlicensees. Such records shall be maintained in the form and manner as prescribed by §§ 178.124 and 178.125 in regard to firearms transaction records and records of acquisition and disposition of firearms.

(Approved by the Office of Management and Budget under control number 1512-0369)

[T.D. ATF-270, 53 FR 10501, Mar. 31, 1988]

§ 178.124 Firearms transaction record.

(a) A licensed importer, licensed manufacturer, or licensed dealer shall not sell or otherwise dispose, temporarily or permanently, of any firearm to any person, other than another licensee, unless the licensee records the transaction on a firearms transaction record, Form 4473: **Provided**, That a firearms transaction record, Form 4473, shall not be required to record the disposition made of a firearm delivered to a licensee for the sole purpose of repair or customizing when such firearm or a replacement firearm is returned to the person from whom received.

(b) A licensed manufacturer, licensed importer, or licensed dealer shall retain in alphabetical (by name of purchaser), chronological (by date of disposition), or numerical (by transaction serial number) order, and as a part of the required records, each Form 4473 obtained in the course of transferring custody of the firearms.

(c) (1) Prior to making an over-the -counter transfer of a firearm to a nonlicensee who is a resident of the State in which the licensee's business premises is located, the licensed importer, licensed manufacturer, or licensed dealer so transferring the firearm shall obtain a Form 4473 from the transferee showing the transferee's name, sex, residence address (including county or similar political subdivision), date and place of birth; height, weight and race of the transferee; whether the transferee is a citizen of the United States; the transferee's State of residence; and certification by the transferee that the transferee is not prohibited by the Act from transporting or shipping a firearm in interstate or foreign commerce or receiving a firearm which has been shipped or transported in interstate or foreign commerce or possessing a firearm in or affecting commerce.

(2) In order to facilitate the transfer of a firearm and enable NICS to verify the identity of the person acquiring the firearm, ATF Form 4473 also requests certain optional information. This information includes the transferee's social security number and alien registration number (if applicable). Such information may help avoid the possibility of the transferee being misidentified as a felon or other prohibited person.

(3) After the transferee has executed the Form 4473, the licensee:

(i) Shall verify the identity of the transferee by examining the identification document (as defined in § 178.11) presented, and shall note on the Form 4473 the type of identification used;

(ii) Shall, in the case of a transferee who is an alien legally in the United States, cause the transferee to present documentation establishing that the transferee is a resident of the State (as defined in § 178.11) in which the licensee's business premises is located, and shall note on the form the documentation used. Examples of acceptable documentation include utility bills or a lease agreement which show that the transferee has resided in the State continuously for at least 90 days prior to the transfer of the firearm; and

(iii) Shall comply with the requirements of § 178.102 and record on the form the date on which the licensee contacted the NICS, as well as any response provided by the system, including any identification number provided by the system.

(4) The licensee shall identify the firearm to be transferred by listing on the Form 4473 the name of the manufacturer, the name of the importer (if any), the type, model, caliber or gauge, and the serial number of the firearm.

(5) The licensee shall sign and date the form if the licensee does not know or have reasonable cause to believe that the transferee is disqualified by law from receiving the firearm and transfer the firearm described on the Form 4473.

(d) Prior to making an over-th -counter transfer of a shotgun or rifle under the provisions contained in § 178.96(c) to a nonlicensee who is not a resident of the State in which the licensee's business premises is located, the licensee so transferring the shotgun or rifle, and such transferee, shall comply with the requirements of paragraph (c) of this section **Provided,** That in the case of a transferee who is an alien legally in the United States, the documentation required by paragraph (c)(3)(ii) of this section need only establish that the transferee is a resident of any State and has resided in such State continuously for at least 90 days prior to the transfer of the firearm. Examples of acceptable documentation include utility bills or a lease agreement. The licensee shall note on the form the documentation used.

(e) Prior to making a transfer of a firearm to any nonlicensee who is not a resident of the State in which the licensee's business premises is located, and such nonlicensee is acquiring the firearm by loan or rental from the licensee for temporary use for lawful sporting purposes, the licensed importer, licensed manufacturer, or licensed dealer so furnishing the firearm, and such transferee, shall comply with the provisions of paragraph (c) of this section, except for the provisions of paragraph (c)(3)(ii).

(f) Form 4473 shall be submitted, in duplicate, to a licensed importer, licensed manufacturer, or licensed dealer by a transferee who is purchasing or otherwise acquiring a firearm by other than an over-th -counter transaction, who is not subject to the provisions of § 178.102(a), and who is a resident of the State in which the licensee's business premises are located. The Form 4473 shall show the name, address, date and place of birth, height, weight, and race of the transferee; and the title, name, and address of the principal law enforcement officer of the locality to which the firearm will be delivered The transferee also must date and execute the sworn statement contained on the form showing, in case the firearm to be transferred is a firearm other than a shotgun or rifle, the transferee is 21 years or more of age; in case the firearm to be transferred is a shotgun or rifle, the transferee is 18 years or more of age; whether the transferee is a citizen of the United States; the transferee's State of residence, and in the case of a transferee who is an alien legally in the United States, the transferee has resided in that State continuously for at least 90 days prior to the transfer of the firearm; the transferee is not prohibited by the provisions of the Act from shipping or transporting a firearm in interstate or foreign commerce or receiving a firearm which has been shipped or transported in interstate or foreign commerce or possessing a firearm in or affecting commerce; and the transferee's receipt of the firearm would not be in violation of any statute of the State or published ordinance applicable to the locality in which the transferee resides. Upon receipt of such Forms 4473, the licensee shall identify the firearm to be transferred by listing in the Forms 4473 the name of the manufacturer, the name of the importer (if any), the type, model, caliber or gauge, and the serial number of the firearm to be transferred The licensee shall prior to shipment or delivery of the firearm to such transferee, forward by registered or certified mail (return receipt requested) a copy of the Form 4473 to the principal law enforcement officer named in the Form 4473 by the transferee, and shall delay shipment or delivery of the firearm to the transferee for a period of at least 7 days following receipt by the licensee of the return receipt evidencing delivery of the copy of the Form 4473 to such principal law enforcement officer, or the return of the copy of the Form 4473 to the licensee due to the refusal of such principal law enforcement officer to accept same in accordance with U.S. Postal Service regulations. The original Form 4473, and evidence of receipt or rejection of delivery of the copy of the Form 4473 sent to the principal law enforcement officer, shall be retained by the licensee as a part of the records required to be kept under this subpart.

(g) A licensee who sells or otherwise disposes of a firearm to a nonlicensee who is other than an individual, shall obtain from the transferee the information required by this section from an individual authorized to act on behalf of the transferee In addition, the licensee shall obtain from the individual acting on behalf of the transferee a written statement, executed under the penalties of perjury, that the firearm is being acquired for the use of and will be the property of the transferee, and showing the name and address of that transferee.

(h) The requirements of this section shall be in addition to any other recordkeeping requirement contained in this part.

(i) A licensee may obtain, upon request, an emergency supply of Forms 4473 from any Director of Industry Operations. For normal usage, a licensee should request a year's supply from the ATF Distribution Center, 7943 Angus Court, Springfield, Virginia 22153.

(Paragraph (f) approved by the Office of Management and Budget under control number 1512-0130; all other recordkeeping approved by the Office of Management and Budget under control number 1512-0129)

[33 FR 18555, Dec. 14, 1968, as amended by T.D. ATF -172, 49 FR 14942, Apr. 16, 1984; T.D. ATF-241, 51 FR 39625, Oct. 29, 1986; T.D. ATF-270, 53 FR 10502, Mar. 31, 1988; T.D. ATF-389, 62 FR 19442, Apr. 21, 1997; T.D. ATF-415, 63 FR 58272, Oct. 29, 1998]

§ 178.124a Firearms transaction record in lieu of record of receipt and disposition.

(a) A licensed dealer acquiring firearms after August 1, 1988 and contemplating the disposition of not more than 50 firearms within a succeeding 21 -month period to licensees o nonlicensees may maintain a record of the acquisition and disposition of such firearms on a firearms transaction record, Form 4473(LV), Part I or II, in lieu of the records prescribed by § 178.125. Such 12-month period shall commence from the date the licensed dealer first records the purchase or other acquisition of a firearm on Form 4473(LV) pursuant to this section. A licensed dealer who maintains records pursuant to this section, but whose firearms dispositions exceed 50 firearms within such 12-month period, shall make and maintain the acquisition and disposition records required by §178.125 with respect to each firearm exceeding 50.

(b) Each licensed dealer maintaining firearms acquisition and disposition records pursuant to this section shall record the purchase or other acquisition of a firearm on Form 4473(LV), Part I or II, in accordance with the instructions on the form not later than the close of the next business day following the date of such purchase or acquisition. However, when disposition is made of a firearm before the close of the next business day after the receipt of that firearm, the licensed dealer making such disposition shall enter all required acquisition information regarding the firearm on the Form 4473(LV) at the time such transfer or disposition is made. The record on Form 4473(LV) shall show the date of receipt, the name and address or the name and license number of the person from whom received, the name of the manufacturer and importer (if any), the model, serial number, type, and caliber or gauge of the firearm.

(c) Each licensed dealer maintaining firearms acquisition and disposition records pursuant to this section shall retain Form 4473(LV), Part I or II, reflecting firearms possessed by such business in chronological (by date of receipt) or numerical (by transaction serial number) order. Forms 4473(LV) reflecting the licensee's sale or disposition of firearms shall be retained in alphabetical (by name of purchaser), chronological (by date of disposition) or numerical (by transaction serial number) order.

(d) A licensed dealer maintaining records pursuant to this section shall record the sale or other disposition of a firearm to another licensee by entering on the Form 4473(LV), Part I, associated with such firearm, the name and license number of the person to whom transferred and by signing and dating the form.

(e) A licensed dealer shall obtain the Form 4473(LV), Part I, associated with the firearm in lieu of a Form 4473 and comply with the requirements specified in § 178.124(c) prior to making an over-the-counter transfer of a firearm to a nonlicensee.

(1) Who is a resident of the State in which the licensee's business premises is located,

(2) Who is not a resident of the State in which the licensee's business premises is located and the firearm is a shotgun or rifle and the transfer is under the provisions of § 178.96(c), or

(3) Who is not a resident of the State in which the licensee's business premises is located and who is acquiring the firearm by loan or rental for temporary use for lawful sporting purposes.

(f) A licensed dealer shall obtain the Form 4473(LV), Part II, associated with the firearm in lieu of a Form 4473 and comply with the requirements specified in § 178.124(f) prior to making a disposition of a firearm to a nonlicensee who is purchasing or otherwise acquiring a firearm by other than an over-the-counter transaction and who is a resident of the State in which the licensee's business premises is located If the licensee's record of the acquisition of the firearm is, at the time of the disposition, being maintained on a Form 4473(LV), Part I, for over-the-counter transactions, the licensee shall transfer the information relative to the receipt of the firearm, as required by paragraph (b) of this section, to Form 4473(LV), Part II. The corresponding form 4473(LV), Part I, may then be destroyed.

[T.D. ATF-273, 53 FR 24687, June 30, 1988; T.D. ATF-415, 63 FR 58272, Oct. 29, 1998]

§ 178.125 Record of receipt and disposition.

(a) **Armor piercing ammunition sales by licensed collectors to nonlicensees.** The sale or other disposition of armor piercing ammunition by licensed collectors shall be recorded in a bound record at the time a transaction is made The bound record shall be maintained in chronological order by date of sale or disposition of the armor piercing ammunition, and shall be retained on the licensed premises of the licensee for a period not less than two years following the date of the recorded sale or disposition of the armor piercing ammunition. The bound record entry shall show:

(1) The date of the transaction;

(2) The name of the manufacturer;

(3) The caliber or gauge;

(4) The quantity of projectiles;

(5) The name, address, and date of birth of the nonlicensee; and

(6) The method used to establish the identity of the armor piercing ammunition purchaser.

The format required for the bound record is as follows:

Disposition Record of Armor Piercing Ammunition

Date	Manufacturer	Caliber or gauge	Quantity of projectiles	Purchaser		Enter a (x) in the "known" column if purchaser is personally known to you. Otherwise, establish the purchaser's identification		
				Name and Address	Date of Birth	Known	Driver's license	Other type (specify)

However, when a commercial record is made at the time a transaction is made, a licensee may delay making an entry into the bound record if the provisions of paragraph (d) of this section are complied with.

(b) Armor piercing ammunition sales by licensed collectors to licensees. Sales or other dispositions of armor piercing ammunition from a licensed collector to another licensee shall be recorded and maintained in the manner prescribed in §178.122(b) for importers: **Provided,** That the license number of the transferee may be recorded in lieu of the transferee's address.

(c) Armor piercing ammunition sales by licensed dealers to governmental entities. A record of armor piercing ammunition disposed of by a licensed dealer to a governmental entity pursuant to § 178.99(e) shall be maintained by the licensed dealer on the licensed premises and shall show the name of the manufacturer, the caliber or gauge, the quantity, the name and address of the entity to which the armor piercing ammunition was transferred, and the date of the transaction. Such information shall be recorded under the format prescribed by § 178.122(b). Each licensed dealer disposing of armor piercing ammunition pursuant to § 178.99(e) shall also maintain a record showing the date of acquisition of such ammunition which shall be filed in an orderly manner separate from other commercial records maintained and be readily available for inspection. The records required by this paragraph shall be retained on the licensed premises of the licensee for a period not less than two years following the date of the recorded sale or disposition of the armor piercing ammunition.

(d) Commercial records of armor piercing ammunition transactions. When a commercial record is made at the time of sale or other disposition of armor piercing ammunition, and such record contains all information required by the bound record prescribed by paragraph (a) of this section, the licensed collector transferring the armor piercing ammunition may, for a period not exceeding 7 days following the date of such transfer, delay making the required entry into such bound record: **Provided,** That the commercial record pertaining to the transfer is:

(1) Maintained by the licensed collector separate from other commercial documents maintained by such licensee, and

(2) Is readily available for inspection on the licensed premises until such time as the required entry into the bound record is made.

(e) Firearms receipt and disposition by dealers. Except as provided in § 178.124a with respect to alternate records for the receipt and disposition of firearms by dealers, each licensed dealer shall enter into a record each receipt and disposition of firearms. In addition, before commencing or continuing a firearms business, each licensed dealer shall inventory the firearms possessed for such business and shall record same in the record required by this paragraph. The record required by this paragraph shall be maintained in bound form under the format prescribed below. The purchase or other acquisition of a firearm shall, except as provided in paragraph (g) of this section, be recorded not later than the close of the next business day following the date of such purchase or acquisition. The record shall show the date of receipt, the name and address or the name and license number of the person from whom received, the name of the manufacturer and importer (if any), the model, serial number, type, and the caliber or gauge of the efirearm. Th sale or other disposition of a firearm shall be recorded by the licensed dealer not later than 7 days following the date of such transaction. When such disposition is made to a nonlicensee, th firearms transaction record, Form 4473, obtained by the licensed dealer shall be retained, until the transaction is recorded, separate from the licensee's Form 4473 file and be readily available for inspection. When such disposition is made to a licensee, the commercial record of the transaction shall be retained, until the transaction is recorded, separate from other commercial documents maintained by the licensed dealer, and be readily available for inspection The record shall show the date of the sale or other disposition of each firearm, the name and address of the person to whom the firearm is transferred, or the name and license number of the person to whom transferred if such person is a licensee, or the firearms transaction record, Form 4473, serial number if the licensed dealer transferring the firearm serially numbers the Forms 4473 and files them numerically. The format required for the record of receipt and disposition of firearms is as follows:

FIREARMS ACQUISTION AND DISPOSTION RECORDS

Description of firearm					Receipt		Disposition		
Manufacturer and/or importer	Model	Serial No.	Type	Caliber or gauge	Date	Name and address or name and license No.	Date	Name	Address or license No. if licensee, or Form 4473 Serial No. if Forms 4473 filed numerically

(f) Firearms receipt and disposition by licensed collectors. Each licensed collector shall enter into a record each receipt and disposition of firearms curios or relics. The record required by this paragraph shall be maintained in bound form under the format prescribed below. The purchase or other acquisition of a curio or relic shall, except as provided in paragraph (g) of this section, be recorded not later than the close of the next business day following the date of such purchase or other acquisition. The record shall show the date of receipt, the name and address or the name and license number of the person from whom received, the name of the manufacturer and importer (if any), the model, serial number, type, and the caliber or gauge of the firearm curio or relic. eTh sale or other disposition of a curio or relic shall be recorded by the licensed collector not later than 7 days following the date of such transaction. When such disposition is made to a licensee, the commercial record of the transaction shall be retained, until the transaction is recorded, separate from other commercial documents maintained by the licensee, and be readily available .for inspection The record shall show the date of the sale or other disposition of each firearm curio or relic, the name and address of the person to whom the firearm curio or relic is transferred, or the name and license number of the person to whom transferred if such person is a licensee, and the date of birth of the transferee if other than a licensee. In addition, the licensee shall—

(1) Cause the transferee, if other than a licensee, to be identified in any manner customarily used in commercial transactions (e.g., a driver's license), and note on the record the method used, and

(2) In the case of a transferee who is an alien legally in the United States and who is other than a licensee—

(i) Verify the identity of the transferee by examining an identification document (as defined in §178.11), and

(ii) Cause the transferee to present documentation establishing that the transferee is a resident of the State (as defined in § 178.11) in which the licensee's business premises is located if the firearm curio or relic is other than a shotgun or rifle, and note on the record the documentation used or is a resident of any State and has resided in such State continuously for at least 90 days prior to the transfer of the firearm if the firearm curio or relic is a shotgun or rifle and shall note on the record the documentation used. Examples of acceptable documentation include utility bills or a lease agreement which show that the transferee has resided in the State continuously for at least 90 days prior to the transfer of the firearm curio or relic.

(3) The format required for the record of receipt and disposition of firearms by collectors is as follows:

Firearms Collectors Acquisition and Disposition Record

Description of firearm					Receipt		Disposition			
Manufacturer and/or importer	Model	Serial No.	Type	Caliber or gauge	Date	Name and address or name and license No.	Date	Name and address or name and license No.	Date of birth if non-licensee	Driver's license No. or other identification if non-licensee

(g) Commercial records of firearms received. When a commercial record is held by a licensed dealer or licensed collector showing the acquisition of a firearm or firearm curio or relic, and such record contains all acquisition information required by the bound record prescribed by paragraphs (e) and (f) of this section, the licensed dealer or licensed collector acquiring such firearm or curio or relic, may, for a period not exceeding 7 days following the date of such acquisition, delay making the required entry into such bound record **Provided,** That the commercial record is, until such time as the required entry into the bound record is made **, (1)** maintained by the licensed dealer or licensed collector separate from other commercial documents maintained by such licensee, an **(2)** readily available for inspection on the licensed premises **Provided further,** That when disposition is made of a firearm or firearm curio or relic not entered in the bound record under the provisions of this paragraph, the licensed dealer or licensed collector making such disposition shall enter all required acquisition information regarding the firearm or firearm curio or relic in the bound record at the time such transfer or disposition is made.

(h) Alternate records. Notwithstanding the provisions of paragraphs (a), (e), and (f) of this section, the Director of Industry Operations may authorize alternate records to be maintained by a licensed dealer or licensed collector to record the acquisition and disposition of firearms or curios or relics and the disposition of armor piercing ammunition when it is shown by the licensed dealer or the licensed collector that such alternate records will accurately and readily disclose the required information. A licensed dealer or licensed collector who proposes to use alternate records shall submit a letter application, in duplicate, to the Director of Industry Operations and shall describe the proposed alternate records and the need therefor Such alternate records shall not be employed by the licensed dealer or licensed collector until approval in such regard is received from the Director of Industry Operations.

(i) **Requirements for importers and manufacturers.** Each licensed importer and licensed manufacturer selling or otherwise disposing of firearms or armor piercing ammunition t nonlicensees shall maintain such records of such transactions as are required of licensed dealers by this section.

(Approved by the Office of Management and Budget under control number 1512-0387)

[T.D. ATF-270, 53 FR 10503, Mar. 31, 1988, as amended by T.D. ATF-273, 53 FR 24687, June 30, 1988; T.D. ATF-313, 56 FR 32508, July 17, 1991; T.D. ATF-389, 62 FR 19442, Apr. 21, 1997]

§ 178.125a Personal firearms collection.

(a) Notwithstanding any other provision of this subpart, a licensed manufacturer, licensed importer, or licensed dealer is not required to comply with the provisions of § 178.102 or record on a firearms transaction record, Form 4473, the sale or other disposition of a firearm maintained as part of the licensee's personal firearms collection: **Provided**, That (1) the licensee has maintained the firearm as part of such collection for 1 year from the date the firearm was transferred from the business inventory into the personal collection or otherwise acquired as a personal firearm, (2) the licensee recorded in the bound record prescribed by § 178.125(e) the receipt of the firearm into the business inventory or other acquisition, (3) the licensee recorded the firearm as a disposition in the bound record prescribed by § 178.125(e) when the firearm was transferred from the business inventory into the personal; firearms collection or otherwise acquired as a personal firearm, and (4) the licensee enters the sale or other disposition of the firearm from the personal firearms collection into a bound record, under the format prescribed below, identifying the firearm transferred by recording the name of the manufacturer and importer (if any), the model, serial number, type, and the caliber or gauge, and showing the date of the sale or other disposition, the name and address of the transferee, or the name and business address of the transferee if such person is a licensee, and the date of birth of the transferee if other than a licensee. In addition, the licensee shall cause the transferee, if other than a licensee, to be identified in any manner customarily used in commercial transactions (e.g., a driver's license). The format required for the disposition record of personal firearms is as follows:

Disposition Record of Personal Firearms

Description of firearm					Disposition		
Manufacturer and/or importer	Model	Serial No.	Type	Caliber or gauge	Date	Name and address (business address if licensee)	Date of birth of nonlicensee

(b) Any licensed manufacturer, licensed importer, or licensed dealer selling or otherwise disposing of a firearm from the licensee's personal firearms collection under this section shall be subject to the restrictions imposed by the Act and this part on the dispositions of firearms by persons other than licensed manufacturers, licensed importers, and licensed dealers.

(Approved by the Office of Management and Budget under control number 1512-0387)

[T.D. ATF-270, 53 FR 10504, Mar. 31, 1988, as amended by T.D. ATF-313, 56 FR 32509, July 17, 1991; T.D. ATF-415, 63 FR 58272, Oct. 29, 1998]

§ 178.126 Furnishing transaction information.

(a) Each licensee shall, when required by letter issued by the Director of Industry Operations, and until notified to the contrary in writing by such officer, submit on Form 4483, Report of Firearms Transactions, for the periods and at the times specified in the letter issued by the Director of Industry Operations, all record information required by this subpart, or such lesser record information as the Director of Industry Operations in his letter may specify.

(b) The Director of Industry Operations may authorize the information to be submitted in a manner other than that prescribed in paragraph (a) of this section when it is shown by a licensee that an alternate method of reporting is reasonably necessary and will not unduly hinder the effective administration of this part. A licensee who proposes to use an alternate method of reporting shall submit a letter application, in duplicate, to the Director of Industry Operations and shall describe the proposed alternate method of reporting and the need therefor An alternate method of reporting shall not be employed by the licensee until approval in such regard is received from the Director of Industry Operations.

(Approved by the Office of Management and Budget under control number 1512-0387)

[33 FR 18555, Dec. 14,. 1968 Redesignated at 40 FR 16835, Apr. 15, 1975, and amended by T.D. ATF-172, 49 FR 14942, Apr. 16, 1984]

§ 178.126a Reporting multiple sales or other disposition of pistols and revolvers.

Each licensee shall prepare a report of multiple sales or other disposition whenever the licensee sells or otherwise disposes of, at one time or during any five consecutive business days, two or more pistols, or revolvers, or any combination of pistols and revolvers totaling two or more, to an unlicensed person: **Provided**, That a report need not be made where pistols or revolvers, or any combination thereof, are returned to the same person from whom they were received. The report shall be prepared on Form 3310.4, Report of Multiple Sale or Other Disposition of Pistols and Revolvers Not later than the close of business on the day that the multiple sale or other disposition occurs, the licensee shall forward two copies of Form 3310.4 to the ATF office specified thereon and one copy to the State police or to the local law enforcement agency in which the sale or other disposition took place. Where the State or local law enforcement officials have notified the licensee that a particular official has been designated to receive Forms 3310.4, the licensee shall forward such forms to that designated official. The licensee shall retain one copy of Form 3310.4 and attacheit to th firearms transaction record, Form 4473, executed upon delivery of the pistols or revolvers.

Example 1. A licensee sells a pistol and revolver in a single transaction to an unlicensed person. This is a multiple sale and must be reported not later than the close of business on the date of the transaction.

Example 2. A licensee sells a pistol on Monday and sells a revolver on the following Friday to the same unlicensed person. This is a multiple sale and must be reported not later than the close of business on Friday. If the licensee sells the same unlicensed person another pistol or revolver on the following Monday, this would constitute an additional multiple sale and must also be reported.

   Example 3. A licensee maintaining business hours on Monday through Saturday sells a revolver to an unlicensed person on Monday and sells another revolver to the same person on the following Saturday. This does not constitute a multiple sale and need not be reported since the sales did not occur during five consecutive business days.

(Approved by the Office of Management and Budget under control number 1512-0006)

[T.D. ATF-16, 40 FR 19202, May 2, 1975, as amended by T.D. ATF-172, 49 FR 14942, Apr. 16, 1984; T.D. ATF-270, 53 FR 10505, Mar. 31, 1988; T.D. ATF-354, 59 FR 7113, Feb. 14, 1994; T.D. ATF-361, 60 FR 10787, Feb. 27, 1995]

§ 178.127 Discontinuance of business.

   Where a licensed business is discontinued and succeeded by a new licensee, the records prescribed by this subpart shall appropriately reflect such facts and shall be delivered to the successor. Where discontinuance of the business is absolute, the records shall be delivered within 30 days following the business discontinuance to the ATF Out-of-Business Records Center, Spring Mills Office Park, 2029 Stonewall Jackson Drive, Falling Waters, West Virginia 25419, or any ATF office in the division in which the business was located: **Provided, however,** Where State law or local ordinance requires the delivery of records to other responsible authority, the Chief, National Licensing Center may arrange for the delivery of the records required by this subpart to such authority **Provided further,** That where a licensed business is discontinued and succeeded by a new licensee, the records may be delivered within 30 days following the business discontinuance to the ATF Out-of-Business Records Center or to any ATF office in the division in which the business was located.

[T.D. ATF-290, 54 FR 53055, Dec. 27, 1989; as amended by T.D. ATF-363, 60 FR 17455, April 6, 1995]

§ 178.128 False statement or representation.

   (a) Any person who knowingly makes any false statement or representation in applying for any license or exemption or relief from disability, under the provisions of the Act, shall be fined not more than $5,000 or imprisoned not more than 5 years, or both.

   (b) Any person other than a licensed manufacturer, licensed importer, licensed dealer, or licensed collector who knowingly makes any false statement or representation with respect to any information required by the provisions of the Act or this part to be kept in the records of a person licensed under the Act or this part shall be fined not more than $5,000 or imprisoned not more than 5 years, or both.

   (c) Any licensed manufacturer, licensed importer, licensed dealer, or licensed collector who knowingly makes any false statement or representation with respect to any information required by the provisions of the Act or this part to be kept in the records of a person licensed under the Act or this part shall be fined not more than $1,000 or imprisoned not more than 1 year, or both.

[T.D. ATF-270, 53 FR 10505, Mar. 31, 1988]

§ 178.129 Record retention.

   (a) Records prior to Act. Licensed importers and licensed manufacturers may dispose of records of sale or other disposition of firearms prior to December 16, 1968. Licensed dealers and licensed collectors may dispose of all records o firearms transactions that occurred prior to December 16, 1968.

(b) Firearms transaction record. Licensees shall retain each Form 4473 and Form 4473(LV) for a period of not less than 20 years after the date of sale or disposition Where a licensee has initiated a NICS check for a proposed firearms transaction, but the sale, delivery, or transfer of the firearm is not made, the licensee shall record any transaction number on the Form 4473, and retain the Form 4473 for a period of not less than 5 years after the date of the NICS inquiry. Forms 4473 shall be retained in the licensee's records as provided in § 178.124(b) **Provided**, That Forms 4473 with respect to which a sale, delivery or transfer did not take place shall be separately retained in alphabetical (by name of transferee) or chronological (by date of transferee's certification) order.

(c) Statement of intent to obtain a handgun, reports of multiple sales or other disposition of pistols and revolvers, and reports of theft or loss of firearms. Licensees shall retain each Form 5300.35 (Statement of Intent to Obtain a Handgun(s)) for a period of not less than 5 years after notice of the intent to obtain the handgun was forwarded to the chief law enforcement officer, as defined in § 178.150(c). Licensees shall retain each copy of Form 3310.4 (Report of Multiple Sale or Other Disposition of Pistols and Revolvers) for a period of not less than 5 years after the date of sale or other disposition. Licensees shall retain each copy of Form 3310.11 (Federal Firearms Licensee Theft/Loss Report) for a period of not less than 5 years after the date the theft or loss was reported to ATF.

(d) Records of importation and manufacture. Licensed importers and licensed manufacturers shall maintain permanent records of the importation, manufacture or other acquisition of firearms. Licensed importers' records and licensed manufacturers' records of the sale or other disposition of firearms after December 15, 1968, shall be retained through December 15, 1988, after which records of transactions over 20 years of age may be discarded.

(e) Records of dealers and collectors under the Act. The records prepared by licensed dealers and licensed collectors under the Act of the sale or other disposition of firearms and the corresponding record of receipt of such firearms shall be retained through December 15, 1988, after which records of transactions over 20 years of age may be discarded.

(f) Retention of records of transactions in semiautomatic assault weapons. The documentation required by §§ 178.40(c) and 178.132 shall be retained in the licensee's permanent records for a period of not less than 5 years after the date of sale or other disposition.

(Paragraph (b) approved by the Office of Management and Budget under control number 1512-0544; Paragraph (c) approved by the Office of Management and Budget under control numbers 1512-0520, 1512-0006, and 1512-0524; Paragraph (f) approved by the Office of Management and Budget under control number 1512-0526; all other recordkeeping approved by the Office of Management and Budget under control number 1512-0129)

[T.D. ATF-208, 50 FR 26704, June 28, 1985 and correctly designated at 50 FR 35081, Aug. 29, 1985, as amended by T.D. ATF-273, 53 FR 24687, June 30, 1988; T.D. ATF-361, 60 FR 10787, Feb. 27, 1995; T.D. ATF-363, 60 FR 17455, April 6, 1995; T.D. ATF-415, 63 FR 58272, Oct. 29, 1998]

§ 178.130 [Removed]

§ 178.131 Firearms transactions not subject to a NICS check.

(a)(1) A licensed importer, licensed manufacturer, or licensed dealer whose sale, delivery, or transfer of a firearm is made pursuant to the alternative provisions of § 178.102(d) and is not subject to the NICS check prescribed by § 178.102(a) shall maintain the records required by paragraph (a) of this section.

(2) If the transfer is pursuant to a permit or license in accordance with § 178.102(d)(1), the licensee shall either retain a copy of the purchaser's permit or license and attach it to the firearms transaction record, Form 4473, or record on the firearms transaction record, Form 4473, any identifying number, the date of issuance, and the expiration date (if provided) from the permit or license.

(3) If the transfer is pursuant to a certification by ATF in accordance with §§ 178.102(d)(3) and 178.150, the licensee shall maintain the certification as part of the records required to be kept under this subpart and for the period prescribed for the retention of Form 5300.35 in § 178.129(c).

(b) The requirements of this section shall be in addition to any other recordkeeping requirements contained in this part.

(Approved by the Office of Management and Budget under control number 1512-0544)

[T.D. ATF-361, 60 FR 10788, Feb. 27, 1995; T.D. ATF-415, 63 FR 58272, Oct. 29, 1998]

§ 178.132 Dispositions of semiautomatic assault weapons and large capacity ammunition feeding devices to law enforcement officers for official use and to employees or contractors of nuclear facilities.

Licensed manufacturers, licensed importers, and licensed dealers in semiautomatic assault weapons, as well as persons who manufacture, import, or deal in large capacity ammunition feeding devices, may transfer such weapons and devices manufactured after September 13, 1994, to law enforcement officers and to employees or contractors of nuclear facilities with the following documentation:

(a) Law enforcement officers. (1) A written statement from the purchasing officer, under penalty of perjury, stating that the weapon or device is being purchased for use in performing official duties and that the weapon or device is not being acquired for personal use or for purposes of transfer or resale; and

(2) A written statement from a supervisor of the purchasing officer, on agency letterhead, under penalty of perjury, stating that the purchasing officer is acquiring the weapon or device for use in official duties, that the firearm is suitable for use in performing official duties, and that the weapon or device is not being acquired for personal use or for purposes of transfer or resale.

(b) Employees or contractors of nuclear facilities. (1) Evidence that the employee is employed by a nuclear facility licensed pursuant to 42 U.S.C. 2133 or evidence that the contractor has a valid contract with such a facility.

(2) A written statement from the purchasing employee or contractor under penalty of perjury, stating that the weapon or device is being purchased for one of the purposes authorized in §§ 178.40(b)(7) and 178.40a(b)(3), i.e., on-site physical protection, on-site or off-site training, or off-site transportation of nuclear materials.

(3) A written statement from a supervisor of the purchasing employee or contractor, on agency or company letterhead, under penalty of perjury, stating that the purchasing employee or contractor is acquiring the weapon or device for use in official duties, and that the weapon or device is not being acquired for personal use or for purposes of transfer or resale.

(Approved by the Office of Management and Budget under control number 1512-0526)

[T.D. ATF-363, 60 FR 17455, April 6, 1995; [T.D. ATF-396, 63 FR 12643, March 16, 1998]]

Editor's Note

Dispositions of semiautomatic assault weapons to law enforcement officers must also comply with section 178.134. Therefore, the written statement of the purchasing officer's supervisor required by section 178.132(a)(2) must also state that a records check revealed that the purchasing officer has no convictions for misdemeanor crimes of domestic violence.

§ 178.133 Records of transactions in semiautomatic assault weapons.

The evidence specified in § 178.40(c), relating to transactions in semiautomatic assault weapons, shall be retained in the permanent records of the manufacturer or dealer and in the records of the licensee to whom the weapons are transferred.

(Approved by the Office of Management and Budget under control number 1512-0526)

[T.D. ATF-363, 60 FR 17455, April 6, 1995]

§ 178.134 Sale of firearms to law enforcement officers.

(a) Law enforcement officers purchasing firearms for official use who provide the licensee with a certification on agency letterhead, signed by a person in authority within the agency (other than the officer purchasing the firearm), stating that the officer will use the firearm in official duties and that a records check reveals that the purchasing officer has no convictions for misdemeanor crimes of domestic violence are not required to complete Form 4473 or Form 5300.35. The law enforcement officer purchasing the firearm may purchase a firearm from a licensee in another State, regardless of where the officer resides or where the agency is located.

(b)(1) The following individuals are considered to have sufficient authority to certify that law enforcement officers purchasing firearms will use the firearms in the performance of official duties:

(i) In a city or county police department, the director of public safety or the chief or commissioner of police.

(ii) In a sheriff's office, the sheriff.

(iii) In a State police or highway patrol department, the superintendent or the supervisor in charge of the office to which the State officer or employee is assigned.

(iv) In Federal law enforcement offices, the supervisor in charge of the office to which the Federal officer or employee is assigned.

(2) An individual signing on behalf of the person in authority is acceptable, provided there is a proper delegation of authority.

(c) Licensees are not required to prepare a Form 4473 or Form 5300.35 covering sales of firearm made in accordance with paragraph (a) of this section to law enforcement officers for official use. However, disposition to the officer must be entered into the licensee's permanent records, and the certification letter must be retained in the licensee's files.

[T.D. ATF-401, 63 FR 35520, June 30, 1998]

Subpart I-Exemptions, Seizures, and Forfeitures

§ 178.141 General.

With the exception of §§ 178.32(a)(9) and (d)(9) and 178.99(c)(9), the provisions of this part shall not apply with respect to:

(a) The transportation, shipment, receipt, possession, or importation of any firearm or ammunition imported for, sold or shipped to, or issued for the use of, the United States or any department or agency thereof or any State or any department, agency, or political subdivision thereof.

(b) The shipment or receipt of firearms or ammunition when sold or issued by the Secretary of the Army pursuant to section 4308 of Title 10, U.S.C., and the transportation of any such firearm or ammunition carried out to enable a person, who lawfully received such firearm or ammunition from the Secretary of the Army, to engage in military training or in competitions.

(c) The shipment, unless otherwise prohibited by the Act or any other Federal law, by a licensed importer, licensed manufacturer, or licensed dealer to a member of the U.S. Armed Forces on active duty outside the United States or to clubs, recognized by the Department of Defense, whose entire membership is composed of such members of the U.S. Armed Forces, and such members or clubs may receive a firearm or ammunition determined by the Director to be generally recognized as particularly suitable for sporting purposes and intended for the personal use of such member or club. Before making a shipment of firearms or ammunition under the provisions of this paragraph, a licensed importer, licensed manufacturer, or licensed dealer may submit a written request, in duplicate, to the Director for a determination by the Director whether such shipment would constitute a violation of the Act or any other Federal law, or whether the firearm or ammunition is considered by the Director to be generally recognized as particularly suitable for sporting purposes.

(d) The transportation, shipment receipt, possession, or importation of any antique firearm.

[33 FR 18555, Dec. 14,. 1968 Redesignated at 40 FR 10835, Apr. 15, 1975, and amended by T.D. ATF-241, 51 FR 39628, Oct. 29, 1986; T.D. ATF-270, 53 FR 10505, Mar. 31, 1988; T.D. ATF-313, 56 FR 32509, July 17, 1991; T.D. ATF-401, 63 FR 35520, June 30, 1998]

§ 178.142 Effect of pardons and expunctions of convictions.

(a) A pardon granted by the President of the United States regarding a Federal conviction for a crime punishable by imprisonment for a term exceeding 1 year shall remove any disability which otherwise would be imposed by the provisions of this part with respect to that conviction.

(b) A pardon granted by the Governor of a State or other State pardoning authority or by the pardoning authority of a foreign jurisdiction with respect to a conviction, or any expunction, reversal, setting aside of a conviction, or other proceeding rendering a conviction nugatory, or a restoration of civil rights shall remove any disability which otherwise would be imposed by the provisions of this part with respect to the conviction, unless:

(1) The pardon, expunction, setting aside, or other proceeding rendering a conviction nugatory, or restoration of civil rights expressly provides that the person may not ship, transport, possess or receive firearms; or

(2) The pardon, expunction, setting aside, or other proceeding rendering a convictio nugatory, or restoration of civil rights did not fully restore the rights of the person to possess or receive firearms under the law of the jurisdiction where the conviction occurred.

[T.D. ATF-270, 53 FR 10505, Mar. 31, 1988]

§ 178.143 Relief from disabilities incurred by indictment.

A licensed importer, licensed manufacturer, licensed dealer, or licensed collector who is indicted for a crime punishable by imprisonment for a term exceeding 1 year may, notwithstanding any other provision of the Act, continue operations pursuant to his existing license during the term of such indictment and until any conviction pursuant to the indictment becomes final: **Provided**, That if the term of the license expires during the period between the date of the indictment and the date the conviction thereunder becomes final, such importer, manufacturer, dealer, or collector must file a timely application for the renewal of his license in order to continue operations. Such application shall show that the applicant is under indictment for a crime punishable by imprisonment for a term exceeding 1 year.

§ 178.144 Relief from disabilities under the Act.

(a) Any person may make application for relief from the disabilities under sections 922(g) and (n) of the Act (see § 178.32).

(b) An application for such relief shall be filed, in triplicate, with the Director. It shall include the information required by this section and such other supporting data as the Director and the applicant deem appropriate.

(c) Any record or document of a court or other government entity or official required by this paragraph to be furnished by an applicant in support of an application for relief shall be certified by the court or other government entity or official as a true copy. An application shall include:

(1) In the case of an applicant who is an individual, a written statement from each of 3 references, who are not related to the applicant by blood or marriage and have known the applicant for at least 3 years, recommending the granting of relief;

(2) Written consent to examine and obtain copies of records and to receive statements and information regarding the applicant's background, including records, statements and other information concerning employment, medical history, military service, and criminal record;

(3) In the case of an applicant under indictment, a copy of the indictment or information;

(4) In the case of an applicant having been convicted of a crime punishable by imprisonment for a term exceeding 1 year, a copy of the

indictment or information on which the applicant was convicted, the judgment of conviction or record of any plea o nolo contendere or plea of guilty or finding of guilt by the court, and any pardon, expunction, setting aside or other record purporting to show that the conviction was rendered nugatory or that civil rights were restored;

(5) In the case of an applicant who has been adjudicated a mental defective or committed to a mental institution, a copy of the order of a court, board, commission, or other lawful authority that made the adjudication or ordered the commitment, any petition that sought to have the applicant so adjudicated or committed, any medical records reflecting the reasons for commitment and diagnoses of the applicant, and any court order or finding of a court, board, commission, or other lawful authority showing the applicant's discharge from commitment, restoration of mental competency and the restoration of rights;

(6) In the case of an applicant who has been discharged from the Armed Forces under dishonorable conditions, a copy of the applicant's summary of service record (Department of Defense Form 214), charge sheet (Department of Defense Form 458), and final court martial order;

(7) In the case of an applicant who, having been a citizen of the United States, has renounced his or her citizenship, a copy of the formal renunciation of nationality before a diplomatic or consular officer of the United States in a foreign state or before an officer designated by the Attorney General when the United States was in a state of war (See 8 U.S.C. 1481(a)(5) and (6)); and

(8) In the case of an applicant who has been convicted of a misdemeanor crime of domestic violence, a copy of the indictment or information on which the applicant was convicted, the judgment of conviction or record of anyfplea o nolo contendere or plea of guilty or finding of guilt by the court, and any pardon, expunction, setting aside or other record purporting to show that the conviction was rendered nugatory or that civil rights were restored.

(d) The Director may grant relief to an applicant if it is established to the satisfaction of the Director that the circumstances regarding the disability, and the applicant's record and reputation, are such that the applicant will not be likely to act in a manner dangerous to public safety, and that the granting of the relief would not be contrary to the public interest. The Director will not ordinarily grant relief if the applicant has not been discharged from parole or probation for a period of at least 2 years. Relief will not be granted to an applicant who is prohibited from possessing all types of firearms by the law of the State where such applicant resides.

(e) In addition to meeting the requirements of paragraph (d) of this section, an applicant who has been adjudicated a mental defective or committed to a mental institution will not be granted relief unless the applicant was subsequently determined by a court, board, commission, or other lawful authority to have been restored to mental competency, to be no longer suffering from a mental disorder, and to have had all rights restored.

(f) Upon receipt of an incomplete or improperly executed application for relief, the applicant shall be notified of the deficiency in the application. If the application is not corrected and returned within 30 days following the date of notification, the application shall be considered as having been abandoned.

(g) Whenever the Director grants relief to any person pursuant to this section, a notice of such action shall be promptly published in the FEDERAL REGISTER, together with the reasons therefor.

(h) A person who has been granted relief under this section shall be relieved of any disabilities imposed by the Act with respect to the acquisition, receipt, transfer, shipment, transportation, or possession of firearms or ammunition and incurred by reason of such disability.

(i)(1) A licensee who incurs disabilities under the Act (See § 178.32(a)) during the term of a current license or while the licensee has pending a license renewal application, and who files an application for removal of such disabilities, shall not be barred from licensed operations for 30 days following the date on which the applicant was first subject to such disabilities (or 30 days after the date upon which the conviction for a crime punishable by imprisonment for a term exceeding 1 year becomes final), and if the licensee files the application for relief as provided by this section within such 30-day period, the licensee may further continue licensed operations during th pendency of the application A licensee who does not file such application within such 30-day period shall not continue licensed operations beyond 30 days following the date on which the licensee was first subject to such disabilities (or 30 days from the date the conviction for a crime punishable by imprisonment for a term exceeding 1 year becomes final).

(2) In the event the term of a license of a person expires during the 30-day period specified in paragraph (i)(1) of this section, or during the pendency of the application for relief, a timely application for renewal of the license must be filed in order to continue licensed operations. Such license application shall show that the applicant is subject to Federal firearms disabilities, shall describe the event giving rise to such disabilities, and shall state when the disabilities were incurred.

(3) A licensee shall not continue licensed operations beyond 30 days following the date the Director issues notification that the licensee's applications for removal of disabilities has been denied.

(4) When as provided in this paragraph a licensee may no longer continue licensed operations, any application for renewal of license filed by the licensee during th pendency of the application for removal of disabilities shall be denied by the Director of Industry Operations.

[T.D. ATF-270, 53 FR 10506, Mar. 31, 1988, as amended by T.D. ATF-313, 56 FR 32509, July 17, 1991; 56 FR 43649, Sept. 3, 1991; T.D. ATF-401, 63 FR 35520, June 30, 1998]

§ 178.145 Research organizations.

The provisions of § 178.98 with respect to the sale or delivery of destructive devices, machine guns, short-barreled shotguns, and short-barreled rifles shall not apply to the sale or delivery of such devices and weapons to any research organization designated by the Director to receiv same. A research organization desiring such designation shall submit a letter application, in duplicate, to the Director Such application shall contain the name and address of the research organization, the names and addresses of the persons directing or controlling, directly or indirectly, the policies and management of such organization, the nature and purpose of the research being conducted, a description of the devices and weapons to be received, and the identity of the person or persons from whom such devices and weapons are to be received.

[T.D. ATF-270, 53 FR 10507, Mar. 31, 1988]

§ 178.146 Deliveries by mail to certain persons.

The provisions of this part shall not be construed as prohibiting a licensed importer, licensed manufacturer, or licensed dealer from depositing a firearm for conveyance in the mails to any officer, employee, agent, or watchman who, pursuant to the provisions of section 1715 of title 18, U.S.C., is eligible to receive through the mails pistols, revolvers, and other firearms capable of being concealed on the person, for use in connection with his official duties.

§ 178.147 Return of firearm.

A person not otherwise prohibited by Federal, State or local law may ship a firearm to a licensed importer, licensed manufacturer, or licensed dealer for any lawful purpose, and, notwithstanding any other provision of this part, the licensed manufacturer, licensed importer, or licensed dealer may return in interstate or foreign commerce to that person the firearm or a replacement firearm of the same kind and type. See § 178.124(a) for requirements of a Form 4473 prior to return A person not otherwise prohibited by Federal, State or local law may ship a firearm curio or relic to a licensed collector for any lawful purpose, and, notwithstanding any other provision of this part, the licensed collector may return in interstate or foreign commerce to that person the firearm curio or relic.

[T.D. ATF-270, 53 FR 10507, Mar. 31, 1988]

§ 178.148 Armor piercing ammunition intended for sporting or industrial purposes.

The Director may exempt certain armor piercing ammunition from the requirements of this part. A person who desires to obtain an exemption under this section for any such ammunition which is primarily intended for sporting purposes or intended for industrial purposes, including charges used in oil and gas well perforating devices, shall submit a written request to the Director Each request shall be executed under the penalties of perjury and contain a complete and accurate description of the ammunition, the name and address of the manufacturer or importer, the purpose of and use for which it is designed and intended, and any photographs, diagrams, or drawings as may be necessary to enable the Director to make a determination. The Director may require that a sample of the ammunition be submitted for examination and evaluation.

[T.D. ATF-270, 53 FR 10507, Mar. 31, 1988]

§ 178.149 Armor piercing ammunition manufactured or imported for the purpose of testing or experimentation.

The provisions of §§ 178.37 and 178.99(d) with respect to the manufacture or importation of armor piercing ammunition and the sale or delivery of armor piercing ammunition by manufacturers and importers shall not apply to the manufacture, importation, sale or delivery of armor piercing ammunition for the purpose of testing or experimentation as authorized by the Director. A person desiring such authorization to receive armor piercing ammunition shall submit a letter application, in duplicate, to the Director Such application shall contain the name and addresses of the persons directing or controlling, directly or indirectly, the policies and management of the applicant, the nature or purpose of the testing or experimentation, a description of the armor piercing ammunition to be received, and the identity of the manufacturer or importer from whom such ammunition is to be received. The approved application shall be submitted to the manufacturer or importer who shall retain a copy as part of the records required by subpart H of this part.

[T.D. ATF-270, 53 FR 10507, Mar. 31, 1988]

§ 178.150   Alternative to NICS in certain geographical locations.

(a) The provisions of § 178.102(d)(3) shall be applicable when the Director has certified that compliance with the provisions of § 178.102(a)(1) is impracticable because:

(1) The ratio of the number of law enforcement officers of the State in which the transfer is to occur to the number of square miles of land area of the State does not exceed 0.0025;

(2) The business premises of the licensee at which the transfer is to occur are extremely remote in relation to the chief law enforcement officer; and

(3) There is an absence of telecommunications facilities in the geographical area in which the business premises are located.

(b) A licensee who desires to obtain a certification under this section shall submit a written request to the Director. Each request shall be executed under the penalties of perjury and contain information sufficient for the Director to make such certification Such information shall include statistical data, official reports, or other statements of government agencies pertaining to the ratio of law enforcement officers to the number of square miles of land area of a State and statements of government agencies and private utility companies regarding the absence of telecommunications facilities in the geographical area in which the licensee's business premises are located.

(c) For purposes of this section and § 178.129(c), the "chief law enforcement officer" means the chief of police, the sheriff, or an equivalent officer or the designee of any such individual.

(Approved by the Office of Management and Budget under control number 1512-0544)

[T.D. ATF-354, 59 FR 7114, Feb. 14, 1994, as amended by T.D. ATF-361, 60 FR 10788, Feb. 27, 1995; T.D. ATF-415, 63 FR 58272, Oct. 29, 1998]

§ 178.151 Semiautomatic rifles or shotguns for testing or experimentation.

(a) The provisions of § 178.39 shall not apply to the assembly of semiautomatic rifles or shotguns for the purpose of testing or experimentation as authorized by the Director.

(b) A person desiring authorization to eassembl nonsporting semiautomatic rifles or shotguns shall submit a written request, in duplicate, to the Director. Each such request shall be executed under the penalties of perjury and shall contain a complete and accurate description of the firearm eto b assembled, and such diagrams or drawings as may be necessary to enable the Director to make a determination. The Director may require the submission of the firearm parts for examination and evaluation. If the submission of the firearm parts is impractical, the person requesting the authorization shall so advise the Director and designate the place where the firearm parts will be available for examination and evaluation.

[T.D. ATF-346, 58 FR 40590, July 29, 1993]

§ 178.152 Seizure and forfeiture.

(a) Any firearm or ammunition involved in or used in any knowing violation of subsections (a)(4), (a)(6), (f), (g), (h), (i), (j), or (k) of section 922 of the Act, or knowing importation or bringing into the United States or any possession thereof any firearm or ammunition in violation of section 922(l) of the Act, or knowing violation of section 924 of the Act, or willful violation of any other provision of the Act or of this part, or any violation of any other criminal law of the United States, or any firearm or ammunition intended to be used in any offense referred to in paragraph (c) of this section, where such intent is demonstrated by clear and convincing evidence, shall be subject to seizure and forfeiture, and all provisions of the Internal Revenue Code of 1986 relating to the seizure, forfeiture, and disposition of firearms, as defined in section 5845(a) of that Code, shall, so far as applicable, extend to seizures and forfeitures under the provisions of the Act: **Provided**, That upon acquittal of the owner or possessor, or dismissal of the charges against such person other than upon motion of the Government prior to trial, or lapse of or court termination of the restraining order to which he is subject, the seized or relinquished firearms or ammunition shall be returned forthwith to the owner or possessor or to a person delegated by the owner or possessor unless the return of the firearms or ammunition would place the owner or possessor or the delegate of the owner or possessor in violation of law. Any action or proceeding for the forfeiture of firearms or ammunition shall be commenced within 120 days of such seizure.

(b) Only those firearms or quantities of ammunition particularly named and individually identified as involved in or used in any violation of the provisions of the Act or this part, or any other criminal law of the United States or as intended to be used in any offense referred to in paragraph (c) of this section, where such intent is demonstrated by clear and convincing evidence, shall be subject to seizure, forfeiture and disposition.

(c) The offenses referred to in paragraphs (a) and (b) of this section for which firearms and ammunition intended to be used in such offenses are subject to seizure and forfeiture are:

(1) Any crime of violence, as that term is defined in section 924(c)(3) of the Act;

(2) Any offense punishable under the Controlled Substances Act (21 U.S.C. 801 et seq.) or the Controlled Substances Import and Export Act (21 U.S.C. 951 et seq.);

(3) Any offense described in section 922(a)(1), 922(a)(3), 922(a)(5), or 922(b)(3) of the Act, where the firearm or ammunition intended to be used in such offense is involved in a pattern of activities which includes a violation of any offense described in section 922(a)(1), 922(a)(3), 922(a)(5), or 922(b)(3) of the Act;

(4) Any offense described in section 922(d) of the Act where the firearm or ammunition is intended to be used in such offense by the transferor of such firearm or ammunition;

(5) Any offense described in section 922(i), 922(j), 922(l), 922(n), or 924(b) of the Act; and

(6) Any offense which may be prosecuted in a court of the United States which involves the exportation of firearms or ammunition.

[T.D. ATF-270, 53 FR 10507, Mar. 31, 1988 Redesignated at T.D. ATF-354, 59 FR 7114, Feb. 14, 1994, and furthe redesignated by T.D. ATF-361, 60 FR 10788, Feb. 27, 1995; T.D. ATF-363, 60 FR 17455, April 6, 1995]

§ 178.153 Semiautomatic assault weapons and large capacity ammunition feeding devices manufactured or imported for the purposes of testing or experimentation.

The provisions of § 178.40 with respect to the manufacture, transfer, or possession of a semiautomatic assault weapon, and § 178.40a with respect to large capacity ammunition feeding devices, shall not apply to the manufacture, transfer, or possession of such weapons or devices by a manufacturer or importer for the purposes of testing or experimentation as authorized by the Director. A person desiring such authorization shall submit a letter application, in duplicate, to the Director. Such application shall contain the name and addresses of the persons directing or controlling, directly or indirectly, the policies and management of the applicant, the nature or purpose of the testing or experimentation, a description of the weapons or devices to be manufactured or imported, and the source of the weapons or devices. The approved application shall be retained as part of the records required by Subpart H of this part.

[T.D. ATF-363, 60 FR 17456, April 6, 1995]

Subpart J - [Reserved]

Subpart K-Exportation

§ 178.171 Exportation.

Firearms and ammunition shall be exported in accordance with the applicable provisions of section 38 of the Arms Export Control Act (22 U.S.C. 2778) and regulations thereunder. However, licensed manufacturers, licensed importers, and licensed dealers exporting firearms shall maintain records showing the manufacture or acquisition of the firearms as required by this part and records showing the name and address of the foreign consignee of the firearms and the date the firearms were exported. Licensed manufacturers and licensed importers exporting armor piercing ammunition and semiautomatic assault weapons manufactured after September 13, 1994, shall maintain records showing the name and address of the foreign consignee and the date the armor piercing ammunition or semiautomatic assault weapons were exported.

[T.D. ATF-270, 53 FR 10507, Mar. 31, 1988; T. D. ATF-363, 60 FR 17456, April 6, 1995]

PART 179-MACHINE GUNS, DESTRUCTIVE DEVICES, AND CERTAIN OTHER FIREARMS

Subpart A-Scope of Regulations

Subpart H-Importation and Exportation

Importation

Exportation

Arms Export Control Act

Subpart I-Records and Returns

Subpart J-Stolen or Lost Firearms or Documents

Subpart K-Examination of Books and Records

Subpart L-Distribution and Sale of Stamps

Subpart M-Redemption of or Allowance for Stamps or Refunds

Subpart N-Penalties and Forfeitures

Subpart O-Other Laws Applicable

Authority: 26 U.S.C. 7805.

Source: 36 FR 14256, Aug. 3, 1971, unless otherwise noted Redesignated at 40 FR 16835, Apr. 15, 1975.

Subpart A-Scope of Regulations

§ 179.1 General.

This part contains the procedural and substantive requirements relative to e th importation, manufacture, making, exportation, identification and registration of, and the dealing in, machine guns, destructive devices and certain other firearms under the provisions of the National Firearms Act (26 U.S.C. Chapter 53).

[36 FR 14256, Aug. 3, .1971 Redesignated at 40 FR 16835, Apr. 15, 1975, and amended by T.D. ATF-48, 44 FR 55842, Sept. 28, 1979]

Subpart B-Definitions

§ 179.11 Meaning of terms.

When used in this part and in forms prescribed under this part, where not otherwise distinctly expressed or manifestly incompatible with the intent thereof, terms shall have the meanings ascribed in this section. Words in the plural form shall include the singular, and vice versa, and words importing the masculine gender shall include the feminine. The terms "includes" and "including" do not exclude other things not enumerated which are in the same general class or are otherwise within the scope thereof.

Antique firearm. Any firearm not designed or redesigned for using rim fire or conventional center fire ignition with fixed ammunition and manufactured in or before 1898 (including any matchlock, flintlock, percussion cap, or similar type of ignition system or replica thereof, whether actually manufactured before or after the year 1898) and also any firearm using fixed ammunition manufactured in or before 1898, for which ammunition is no longer manufactured in the United States and is not readily available in the ordinary channels of commercial trade.

Any other weapon. Any weapon or device capable of being concealed on the person from which a shot can be discharged through the energy of an explosive, a pistol or revolver having a barrel with a smooth bore designed or redesigned to fire a fixed shotgun shell, weapons with combination shotgun and rifle barrels 12 inches or more, less than 18 inches in length, from which only a single discharge can be made from either barrel without manual reloading, and shall include any such weapon which may be readily restored to fire. Such term shall not include a pistol or a revolver having a rifled bore, or rifled bores, or weapons designed, made, or intended to be fired from the shoulder and not capable of firing fixed ammunition.

ATF officer. An officer or employee of the Bureau of Alcohol, Tobacco and Firearms (ATF) authorized to perform any function relating to the administration or enforcement of this part.

Customs officer. Any officer of the Customs Service or any commissioned, warrant, or petty officer of the Coast Guard, or any agent or other person authorized by law or designated by the Secretary of the Treasury to perform any duties of an officer of the Customs Service.

Dealer. Any person, not a manufacturer or importer, engaged in the business of selling, renting, leasing, or loaning firearms and shall include pawnbrokers who accept firearms as collateral for loans.

Destructive device. (a) Any explosive, incendiary, or psison ga (1) bomb , (2) grenade (3) rocket having a propellant charge of more than 4 ounces, **(4)** missile having an explosive or incendiary charge of more than one-quarter ounce **(5)** mine, ro **(6)** similar device **(b)** any type of weapon by whatever name known which will, or which may be readily converted to, expel a projectile by the action of an explosive or other propellant, the barrel or barrels of which have a bore of more than one-half inch in diameter, except a shotgun or shotgun shell which the Director finds is generally recognized as particularly suitable for sporting purposes; and **(c)** any combination of parts either designed or intended for use in converting any device into a destructive device as described in paragraphs (a) and (b) of this definition and from which a destructive device may be readily assembled The term shall not include any device which is neither designed or redesigned for use as a weapon; any device, although originally designed for use as a weapon, which is redesigned for use as a signaling, pyrotechnic, line throwing, safety, or similar device; surplus ordnance sold, loaned, or given by the Secretary of the Army under 10 U.S.C. 4684(2), 4685, or 4686, or any device which the Director finds is not likely to be used as a weapon, or is an antique or is a rifle which the owner intends to use solely for sporting purposes.

Director. The Director, Bureau of Alcohol, Tobacco, and Firearms, the Department of the Treasury, Washington, DC.

Director of the Service Center. A director of an Internal Revenue Service Center in an internal revenue region.

District director. A district director of the Internal Revenue Serviceni an internal revenue district.

Executed under penalties of perjury. Signed with the prescribed declaration under the penalties of perjury as provided on or with respect to the return, form, or other document or, where no form of declaration is prescribed, with the declaration:

"I declare under the penalties of perjury that this - (insert type of document, such as, statement, application, request, certificate), including the documents submitted in support thereof, has been examined by me and, to the best of my knowledge and belief, is true, correct, and complete."

Exportation. The severance of goods from the mass of things belonging to this country with the intention of uniting them to the mass of things belonging to some foreign country.

Exporter. Any person who exports firearms from the United States.

Firearm. **(a)** A shotgun having a barrel or barrels of less than 18 inches in length **(b)** a weapon made from a shotgun if such weapon as modified has an overall length of less than 26 inches or a barrel or barrels of less than 18 inches in length **(c)** a rifle having a barrel or barrels of less than 16 inches in length **(d)** a weapon made from a rifle if such weapon as modified has an overall length of less than 26 inches or a barrel or barrels of less than 16 inches in length **(e)** any other weapon, as defined in this subpart **(f)** a machine gun **(g)** a muffler or a silencer for any firearm whether or not such firearm is included within this definition; an **(h)** a destructive device The term shall not include an antique firearm or any device (other than a machine gun or destructive device) which, although designed as a weapon, the Director finds by reason of the date of its manufacture, value, design, and other characteristics is primarily a collector's item and is not likely to be used as a weapon. For purposes of this definition, the length of the barrel having an integral chamber(s) on a shotgun or rifle shall be determined by measuring the distance between the muzzle and the face of the bolt, breech, or breech block when closed and when the shotgun or rifle is cocked. The overall length of a weapon made from a shotgun or rifle is the distance between the extreme ends of the weapon measured along a line parallel to the center line of the bore.

Fixed ammunition. That self-contained unit consisting of the case, primer, propellant charge, and projectile or projectiles.

Frame or receiver. That part of a firearm which provides housing for the hammer, bolt o breechblock and firing mechanism, and which is usually threaded at its forward portion to receive the barrel.

Importation. The bringing of a firearm within the limits of the United States or any territory under its control or jurisdiction, from a place outside thereof (whether such place be a foreign country or territory subject to the jurisdiction of the United States), with intent to unlade Except that, bringing a firearm from a foreign country or a territory subject to the jurisdiction of the United States into a foreign trade zone for storage pending shipment to a foreign country or subsequent importation into this country, under Title 26 of the United States Code, and this part, shall not be deemed importation.

Importer. Any person who is engaged in the business of importing or bringing firearms into the United States.

Machine gun. Any weapon which shoots, is designed to shoot, or can be readily restored to shoot, automatically more than one shot, without manual reloading, by a single function of .the trigger The term shall also include the frame or receiver of any such weapon, any part designed and intended solely and exclusively, or combination of parts designed and intended, for use in converting a weapon into a machine gun, and any combination of parts from which a machine gun can be assembled if such parts are in the possession or under the control of a person.

Make. This term and the various derivatives thereof shall include manufacturing (other than by one qualified to engage in such business under this part), putting together altering, any combination of these, or otherwise producing a firearm.

Manual reloading. The inserting of a cartridge or shell into the chamber of a firearm either with the hands or by means of a mechanical device controlled and energized by the hands.

Manufacturer. Any person who is engaged in the business of manufacturing firearms.

Muffler or silencer. Any device for silencing, muffling, or diminishing the report of a portable firearm, including any combination of parts, designed or redesigned, and intended for the use in assembling or fabricating a firearm silencer or firearm muffler, and any part intended only for use in such assembly or fabrication.

Person. A partnership, company, association, trust, estate, or corporation, as well as a natural person.

Pistol. A weapon originally designed, made, and intended to fire a projectile (bullet) from one or more barrels when held in one hand, and having (a) a chamber(s) as an integral part(s) of, or permanently aligned with, the bore(s); and (b) a short stock designed to be gripped by one hand and at an angle to and extending below the line of the bore(s).

Regional director (compliance). The principal ATF regional official responsible for administering regulations in this part.

Revolver. A projectile weapon, of the pistol type, having a breechloading chambered cylinder so arranged that the cocking of the hammer or movement of the trigger rotates it and brings the next cartridge in line with the barrel for firing.

Rifle. A weapon designed or redesigned, made or remade, and intended to be fired from the shoulder and designed or redesigned and made or remade to use the energy of the explosive in a fixed cartridge to fire only a single projectile through a rifled bore for each single pull of the trigger, and shall include any such weapon which may be readily restored to fire a fixed cartridge.

Shotgun. A weapon designed or redesigned, made or remade, and intended to be fired from the shoulder and designed or redesigned and made or remade to use the energy of the explosive in a fixed shotgun shell to fire through a smooth bore either a number of projectiles (ball shot) or a single projectile for each pull of the trigger, and shall include any such weapon which may be readily restored to fire a fixed shotgun shell.

Transfer. This term and the various derivatives thereof shall include selling, assigning, pledging, leasing, loaning, giving away, or otherwise disposing of.

United States. The States and the District of Columbia.

U.S.C. The United States Code.

Unserviceable firearm. A firearm which is incapable of discharging a shot by means of an explosive and incapable of being readily restored to a firing condition.

(26 U.S.C. 7805A (68 Stat. 917), 27 U.S.C. 9205 (4 Stat. 981 as amended), 18 U.S.C. 9262 (8 Stat. 959), and sec. 38, Arms Export Control Act (22 U.S.C. 2778Q 9 Stat. 744))

[T.D. ATF-48, 43 FR 13538, Mar. 31, 1978; 44 FR 55842, Sept. 28, 1979; T.D. ATF-241, 51 FR 39630, Oct. 29, 1986; T.D. ATF-270, 53 FR 10492, Mar. 31, 1988; T.D. ATF-396, 63 FR 12647, Mar. 16, 1998]

Subpart C-Administrative and Miscellaneous Provisions

§ 179.21   Forms prescribed.

(a) The Director is authorized to prescribe all forms required by this part. All of the information called for in each form shall be furnished as indicated by the headings on the form and the instructions on or pertaining to the form. In addition, information called for in each form shall be furnished as required by this part. Each form requiring that it be executed under penalties of perjury shall be executed under penalties of perjury.

(b) Requests for forms should be mailed to the ATF Distribution Center, 7943 Angus Court, Springfield Virginia 22153.

(5 U.S.C. 552(a); 8 Stat. 383, as amended)

[T.D. ATF-92, 46 FR 46916, Sept. 23, 1981, as amended by T.D. ATF-241, 51 FR 39630, Oct. 29, 1986; T.D. ATF-270, 53 FR 10508, Mar. 31, 1988; T.D. 3-61 FR 20725, May 8, 1996]

§ 179.22   Right of entry and examination.

Any ATF officer or employee of the Bureau of Alcohol, Tobacco and

Firearms duly authorized to perform any function relating to the administration or enforcement of this part may enter during business hours the premises (including places of storage) of any importer or manufacturer of or dealer in firearms, to examine any books, papers, or records required to be kept pursuant to this part, and any firearms kept by such importer, manufacturer or dealer on such premises, and may require the production of any books, papers, or records necessary to determine any liability for tax under 26 U.S.C. Chapter 53, or the observance of 26 U.S.C. Chapter 53, and this part.

[36 FR 14256, Aug. 3, 1971 Redesignated at 40 FR 16835, Apr. 15, 1975, and amended by T.D. ATF-48, 44 FR 55842, Sept. 28, 1979]

§ 179.23 Restrictive use of required information.

No information or evidence obtained from an application, registration, or record required to be submitted or retained by a natural person in order to comply with any provision of 26 U.S.C. Chapter 53, or this part or section 207 of the Gun Control Act of 1968 shall be used, directly or indirectly, as evidence against that person in a criminal proceeding with respect to a violation of law occurring prior to or concurrently with the filing of the application or registration, or the compiling of the record containing the information: or evidence **Provided, however,** That the provisions of this section shall not preclude the use of any such information or evidence in a prosecution or other action under any applicable provision of law with respect to the furnishing of false information.

[36 FR 14256, Aug. 3, 1971 Redesignated at 40 FR 16835, Apr. 15, 1975, and amended by T.D. ATF-48, 44 FR 55842, Sept. 28, 1979]

§ 179.24 Destructive device determination.

The Director shall determine in accordance with 26 U.S.C. 5845(f), whether a device is excluded from the definition of a destructive device. A person who desires to obtain a determination under that provision of law for any device which he believes is not likely to be used as a weapon shall submit a written request, in triplicate, for a ruling thereon to the Director Each such request shall be executed under the penalties of perjury and contain a complete and accurate description of the device, the name and address of the manufacturer or importer thereof, the purpose of and use for which it is intended, and such photographs, diagrams, or drawings as may be necessary to enable the Director to make his determination. The Director may require the submission to him, of a sample of such device for examination and evaluation. If the submission of such device is impracticable, the person requesting the ruling shall so advise the Director and designate the place where the device will be available for examination and evaluation.

[36 FR 14256, Aug. 3, 1971 Redesignated at 40 FR 16835, Apr. 15, 1975, and amended by T.D. ATF-48, 44 FR 55842, Sept. 28, 1979]

§ 179.25 Collector's items.

The Director shall determine in accordance with 26 U.S.C. 5845(a), whether a firearm or device, which although originally designed as a weapon, is by reason of the date of its manufacture, value, design, and other characteristics primarily a collector's item and is not likely to be used as a weapon A person who desires to obtain a determination under that provision of law shall follow the procedures prescribed in § 179.24 relating to destructive device determinations, and shall include information as to date of manufacture, value, design and other characteristics which would sustain a finding that the firearm or device is primarily a collector's item and is not likely to be used as a weapon.

[36 FR 14256, Aug. 3, .1971 Redesignated at 40 FR 16835, Apr. 15, 1975, and amended by T.D. ATF-48, 44 FR 55842, Sept. 28, 1979]

§ 179.26 Alternate methods or procedures; emergency variations from requirements.

(a) **Alternate methods or procedures.** Any person subject to the provisions of this part, on specific approval by the Director as provided in this paragraph, may use an alternate method or procedure in lieu of a method or procedure specifically prescribed in this part. The Director may approve an alternate method or procedure, subject to stated conditions, when it is found that:

(1) Good cause is shown for the use of the alternate method or procedure;

(2) The alternate method or procedure is within the purpose of, and consistent with the effect intended by, the specifically prescribed method or procedure and that the alternate method or procedure is substantially equivalent to that specifically prescribed method or procedure; and

(3) The alternate method or procedure will not be contrary to any provision of law and will not result in an increase in cost to the Government or hinder the effective administration of this part. Where such person desires to employ an alternate method or procedure, a written application shall be submitted to the appropriate regional director (compliance), for transmittal to the Director. The application shall specifically describe the proposed alternate method or procedure and shall set forth the reasons for it. Alternate methods or procedures may not be employed until the application is approved by the Director. Such person shall, during the period of authorization of an alternate method or procedure, comply with the terms of the approved application. Authorization of any alternate method or procedure may be withdrawn whenever, in the judgment of the Director, the effective administration of this part is hindered by the continuation of the authorization.

(b) **Emergency variations from requirements.** The Director may approve a method of operation other than as specified in this part, where it is found that an emergency exists and the proposed variation from the specified requirements are necessary and the proposed variation (1) will not hinder the effective administration of this part; an (2) will not be contrary to any provisions of law. Variations from requirements granted under this paragraph are conditioned on compliance with the procedures, conditions, and limitations set forth in the approval of the application. Failure to comply in good faith with the procedures, conditions, and limitations shall automatically terminate the authority for the variations, and the person granted the varianc shall fully comply with the prescribed requirements of regulations from which the variations were authorized. Authority for any variation may be withdrawn whenever, in the judgment of the Director, the effective administration of this part is hindered by the continuation of the variation. Where a person desires to employ an emergency variation, a written application shall be submitted to the appropriate regional director (compliance) for transmittal to the Director. The application shall describe the proposed variation and set forth the reasons for it. Variations may not be employed until the application is approved.

(c) **Retention of approved variations.** The person granted the varianc shall retain and make available for examination by ATF officers any application approved by the Director under this section.

[T.D. ATF-270, 53 FR 10508 Mar. 31, 1988]

Subpart D-Special (Occupational) Taxes

§ 179.31 Liability for tax.

(a) **General.** Every person who engages in the business of importing, manufacturing, or dealing in (including pawnbrokers) firearms in the United States shall pay a special (occupational) tax at a rate specified by § 179.32. The tax shall be paid on or before the date of commencing the taxable business, and thereafter every year on or before July 1. Special (occupational) tax shall not be prorated. The tax shall be computed for the entire tax year (July 1 through June 30), regardless of the portion of the year during which the taxpayer engages in business. Persons commencing business at any time after July 1 in any year are liable for the special (occupational) tax for the entire tax year.

(b) **Each place of business taxable.** An importer, manufacturer, or dealer in firearms incurs special tax liability at each place of business where an occupation subject to special tax is conducted. A place of business means the entire office, plant or area of the business in any one location under the same proprietorship. Passageways, streets, highways, rail crossings, waterways, or partitions dividing the premises are not sufficient separation to require additional special tax, if the divisions of the premises are otherwise contiguous. See also § 179.38-179.39.

(26 U.S.C. 5143, 5801, 5846)

[T.D. ATF-271, 53 FR 17550, May 17, 1988]

§ 179.32 Special (occupational) tax rates.

(a) Prior to January 1, 1988, the special (occupational) tax rates were as follows:

	Per year or fraction thereof
Class 1-Importer of firearms.	$500
Class 2-Manufacturer of firearms.	$500
Class 3-Dealer in firearms.	$200
Class 4-Importer only of weapons classified as "any other weapon".	$25
Class 5-Manufacturer only of weapons classified as "any other weapon".	$25
Class 6-Dealer only in weapons classified as "any other weapon"	$10

(b) Except as provided in § 179.32a, the special (occupational) tax rates effective January 1, 1988, are as follows:

	Per year or fraction thereof
Class 1-Importer of firearms (including an importer only of weapons classified as "any other weapon")	$1,000
Class 2-Manufacturer of firearms (including a manufacturer only of weapons classified as "any other weapon")	$1,000
Class 3-Dealer in firearms (including a dealer only of weapons classified as "any other weapon")	$500

(c) A taxpayer who was engaged in a business on January 1, 1988, for which a special (occupational) tax was paid for a taxable period which began before January 1, 1988, and included that date, shall pay an increased special tax for the period January 1, 1988, through June 30, 1988. The increased tax shall not exceed one-half the excess (if any) of **(1)** the rate of special tax in effect on January 1, 1988, ove **(2)** the rate of such tax in effect on December 31, 1987. The increased special tax shall be paid on or before April 1, 1988.

[T.D. ATF-271, 53 FR 17550, May 17, 1988]

§ 179.32a Reduced rate of tax for small importers and manufacturers.

(a) General. Effective January 1, 1988, 26 U.S.C. 5801(b) provides for a reduced rate of special tax with respect to any importer or manufacturer whose gross receipts (for the most recent taxable year ending before the first day of the taxable period to which the special tax imposed by §179.32 relates) are less than $500,000. The rate of tax for such an importer or manufacturer is $500 per year or fraction thereof. The "**taxable year**" to be used for determining gross receipts is the taxpayer's income tax year. All gross receipts of the taxpayer shall be include , not just the gross receipts of the business subject to special tax. Proprietors of new businesses that have not yet begun a taxable year, as well as proprietors of existing businesses that have not yet ended a taxable year, who commence a new activity subject to special tax, quality for the reduced special (occupational) tax rate, unless the business is a member aof "**controlled group**"; in that case, the rules of paragraph (b) of this section shall apply.

(b) Controlled group. All persons treated as one taxpayer under 26 U.S.C. 5061(e)(3) shall be treated as one taxpayer for the purpose of determining gross receipts under paragraph (a) of this section. "**Controlled group**" means a controlled group of corporations, as defined in 26 U.S.C. 1563 and implementing regulations in 26 CFR 1.1563-1 through 1.1563-4, except that the words "at least 80 percent" shall be replaced by the words "more than 50 percent" in each place they appear in subsection (a) of 26 U.S.C. 1563, as well as in the implementing regulations. Also, the rules for "**controlled group of corporations**" apply in a similar fashionot groups which include partnerships and/or sole proprietorships. If one entity maintains more than 50% control over a group consisting of corporations and one, or more, partnerships and/or sole proprietorships, all of the members of the controlled group are one taxpayer for the purpose of this section.

(c) Short taxable year. Gross receipts for any taxable year of less than 12 months shall be annualized by multiplying the gross receipts for the short period by 12 and dividing the result by the number of months in the short period, as required by 26 U.S.C. 448(c)(3).

(d) Returns and allowances. Gross receipts for any taxable year shall be reduced by returns and allowances made during that year under 26 U.S.C. 448(c)(3).

(26 U.S.C. 448, 5061, 5801)

[T.D. ATF-271, 53 FR 17550, May 17, 1988]

§ 179.33 Special exemption.

(a) Any person required to pay special (occupational) tax under this part shall be relieved from payment of that tax if he establishes to the satisfaction of the Director that his business is conducted exclusively with, or on behalf of, the United States or any department, independent establishment, or agency thereof. The Director may relieve any person manufacturing firearm for or on behalf of the United States from compliance with any provision of this part in the conduct of the business with respect to such firearms.

(b) The exemption in this section may be obtained by filing with the Director an application, in letter form, setting out the manner in which the applicant conducts his business, the type of firearm to be manufactured, and proof satisfactory to the Director of the existence of the contract with the United States, department, independent establishment, or agency thereof, under which the applicant intends to operate.

§ 179.34 Special tax registration and return.

(a) General. Special tax shall be paid by return. The prescribed return is ATF Form 5630.7, Special Tax Registration and Return. Special tax returns, with payment of tax, shall be filed with ATF in accordance with instructions on the form Properly completing, signing, and timely filing of a return (Form 5630.7) constitutes compliance with 26 U.S.C. 5802.

(b) Preparation of ATF Form 5630.7. All of the information called for on Form 5630.7 shall be provided, including:

(1) The true name of the taxpayer.

(2) The trade name(s) (if any) of th business(es) subject to special tax.

(3) The employer identification number (see § 179.35).

(4) The exact location of the place of business, by name and number of building or street, or if these do not exist, by some description in addition to the post office address. In the case of one return for two or more locations, the address to be shown shall be the taxpayer's principal place of business (or principal office, in the case of a corporate taxpayer).

(5) The class(es) of special tax to which the taxpayer is subject.

(6) Ownership and control information: That is, the name, position, and residence address of every owner of the business and of every person having power to control its management and policies with respect to the activity subject to special tax. "Owner of the business" shall include every partner, if the taxpayer is a partnership, and every person owning 10% or more of its stock, if the taxpayer is a corporation. However, the ownership and control information required by this paragraph need not be stated if the same information has been previously provided to ATF in connection with a license application under Part 178 of this chapter, and if the information previously provided is still current.

(c) Multiple locations and/or classes of tax. A taxpayer subject to special tax for the same period at more than one location or for more than one class of tax shall -

(1) File one special tax return, ATF Form 5630.7, with payment of tax, to cover all such locations and classes of tax; and

(2) Prepare, in duplicate, a list identified with the taxpayer's name, address (as shown on ATF Form 5630.7), employer identification number, and period covered by the return. The list shall show, by States, the name, address, and tax class of each location for which special tax is being paid. The original of the list shall be filed with ATF in accordance with instructions on the return, and the copy shall be retained at the taxpayer's principal place of business (or principal office, in the case of a corporate taxpayer) for not less than 3 years.

(d) **Signing of ATF Forms 5630.7 - (1) Ordinary returns.** The return of an individual proprietor shall be signed by the individual. The return of a partnership shall be signed by a general partner. The return of a corporation shall be signed by any officer. In each case, the person signing the return shall designate his or her capacity as "individual owner," "member of firm," or, in the case of a corporation, the title of the officer.

(2) **Fiduciaries.** Receivers, trustees, assignees, executors, administrators, and other legal representatives who continue the business of a bankrupt, insolvent, deceased person, etc., shall indicate the fiduciary capacity in which they act.

(3) **Agent or attorney in fact.** If a return is signed by an agent or attorney in fact, the signature shall be preceded by the name of the principal and followed by the title of the agent or attorney in fact. A return signed by a person as agent will not be accepted unless there is filed, with the ATF office with which the return is required to be filed, a power of attorney authorizing the agent to perform the act.

(4) **Perjury statement.** ATF Forms 5630.7 shall contain or be verified by a written declaration that the return has been executed under the penalties of perjury.

(e) **Identification of taxpayer.** If the taxpayer is an individual, with the initial return such person shall securely attach to Form 5630.7 a photograph of the individual 2 x 2 inches in size, clearly showing a full front view of the features of the individual with head bare, with the distance from the top of the head to the point of the chin approximately 1 1/4 inches, and which shall have been taken within 6 months prior to the date of completion of the return. The individual shall also attach to the return a properly completed FBI Form FD-258 (Fingerprint Card). The fingerprints must be clear for accurate classification and should be taken by someone properly equipped to take them: **Provided,** That the provisions of this paragraph shall not apply to individuals who have filed with ATF a properly executed Application for License under 18 U.S.C. Chapter 44, Firearms, ATF Form 7 (5310.12)(12-93 edition), as specified in § 178.44(a).

(26 U.S.C. 5142, 5802, 5846, 6061, 6065, 6151)

[T.D. ATF-271, 53 FR 17551, May 17, 1988; T.D. ATF-363, 60 FR 17456, April 6, 1995]

§ 179.35 Employer identification number.

(a) **Requirement.** The employer identification number (defined in 26 CFR 301.7701-12) of the taxpayer who has been assigned such a number shall be shown on each special tax return, including amended returns, filed under this subpart. Failure of the taxpayer to include the employer identification number may result in the imposition of the penalty specified in § 70.113 of this chapter.

(b) **Application for employer identification number.** Each taxpayer who files a special tax return, who has not already been assigned an employer identification number, shall file IRS Form SS-4 to apply for one. The taxpayer shall apply for and be assigned only one employer identification number, regardless of the number of places of business for which the taxpayer is required to file a special tax return. The employer identification number shall be applied for no later than 7 days after the filing of the taxpayer's first special tax return. IRS Form SS-4 may be obtained from the director of an IRS service center or from any IRS district director.

(c) **Preparation and filing of IRS Form SS-4.** The taxpayer shall prepare and file IRS Form SS-4, together with any supplementary statement, in accordance with the instructions on the form or issued in respect to it.

(26 U.S.C. 6109)

[T.D. ATF-271, 53 FR 17551, May 17, 1988; as amended by T.D. ATF-301, 55 FR 47657, Nov. 14, 1990]

§ 179.36 The special tax stamp, receipt for special (occupational) taxes.

Upon filing a properly completed and executed return (Form 5630.7) accompanied by remittance of the full amount due, the taxpayer will be issued a special tax stamp as evidence of payment of the special (occupational) tax.

[36 FR 14256, Aug. 3, 1971 Redesignated at 40 FR 16835, Apr. 15, 1975, and amended by T.D. ATF-251, 52 FR 19334, May 22, 1987; T.D. ATF-363, 60 FR 17456, Apr. 6, 1995]

§ 179.37 Certificates in lieu of stamps lost or destroyed.

When a special tax stamp has been lost or destroyed, such fact should be reported immediately to the regional director (compliance) who issued the stamp. A certificate in lieu of the lost or destroyed stamp will be issued to the taxpayer upon the submission of an affidavit showing to the satisfaction of the regional director (compliance) that the stamp was lost or destroyed.

[36 FR 14256, Aug. 3, 1971 Redesignated at 40 FR 16835, Apr. 15, 1975, and amended by T.D. ATF-251, 52 FR 19334, May 22, 1987]

Editor's Note:

The ATF National Revenue Center, Cincinnati, Ohio handles all issues concerning Special Occupational Tax and the issuance of the tax stamps.

§ 179.38 Engaging in business at more than one location.

A person shall pay th special (occupational) tax for each location where he engages in any business taxable under 26 U.S.C. 5801. However, a person paying a special (occupational) tax covering his principal place of business may utilize other locations solely for storage of firearms without incurring special (occupational) tax liability at such locations. A manufacturer, upon the single payment of the appropriate special (occupational) tax, may sell firearms, if such firearms are of his own manufacture, at the place of manufacture and at his principal office or place of business if no such firearms, except samples, are kept at such office or place of business When a person changes the location of a business for which he has paid the special (occupational) tax, he will be liable for another such tax unless the change is properly registered with the regional director (compliance) for the region in which the special tax stamp was issued, as provided in § 179.46.

[36 FR 14256, Aug. 3, 1971 Redesignated at 40 FR 16835, Apr. 15, 1975, and amended by T.D. ATF-48, 44 FR 55842, Sept. 28, 1979; T.D. ATF-251, 52 FR 19334, May 22, 1987; T.D. ATF-271, 53 FR 17551, May 17, 1988]

§ 179.39 Engaging in more than one business at the same location.

If more than one business taxable unde 26 U.S.C. 5801, is carried on at the same location during a taxable year, the special (occupational) tax imposed on each such business must be paid. This section does not require a qualified manufacturer or importer to qualify as a dealer if such manufacturer or importer also engages in business on his qualified premises as a dealer. However, a qualified manufacturer who engages in business as an importer must also qualify as an importer. Further, a qualified dealer is not entitled to engage in business as a manufacturer or importer.

[36 FR 14256, Aug. 3,.1971 Redesignated at 40 FR 16835, Apr. 15, 1975, and amended by T.D. ATF-48, 44 FR 55842, Sept. 28, 1979; T.D. ATF-271, 53 FR 17551, May 17, 1988]

§ 179.40 Partnership liability.

Any number of persons doing business in partnership at any one location shall be required to pay but one special (occupational) tax.

§ 179.41 Single sale.

A single sale, unattended by circumstances showing the one making the sale to be engaged in business, does not create special (occupational) tax liability.

Change of Ownership

§ 179.42 Changes through death of owner.

Whenever any person who has paid special (occupational) tax dies, the surviving spouse or child, or executors or administrators, or other legal representatives, may carry on this business for the remainder of the term for which tax has been paid and at the place (or places) for which the tax was paid, without any additional payment, subject to the following conditions If the surviving spouse or child, or executor or administrator, or other legal representative of the deceased taxpayer continues the business, such person shall, within 30 days after the date on which the successor begins to carry on the business, file a new return, Form 5630.7, with ATF in accordance with the instructions on the form. The return thus executed shall show the name of the original taxpayer, together with the basis of the succession. (As to liability in case of failure to register, see § 179.49.)

[T.D. ATF-70, 45 FR 33979, May 21, 1980, as amended by T.D. ATF-251, 52 FR 19334, May 22, 1987; T.D. ATF-363, 60 FR 17456, Apr. 6, 1995]

§ 179.43 Changes through bankruptcy of owner.

A receiver or referee in bankruptcy may continue the business under the stamp issued to the taxpayer at the place and for the period for which the tax was paid. An assignee for the benefit of creditors may continue business under his assignor's special tax stamp without incurring additional special (occupational) tax liability. In such cases, the change shall be registered with ATF in a manner similar to that required by § 179.42.

[36 FR 14256, Aug. 3,.1971 Redesignated at 40 FR 16835, Apr. 15, 1975, and amended by T.D. ATF-251, 52 FR 19334, May 22, 1987]

§ 179.44 Change in partnership or unincorporated association.

When one or more members withdraw from a partnership or an unincorporated association, the remaining member, or members, may, without incurring additional special (occupational) tax liability, carry on the same business at the same location for the balance of the taxable period for which special (occupational) tax was paid, provided any such change shall be registered in the same manner as required by § 179.42. Where new member(s) are taken into a partnership or an unincorporated association, the new firm so constituted may not carry on business under the special tax stamp of the old firm. The new firm must file a return, pay the special (occupational) tax and register in the same manner as a person who first engages in business is required to do under § 179.34 even though the name of the new firm may be the same as that of the old. Where the members of a partnership or an unincorporated association, which has paid special (occupational) tax, form a corporation to continue the business, a new special tax stamp must be taken out in the name of the corporation.

§ 179.45 Changes in corporation.

Additional special (occupational) tax is not required by reason of a mere change of name or increase in the capital stock of a corporation if the laws of the State of incorporation provide for such change or increase without the formation of a new corporation. A stockholder in a corporation who after its dissolution continues the business, incurs new special (occupational) tax liability.

Change of Business Location

§ 179.46 Notice by taxpayer.

Whenever during the taxable year a taxpayer intends to remove his business to a location other than specified in his last special (occupational) tax return (see § 179.34), he shall file with AT **(a)** a return, Form 5630.7, bearing the notation "Removal Registry," and showing the new address intended to be used **(b)** his current special tax stamp, and **(c)** a letter application requesting the amendment of his registration. The regional director (compliance), upon approval of the application, shall return the special tax stamp, amended to show the new business location. Firearms operations shall not be commenced at the new business location by the taxpayer prior to the required approval of his application to so change his business location.

[36 FR 14256, Aug. 3,.1971 Redesignated at 40 FR 16835, Apr. 15, 1975, and amended by T.D. ATF-251, 52 FR 19334, May 22, 1987; T.D. ATF-363, 60 FR 17456, Apr. 6, 1995]

Change of Trade Name

§ 179.47 Notice by taxpayer.

Whenever during the taxable year a taxpayer intends to change the name of his business, he shall file with AT **(a)** a return, Form 5630.7, bearing the notation "Amended," and showing the trade name intended to be used, **(b)** his current special tax stamp, an **(c)** a letter application requesting the amendment of his registration. The regional director (compliance), upon approval of the application, shall return the special tax stamp, amended to show the new trade name. Firearms operations shall not be commenced under the new trade name by the taxpayer prior to the required approval of his application to so change the trade name.

[36 FR 14256, Aug. 3,.1971 Redesignated at 40 FR 16835, Apr. 15, 1975, and amended by T.D. ATF-251, 52 FR 19334, May 22, 1987]

Penalties and Interest

§ 179.48 Failure to pay special (occupational) tax.

Any person who engages in a business taxable under 26 U.S.C. 5801, without timely payment of the tax imposed with respect to such business (see § 179.34) shall be liable for such tax, plus the interest and penalties thereon (See 26 U.S.C. 6601 and 6651). In addition, such person may be liable for criminal penalties under 26 U.S.C. 5871.

[36 FR 14256, Aug. 3,.1971 Redesignated at 40 FR 16835, Apr. 15, 1975, and amended by T.D. ATF-48, 44 FR 55842, Sept. 28, 1979]

§ 179.49 Failure to register change or removal.

Any person succeeding to and carrying on a business for which special (occupational) tax has been paid without registering such change within 30 days thereafter, and any taxpayer removing his business with respect to which special (occupational) tax has been paid to a place other than that for which tax was paid without obtaining approval therefor (See § 179.46), will incur liability to an additional payment of the tax, addition to tax and interest, as provided in sections 5801, 6651, and 6601, respectively, I.R.C., for failure to make return (See § 179.50) or pay tax, as well as criminal penalties for carrying on business without payment of special (occupational) tax (see section 5871 I.R.C.).

§ 179.50 Delinquency.

Any person liable for special (occupational) tax under section 5801, I.R.C., who fails to file a return (Form 5630.7), as prescribed, will be liable for a delinquency penalty computed on the amount of tax due unless a return (Form 5630.7) is later filed and failure to file the return timely is shown to the satisfaction of the regional director (compliance), to be due to reasonable cause. The delinquency penalty to be added to the tax is 5 percent if the failure is for not more than 1 month, with an additional 5 percent for each additional month or fraction thereof during which failure continues, not to exceed 25 percent in the aggregate (section 6651, I.R.C.). However, no delinquency penalty is assessed where the 50 percent addition to tax is assessed for fraud (See § 179.51).

[36 FR 14256, Aug. 3,.1971 Redesignated at 40 FR 16835, Apr. 15, 1975, and amended by T.D. ATF-251, 52 FR 19334, May 22, 1987; T.D. ATF-363, 60 FR 17456, Apr. 6, 1995]

§ 179.51 Fraudulent return.

If any part of any underpayment of tax required to be shown on a return is due to fraud, there shall be added to the tax an amount equal to 50 percent of the underpayment, but no delinquency penalty shall be assessed with respect to the same underpayment (section 6653, I.R.C.).

Application of State Laws

§ 179.52 State regulations.

Special tax stamps are merely receipts for the tax. Payment of tax under Federal law confers no privilege to act contrary to State law. One to whom a special tax stamp has been issued may still be punishable under a State law prohibiting or controlling the manufacture, possession or transfer of firearms. On the other hand, compliance with State law confers no immunity under Federal law. Persons who engage in the business of importing, manufacturing or dealing in firearms, in violation of the law of a State are nevertheless required to pay special (occupational) tax as imposed under the internal revenue laws of the United States. For provisions relating to restrictive use of information furnished to comply with the provisions of this part see § 179.23.

Subpart F-Tax on Making Firearms

§ 179.61 Rate of tax.

Except as provided in this subpart, there shall be levied, collected, and paid upon the making of a firearm a tax at the rate of $200 for each firearm made This tax shall be paid by the person making the firearm. Payment of the tax on the making of a firearm shall be represented by a $200 adhesive stamp bearing the words "National Firearms Act." Th stamps are maintained by the Director.

[T.D. ATF-270, 53 FR 10508, Mar. 31, 1988]

Application to Make a Firearm

§ 179.62 Application to make.

No person shall make a firearm unless the person has filed with the Director a written application on Form 1,(Firearms) **Application to Make and Register a Firearm,** in duplicate, executed under the penalties of perjury, to make and register the firearm and has received the approval of the Director to make the firearm which approval shall effectuate registration of the weapon to the applicant. The application shall identify the firearm to be made by serial number, type, model, caliber or gauge, length of barrel, other marks of identification, and the name and address of original manufacturer (if the applicant is not the original manufacturer) The applicant must be identified on the Form 1 (Firearms) by name and address and, if other than a natural person, the name and address of the principal officer or authorized representative and the employer identification number and, if an individual, the identification must include the date and place of birth and the information prescribed in § 179.63. Each applicant shall identify the Federal firearms license and special (occupational) tax stamp issued to the applicant, if any. The applicant shall also show required information evidencing that making or possession of the firearm would not be in violation of law. If the making is taxable, a remittance in the amount of $200 shall be submitted with the application in accordance with the instructions on the form. If the making is taxable and the application is approved, the Director will affix a National Firearms Act stamp to the original application in the space provided therefor and properly cancel the stamp (See § 179.67). The approved application will be returned to the applicant. If the making of the firearm is tax exempt under this part, an explanation of the basis of the exemption shall be attached to the Form 1 (Firearms).

[T.D. ATF-270, 53 FR 10508, Mar. 31, 1988]

§ 179.63 Identification of applicant.

If the applicant is an individual, the applicant shall securely attach to each copy of the Form 1 (Firearms), in the space provided on the form, a photograph of the applicant 2 X 2 inches in size, clearly showing a full front view of the features of the applicant with head bare, with the distance from the top of the head to the point of the chin approximately 1 1/4 inches, and which shall have been taken within 1 year prior to the date of the application. The applicant shall attach two properly completed FBI Forms FD-258 (Fingerprint Card) to the application. The fingerprints must be clear for accurate classification and should be taken by someone properly equipped to take them A certificate of the local chief of police, sheriff of the county, head of the State police, State or local district attorney or prosecutor, or such other person whose certificate may in a particular case be acceptable to the Director, shall be completed on each copy of the Form 1 (Firearms) The certificate shall state that the certifying official is satisfied that the fingerprints and photograph accompanying the application are those of the applicant and that the certifying official has no information indicating that possession of the firearm by the maker would be in violation of State or local law or that the maker will use the firearm for other than lawful purposes.

[T.D. ATF-270, 53 FR 10509, Mar. 31, 1988]

§ 179.64   Procedure for approval of application.

The application to make a firearm, Form 1 (Firearms), must be forwarded directly, in duplicate, by the maker of the firearm to the Director in accordance with the instructions on the form. The Director will consider the application for approval or disapproval. If the application is approved, the Director will return the original thereof to the maker of the firearm and retain the duplicate. Upon receipt of the approved application, the maker is authorized to make the firearm described therein. The maker of the firearm shall not, under any circumstances, make the firearm until the application, satisfactorily executed, has been forwarded to the Director and has been approved and returned by the Director with the National Firearms Act stamp affixed. If the application is disapproved, the original Form 1 (Firearms) and the remittance submitted by the applicant for the purchase of the stamp will be returned to the applicant with the reason for disapproval stated on the form.

[T.D. ATF-270, 53 FR 10509, Mar. 31, 1988]

§ 179.65   Denial of application.

An application to make a firearm shall not be approved by the Director if the making or possession of the firearm would place the person making the firearm in violation of law.

§ 179.66   Subsequent transfer of firearms.

Where a firearm which has been made in compliance with 26 U.S.C. 5821, and the regulations contained in this part, is to be transferred subsequently, the transfer provisions of the firearms laws and regulations must be complied with. (See Subpart F of this part).

[36 FR 14256, Aug. 3,.1971 Redesignated at 40 FR 16835, Apr. 15, 1975, and amended by T.D. ATF-48, 44 FR 55842, Sept. 28, 1979]

§ 179.67   Cancellation of stamp.

The person affixing to a Form 1 (Firearms) a "National Firearms Act" stamp shall cancel it by writing or stamping thereon, in ink, his initials, and the day, month and year, in such manner as to render it unfit for reuse. The cancellation shall not so deface the stamp as to prevent its denomination and genuineness from being readily determined.

Exceptions to Tax on Making Firearms

§ 179.68   Qualified manufacturer.

A manufacturer qualified under this part to engage in such business may make firearms without payment of the making tax. However, such manufacturer shall report and register each firearm made in the manner prescribed by this part.

[36 FR 14256, Aug. 3,.1971 Redesignated at 40 FR 16835, Apr. 15, 1975,
and amended by T.D. ATF-271, 53 FR 17551, May 17, 1988]

§ 179.69 Making a firearm for the United States.

A firearm may be made by, or on behalf of, the United States or any
department, independent establishment, or agency thereof without payment
of the making tax. However, if a firearm is to be made on behalf of the
United States, the maker must file an application, in duplicate, on Form 1
(Firearms) and obtain the approval of the Director in the manner
prescribed in § 179.62.

§ 179.70 Certain government entities.

A firearm may be made without payment of the making tax by, or on
behalf of, any State, or possession of the United States, any political
subdivision thereof, or any official police organization of such a
government entity engaged in criminal investigations. Any person making a
firearm under this exemption shall first file an application, in
duplicate, on Form 1 (Firearms) and obtain the approval of the Director as
prescribed in § 179.62.

Registration

§ 179.71 Proof of registration.

The approval by the Director of an application, Form (Firearms), to make a firearm under this subpart shall effectuate registration of the firearm described in the Form 1 (Firearms) to the person making the firearm. The original Form 1 (Firearms) showing approval by the Director shall be retained by the maker to establish proof of his registration of the firearm described therein, and shall be made available to any ATF officer on request.

Subpart F-Transfer Tax

§ 179.81 Scope of tax.

Except as otherwise provided in this part, each transfer of a firearm in the United States is subject to a tax to be represented by an adhesive stamp of the proper denomination bearing the words "National Firearms Act" to be affixed to the Form 4 (Firearms), Application for Transfer and Registration of Firearm, as provided in this subpart.

§ 179.82 Rate of tax.

The transfer tax imposed with respect to firearms transferred within the United States is at the rate of $200 for each firearm transferred, except that the transfer tax on any firearm classified as "any other weapon" shall be at the rate of $5 for each such firearm transferred. The tax imposed on the transfer of the firearm shall be paid by the transferor.

§ 179.83 Transfer tax in addition to import duty.

The transfer tax imposed by section 5811, I.R.C., is in addition to any import duty.

Application and Order for Transfer of Firearm

§ 179.84 Application to transfer.

Except as otherwise provided in this subpart, no firearm may be transferred in the United States unless an application, Form 4 (Firearms), Application for Transfer and Registration of Firearm, in duplicate, executed under the penalties of perjury to transfer the firearm and register it to the transferee has been filed with and approved by the Director The application, Form 4 (Firearms), shall be filed by the transferor and shall identify the firearm to be transferred by type; serial number; name and address of the manufacturer and importer, if known; model; caliber, gauge or size; in the case of a short-barreled shotgun or a short-barreled rifle, the length of the barrel; in the case of a weapon made from a rifle or shotgun, the overall length of the weapon and the length of the barrel; and any other identifying marks on the firearm. In the event the firearm does not bear a serial number, the applicant shall obtain a serial number from the Regional director (compliance) and shall stamp (impress) or otherwise conspicuously place such serial number on the firearm in a manner not susceptible of being readily obliterated, altered or removed. The application, Form 4 (Firearms), shall identify the transferor by name and address; shall identify the transferor's Federal firearms license and special (occupational) Chapter tax stamp, if any; and if the transferor is other than a natural person, shall show the title or status of the person executing the application. The application also shall identify the transferee by name and address, and, if the transferee is a natural person not qualified as a manufacturer, importer or dealer under this part, he shall be further identified in the manner prescribed in § 179.85. The application also shall identify the special (occupational) tax stamp and Federal firearms license of the transferee, if any. Any tax payable on the transfer must be represented by an adhesive stamp of proper denomination being affixed to the application, Form 4 (Firearms), properly cancelled.

[36 FR 14256, Aug. 3,.1971 Redesignated at 40 FR 16835, Apr. 15, 1975, and amended by T.D. ATF-241, 51 FR 39632, Oct. 29, 1986 T.D. ATF-270, 53 FR 10509, Mar. 31, 1988]

§ 179.85 Identification of transferee.

If the transferee is an individual, such person shall securely attach to each copy of the application, Form 4 (Firearms), in the space provided on the form, a photograph of the applicant 2 x 2 inches in size, clearly showing a full front view of the features of the applicant with head bare, with the distance from the top of the head to the point of the chin approximately 1 1/4 inches, and which shall have been taken within 1 year prior to the date of the application. The transferee shall attach two properly completed FBI Forms FD-258 (Fingerprint Card) to the application. The fingerprints must be clear for accurate classification and should be taken by someone properly equipped to take them   A certificate of the local chief of police, sheriff of the county, head of the State police, State or local district attorney or prosecutor, or such other person whose certificate may in a particular case be acceptable to the Director, shall be completed on each copy of the Form 4 (Firearms)   The certificate shall state that the certifying official is satisfied that the fingerprints and photograph accompanying the application are those of the applicant and that the certifying official has no information indicating that the receipt or possession of the firearm would place the transferee in violation of State or local law or that the transferee will use the firearm for other than lawful purposes.

[T.D. ATF-270, 53 FR 10509, Mar. 31, 1988]

§ 179.86 Action on application.

The Director will consider a completed and properly executed application, Form 4 (Firearms), to transfer a firearm. If the application is approved, the Director will affix the appropriate National Firearms Act stamp, cancel it, and return the original application showing approval to the transferor who may then transfer the firearm to the transferee along with the approved application.   The approval of an application, Form 4 (Firearms), by the Director will effectuate registration of the firearm to the transferee. The transferee shall not take possession of a firearm until the application, Form 4 (Firearms), for th transfer filed by the transferor has been approved by the Director and registration of the firearm is effectuated to the transferee. The transferee shall retain the approved application as proof that the firearm described therein is registered to the transferee, and shall make the approved Form 4 (Firearms) available to any ATF officer on request. If the application, Form 4 (Firearms), to transfer a firearm is disapproved by the Director, the original application and the remittance for purchase of the stamp will be returned to the transferor with reasons for the disapproval stated on the application. An application, Form 4 (Firearms), to transfer a firearm shall be denied if the transfer, receipt, or possession of a firearm would place the transferee in violation of law. In addition to any other records checks that may be conducted to determine whether the transfer, receipt, or possession of a firearm would place the transferee in violation of law, the Director shall contact the National Instant Criminal

Background Check System.

[T.D. ATF-270, 53 FR 10509, Mar. 31, 1988; T.D. ATF-415, 63 FR 58272, Oct. 29, 1998]

§ 179.87 Cancellation of stamp.

The method of cancellation of the stamp required by this subpart as prescribed in § 179.67 shall be used.

Exemptions Relating to Transfers of Firearms

§ 179.88 Special (occupational) taxpayers.

(a) A firearm registered to a person qualified under this part to engage in business as an importer, manufacturer, or dealer may be transferred by that person without payment of the transfer tax to any other person qualified under this part to manufacture, import, or deal in firearms.

(b) The exemption provided in paragraph (a) of this section shall be obtained by the transferor of the firearm filing with the Director an application, Form 3 (Firearms), Application for Tax-exempt Transfer of Firearm and Registration to Special (Occupational) Taxpayer, in duplicate, executed under the penalties of perjury.
The application, Form 3 (Firearms), shal (1) show the name and address of the transferor and of the transferee (2) identify the Federal firearms license and special (occupational) tax stamp of the transferor and of the transferee (3) show the name and address of the manufacturer and the importer of the firearm, if known (4) show the type, model, overall length (if applicable), length of barrel, caliber, gauge or size, serial number, and other marks of identification of the firearm, an (5) contain a statement by the transferor that he is entitled to the exemption because the transferee is a person qualified under this part to manufacture, import, or deal in firearms. If the Director approves an application, Form 3 (Firearms), he shall return the original Form 3 (Firearms) to the transferor with the approval noted thereon. Approval of an application, Form 3 (Firearms), by the Director shall remove registration of the firearm reported thereon from the transferor and shall effectuate the registration of that firearm to the transferee. Upon receipt of the approved Form 3 (Firearms), the transferor shall deliver same with the firearm to the transferee. The transferor shall not transfer the firearm to the transferee until his application, Form 3 (Firearms), has been approved by the Director and the original thereof has been returned to the transferor. If the Director disapproves the application, Form 3 (Firearms), he shall return the original Form 3 (Firearms) to the transferor with the reasons for the disapproval stated thereon.

(c) The transferor shall be responsible for establishing the exempt status of the transferee before making a transfer under the provisions of this section. Therefore, before engaging in transfer negotiations with the transferee, the transferor should satisfy himself as to the claimed exempt status of the transferee and the bona fides of the transaction. If not fully satisfied, the transferor should communicate with the Director, report all circumstances regarding the proposed transfer, and await the Director's advice before making application for the transfer. An unapproved transfer or a transfer to an unauthorized person may subject the transferor to civil and criminal liabilities. (See 26 U.S.C. 5852, 5861, and 5871.)

[36 FR 14256, Aug. 3, 1971 Redesignated at 40 FR 16835, Apr. 15, 1975, and amended by T.D. ATF-48, 44 FR 55842, Sept. 28, 1979; T.D. ATF-271, 53 FR 17551, May 17, 1988]

§ 179.89 Transfers to the United States.

A firearm may be transferred to the United States or any department, independent establishment or agency thereof without payment of the transfer tax. However, the procedures for the transfer of a firearm as provided in § 179.90 shall be followed in a tax-exempt transfer of a firearm under this section, unless the transferor is relieved of such requirement under other provisions of this part.

§ 179.90 Certain government entities.

(a) A firearm may be transferred without payment of the transfer tax to or from any State, possession of the United States, any political subdivision thereof, or any official police organization of such a governmental entity engaged in criminal investigations.

(b) The exemption provided in paragraph (a) of this section shall be obtained by the transferor of the firearm filing with the Director an application, Form 5 (Firearms), Application for Tax-exempt Transfer and Registration of Firearm, in duplicate, executed under the penalties of perjury The application shal **(1)** show the name and address of the transferor and of the transferee **(2)** identify the Federal firearms license and special (occupational) tax stamp, if any, of the transferor and of the transferee **(3)** show the name and address of the manufacturer and the importer of the firearm, if known **(4)** show the type, model, overall length (if applicable), length of barrel, caliber, gauge or size, serial number, and other marks of identification of the firearm, an **(5)** contain a statement by the transferor that the transferor is entitled to the exemption because either the transferor or the transferee is a governmental entity coming within the purview of paragraph (a) of this section. In the case of a transfer of a firearm by a governmental entity to a transferee who is a natural person not qualified as a manufacturer, importer, or dealer under this part, the transferee shall be further identified in the manner prescribed in § 179.85. If the Director approves an application, Form 5 (Firearms), the original Form 5 (Firearms) shall be returned to the transferor with the approval noted thereon. Approval of an application, Form 5 (Firearms), by the Director shall effectuate the registration of that firearm to the transferee. Upon receipt of the approved Form 5 (Firearms), the transferor shall delive same with the firearm to the transferee. The transferor shall not transfer the firearm to the transferee until the application, Form 5 (Firearms), has been approved by the Director and the original thereof has been returned to the transferor.   If the Director disapproves the application, Form 5 (Firearms), the original Form 5 (Firearms) shall be returned to the transferor with the reasons for the disapproval stated thereon.   An application by a governmental entity to transfer a firearm shall be denied if the transfer, receipt, or possession of a firearm would place the transferee in violation of law.

(c) The transferor shall be responsible for establishing the exempt status of the transferee before making a transfer under the provisions of this section.   Therefore, before engaging in transfer negotiations with the transferee, the transferor should satisfy himself of the claimed exempt status of the transferee and the bona fides of the transaction. If not fully satisfied, the transferor should communicate with the Director, report all circumstances regarding the proposed transfer, and await the Director's advice before making application for transfer.   An unapproved transfer or a transfer to an unauthorized person may subject the transferor to civil and criminal liabilities. (See 26 U.S.C. 5852, 5861, and 5871.)

[36 FR 14256, Aug. 3, .1971 Redesignated at 40 FR 16835, Apr. 15, 1975, and amended by T.D. ATF-48, 44 FR 55842, Sept. 28, 1979; T.D. ATF-241, 51 FR 39632, Oct. 29, 1986; T.D. ATF-270, 53 FR 10510, Mar. 31, 1988]

§ 179.91 Unserviceable firearms.

An unserviceable firearm may be transferred as a curio or ornament without payment of the transfer tax However, the procedures for the transfer of a firearm as provided in § 179.90 shall be followed in a tax-exempt transfer of a firearm under this section, except a statement shall be entered on the transfer application, Form 5 (Firearms), by the transferor that he is entitled to the exemption because the firearm to be transferred is unserviceable and is being transferred as a curio or ornament. An unapproved transfer, the transfer of a firearm under the provisions of this section which is in fact not an unserviceable firearm, or the transfer of an unserviceable firearm as something other than a curio or ornament, may subject the transferor to civil and criminal liabilities. (See 26 U.S.C. 5811, 5852, 5861, and 5871.)

[36 FR 14256, Aug. 3,.1971 Redesignated at 40 FR 16835, Apr. 15, 1975, and amended by T.D. ATF-48, 44 FR 55843, Sept. 28, 1979]

§ 179.92 Transportation of firearms to effect transfer.

Notwithstanding any provision of § 178.28 of this chapter, it shall not be required that authorization be obtained from the Director for the transportation in interstate or foreign commerce of a firearm in order to effect the transfer of a firearm authorized under the provisions of this subpart.

[T.D. ATF-270, 53 FR 10510, Mar. 31, 1988]

Other Provisions

§ 179.93 Transfers of firearms to certain persons.

Where the transfer of a destructive device, machine gun, short-barreled shotgun, or short-barreled rifle is to be made by a person licensed under the provisions of Title I of the Gun Control Act of 1968 (82 Stat. 1213) to a person not so licensed, the sworn statement required by § 178.98 of this chapter shall be attached to and accompany the transfer application required by this subpart.

Subpart G-Registration and Identification of Firearms

§ 179.101 Registration of firearms.

(a) The Director shall maintain a central registry of all firearms in the Unite States which are not in the possession of or under the control of the United States. This registry shall be known as the National Firearms Registration and Transfer Record and shall include:

(1) Identification of the firearm as required by this part;

(2) Date of registration; and

(3) Identification and address of person entitled to possession of the firearm as required by this part.

(b) Each manufacturer, importer, and maker shall register each firearm he manufactures, imports, or makes in the manner prescribed by this part. Each firearm transferred shall be registered to the transferee by the transferor in the manner prescribed by this part. No firearm may be registered by a person unlawfully in possession of the firearm except during an amnesty period established under section 207 of the Gun Control Act of 19682(8 Stat. 1235).

(c) A person shown as possessing firearms by the records maintained by the Director pursuant to the National Firearms Act (26 U.S.C. Chapter 53) in force on October 31, 1968, shall be considered to have registered the firearms in his possession which are disclosed by that record as being in his possession on October 31, 1968.

(d) The National Firearms Registration and Transfer Record shall include firearms registered to the possessors thereof under the provisions of section 207 of the Gun Control Act of 1968.

(e) A person possessing a firearm registered to him shall retain proof b registration which shall be made available to any ATF officer upon request.

(f) A firearm not identified as required by this part shall not be registered.

[36 FR 14256, Aug. 3, .1971 Redesignated at 40 FR 16835, Apr. 15, 1975, and amended by T.D. ATF-48, 44 FR 55843, Sept. 28, 1979]

§ 179.102 Identification of firearms.

Each manufacturer, importer, or maker of a firearm shall legibly identify it by engraving, casting, stamping (impressed), or otherwise conspicuously placing or causing to be engraved, cast, stamped (impressed) or placed on the frame or receiver thereof in a manner not susceptible of being readily obliterated, altered, or removed, an individual serial number not duplicating any serial number placed by the manufacturer, importer, or maker on any other firearm, and by engraving, casting, stamping (impressed), or otherwise conspicuously placing or causing to be engraved, cast, stamped (impressed), or placed on the frame, receiver, or barrel thereof in a manner not susceptible of being readily obliterated, altered or removed, the model, if such designation has been made; the caliber or gauge; the name (or recognized abbreviation of same) of the manufacturer, or maker, and also, when applicable, of the importer; in the case of a domestically made firearm, the city and State (or recognized abbreviation thereof) wherein the manufacturer or importer maintains his place of business, or the maker made the firearm; and in the case of an imported firearm, the name of the country in which manufactured and the city and State (or recognized abbreviation thereof) of the importer:
Provided, That the Director may authorize other means of identification of the manufacturer, importer, or maker upon receipt of letter application, in duplicate, from same showing that such other identification is reasonable and will not hinder the effective administration of this part:
Provided, further, That in the case of a destructive device, the Director may authorize other means of identifying that weapon upon receipt of letter application, in duplicate, from the manufacturer, importer, or maker showing that engraving, casting, or stamping (impressing) such a weapon would be dangerous or impracticable A firearm frame or receiver or any other part defined as a machine gun or a muffler or silencer for the purposes of this part which is not a component part of a complete firearm at the time it is sold, shipped, or otherwise disposed of by a manufacturer, importer, or maker shall be identified as required by this section The Director may authorize other means of identification of parts defined as machine guns other than frames or receivers and parts defined as mufflers or silencers upon receipt of a letter application, in duplicate, showing that such other identification is reasonable and will not hinder the effective administration of this part.

[36 FR 14256, Aug. 3, .1971 Redesignated at 40 FR 16835, Apr. 15, 1975, and amended by T.D. ATF-241, 51 FR 39632, Oct. 29, 1986; T.D. ATF-270, 53 FR 10510, Mar. 31, 1988]

§ 179.103 Registration of firearms manufactured.

Each manufacturer qualified under this part shall file with the Director an accurate notice on Form 2 (Firearms), Notice of Firearms Manufactured or Imported, executed under the penalties of perjury, to show his manufacture of firearms The notice shall set forth the name and address of the manufacturer, identify his special (occupational) tax stamp and Federal firearms license, and show the date of manufacture, the type, model, length of barrel, overall length, caliber, gauge or size, serial numbers, and other marks of identification of the firearms he manufactures, and the place where the manufactured firearms will be kept. All firearms manufactured by him during a single day shall be included on one notice, Form 2 (Firearms), filed by the manufacturer no later than the close of the next business day. The manufacturer shall prepare the notice, Form 2 (Firearms), in duplicate, file the original notice as prescribed herein and keep the copy with the records required by Subpart I of this part at the premises covered by his special (occupational) tax stamp. Receipt of the notice, Form 2 (Firearms) by the Director shall effectuate the registration of the firearms listed on that notice. The requirements of this part relating to the transfer of a firearm are applicable to transfers by qualified manufacturers.

§ 179.104 Registration of firearms by certain governmental entities.

Any State, any political subdivision thereof, or any official police organization of such a government entity engaged in criminal investigations, which acquires for official use a firearm not registered to it, such as by abandonment or by forfeiture, will register such firearm with the Director by filing Form 10 (Firearms), Registration of Firearms Acquired by Certain Governmental Entities, and such registration shall become a part of the National Firearms Registration and Transfer Record. The application shall identify the applicant, describe each firearm covered by the application, show the location where each firearm usually will be kept, and, if the firearm is unserviceable, the application shall show how the firearm was made unserviceable. This section shall not apply to a firearm merely being held for use as evidence in a criminal proceeding. The Form 10 (Firearms) shall be executed in duplicate in accordance with the instructions thereon. Upon registering the firearm, the Director shall return the original Form 10 (Firearms) to the registrant with notification thereon that registration of the firearm has been made. The registration of any firearm under this section is for official use only and a subsequent transfer will be approved only to other governmental entities for official use.

[36 FR 14256, Aug. 3, .1971 Redesignated at 40 FR 16835, Apr. 15, 1975, and amended by T.D. ATF-241, 51 FR 39633, Oct. 29, 1986; T.D. ATF-270, 53 FR 10510, Mar. 31, 1988]

Machine guns

§ 179.105 Transfer and possession of machine guns.

(a) General. As provided by 26 U.S.C. 5812 and 26 U.S.C. 5822, an application to make or transfer a firearm shall be denied if the making, transfer, receipt, or possession of the firearm would place the maker or transferee in violation. of law Section 922(o), Title 18, U.S.C., makes it unlawful for any person to transfer or possess a machine gun, except a transfer to or by, or possession by or under the authority of, the United States or any department or agency thereof or a State, or a department, agency, or political subdivision thereof; or any lawful transfer or lawful possession of a machine gun that was lawfully possessed before May 19, 1986. Therefore, notwithstanding any other provision of this part, no application to make, transfer, or import a machine gun will be approved except as provided by this section.

(b) Machine guns lawfully possessed prior to May 19, 1986. A machine gun possessed in compliance with the provisions of this part prior to May 19, 1986, may continue to be lawfully possessed by the person to whom the machine gun is registered and may, upon compliance with the provisions of this part, be lawfully transferred to and possessed by the transferee.

(c) Importation and manufacture. Subject to compliance with the provisions of this part, importers and manufacturers qualified under this part may import and manufacture machine guns on or after May 19, 1986, for sale or distribution to any department or agency of the United States or any State or political subdivision thereof, or for use by dealers qualified under this part as sales samples as provided in paragraph (d) of this section. The registration of such machine guns under this part and their subsequent transfer shall be conditioned upon and restricted to the sale or distribution of such weapons for the official use of Federal, State or local governmental entities. Subject to compliance with the provisions of this part, manufacturers qualified under this part may manufacture machine guns on or after May 19, 1986, for exportation in compliance with the Arms Export Control Act (22 U.S.C. 2778) and regulations prescribed thereunder by the Department of State.

(d) Dealer sales samples. Subject to compliance with the provisions of this part, applications to transfer and register a machine gun manufactured or imported on or after May 19, 1986, to dealers qualified under this part will be approved if it is established by specific information the expected governmental customers who would require a demonstration of the weapon, information as to the availability of the machine gun to fill subsequent orders, and letters from governmental entities expressing a need for a particular model or interest in seeing a demonstration of a particular weapon. Applications to transfer more than one machine gun of a particular model to a dealer must also establish the dealer's need for the quantity of sample sought to be transferred.

(e) The making of machine guns on or after May 19, 1986. Subject to compliance with the provisions of this part, applications to make and register machine guns on or after May 19, 1986, for the benefit of a Federal, State or local governmental entity (e.g., an invention for possible future use of a governmental entity or the making of a weapon in connection with research and development on behalf of such an entity) will be approved if it is established by specific information that the machine gun is particularly suitable for use by Federal, State or local governmental entities and that the making of the weapon is at the request and on behalf of such an entity.

(f) Discontinuance of business. Since section 922(o), Title 18, U.S.C., makes it unlawful to transfer or possess a machine gun except as provided in the law, any qualified manufacturer, importer, or dealer intending to discontinue business shall, prior to going out of business, transfer in compliance with the provisions of this part any machine gun manufactured or imported after May 19, 1986, to a Federal, State or local governmental entity, qualified manufacturer, qualified importer, or, subject to the provisions of paragraph (d) of this section, dealer qualified to possess such, machine gun.

[T.D. ATF-270, 53 FR 10510, Mar. 31, 1988]

Subpart H-Importation and Exportation

Importation

§ 179.111   Procedure.

(a) No firearm shall be imported or brought into the United States or any territory under its control or jurisdiction unless the person importing or bringing in the firearm establishes to the satisfaction of the Director that the firearm to be imported or brought in is being imported or brought in for:

(1) The use of the United States or any department, independent establishment, or agency thereof or any State or possession or any political subdivision thereof; or

(2) Scientific or research purposes; or

(3) Testing,or use as a model by a registered manufacturer or solely for use as a sample by a registered importer or registered dealer.

The burden of proof is affirmatively on any person importing or bringing the firearm into the United States or any territory under its control or jurisdiction to show that the firearm is being imported or brought in under one of the above paragraphs. Any person desiring to import or bring a firearm into the United States under this paragraph shall file with the Director an application on Form 6 (Firearms), Application and Permit for Importation of Firearms, Ammunition and Implements of War, in triplicate, executed under the penalties of perjury. The application shall show the information required by Subpart G of Part 178 of this chapter. A detailed explanation of why the importation of the firearm falls within the standards set out in this paragraph shall be attached to the application. The person seeking to import or bring in the firearm will be notified of the approval or disapproval of his application. If the application is approved, the original Form 6 (Firearms) will be returned to the applicant showing such approval and he will present the approved application, Form 6 (Firearms), to the Customs officer at the port of importation. The approval of an application to import a firearm shall be automatically terminated at the expiration of one year from the date of approval unless, upon request it is further extended by the Director. If the firearm described in the approved application is not imported prior to the expiration of the approval, the Director shall be so notified. Customs officers will not permit release of a firearm from Customs custody, except for exportation, unless covered by an application which has been approved by the Director and which is currently effective. The importation or bringing in of a firearm not covered by an approved application may subject the person responsible to civil and criminal liabilities. (26 U.S.C. 5861, 5871, and 5872.)

(b) Part 178 of this chapter also contains requirements and procedures for the importation of firearms into the United States. A firearm may not be imported into the United States under this part unless those requirements and procedures are also complied with by the person importing the firearm.

(c) The provisions of this subpart shall not be construed as prohibiting the return to the United States or any territory under its control or jurisdiction of a firearm by a person who can establish to the satisfaction of Customs tha (1) the firearm was taken out of the United States or any territory under its control or jurisdiction by such person, (2) the firearm is registered to that person, an (3) if appropriate, the authorization required by Part 178 of this chapter for the transportation of such a firearm in interstate or foreign commerce has been obtained by such person.

[36 FR 14256, Aug. 3,.1971 Redesignated at 40 FR 16835, Apr. 15, 1975, and amended by T.D. ATF-48, 44 FR 55843, Sept. 28, 1979; ATF-325, 57 FR 29787, July 7, 1992]

§ 179.112 Registration of imported firearms.

(a) Each importer shall file with the Director an accurate notice on Form 2 (Firearms), Notice of Firearms Manufactured or Imported, executed under the penalties of perjury, showing the importation of a firearm   The notice shall set forth the name and address of the importer, identify the importer's special (occupational) tax stamp and Federal firearms license, and show the import permit number, the date of release from Customs custody, the type, model, length of barrel, overall length, caliber, gauge or size, serial number, and other marks of identification of the firearm imported, and the place where the imported firearm will be kept.   The Form 2 (Firearms) covering an imported firearm shall be filed by the importer no later than fifteen (15) days from the date the firearm was released from Customs custody.   The importer shall prepare the notice, Form 2 (Firearms), in duplicate, file the original return as prescribed herein, and keep the copy with the records required by subpart I of this part at the premises covered by the special (occupational) tax stamp   The timely receipt by the Director of the notice, Form 2 (Firearms), and the timely receipt by the Director of the copy of Form 6A (Firearms), Release and Receipt of Imported Firearms, Ammunition and Implements of War, required by §178.112 of this chapter, covering the weapon reported on the Form 2 (Firearms) by the qualified importer, shall effectuate the registration of the firearm to the importer.

(b) The requirements of this part relating to the transfer of a firearm are applicable to the transfer of imported firearms by a qualified importer or any other person.

(c) Subject to compliance with the provisions of this part, an application, Form 6 (Firearms), to import a firearm by an importer or dealer qualified under this part, for use as a sample in connection with sales of such firearms to Federal, State or local governmental entities, will be approved if it is established by specific information attached to the application that the firearm is suitable or potentially suitable for use by such entities.   Such information must show why a sales sample of a particular firearm is suitable for such use and the expected governmental customers who would require a demonstration of the firearm.   Information as to the availability of the firearm to fill subsequent orders and letters from governmental entities expressing a need for a particular model or interest in seeing a demonstration of a particular firearm would establish suitability for governmental use.   Applications to import more than one firearm of a particular model for use as a sample by an importer or dealer must also establish th importer's or dealer's need for the quantity of samples sought to be imported.

(d) Subject to compliance with the provisions of this part, an application, Form 6 (Firearms), to import a firearm by an importer or dealer qualified under this part, for use as a sample in connection with sales of such firearms to Federal, State or local governmental entities, will be approved if it is established by specific information attached to the application that the firearm is particularly suitable for use by such entities.   Such information must show why a sales sample of a particular firearm is suitable for such use and the expected governmental customers who would require a demonstration of the firearm.   Information as to the availability of the firearm to fill subsequent orders and letters from governmental entities expressing a need for a particular model or interest in seeing a demonstration of a particular firearm would establish suitability for governmental use.   Applications to import more than one firearm of a particular model for use as a sample by an importer or dealer must also establish th importer's or dealer's need for the quantity of samples sought to be imported.

[36 FR 14256, Aug. 3,.1971 Redesignated at 40 FR 16835, Apr. 15, 1975, and amended by T.D. ATF-241, 51 FR 39633, Oct. 29, 1986; T.D. ATF-270, 53 FR 10511, Mar. 31, 1988]

§ 179.113 Conditional importation.

The Director shall permit the conditional importation or bringing into the United States of any firearm for the purpose of examining and testing the firearm in connection with making a determination as to whether the importation or bringing in of such firearm will be authorized under this subpart. An application under this section shall be filed on Form 6 (Firearms), in triplicate, with the Director The Director may impose conditions upon any importation under this section including a requirement that the firearm be shipped directly from Customs custody to the Director and that the person importing or bringing in the firearm must agree to either export the weapon or destroy it if a final determination is made that it may not be imported or brought in under this subpart. A firearm so imported or brought into the United States may be released from Customs custody in the manner prescribed by th conditional authorization of the Director.

[T.D. ATF-270, 53 FR 10511, Mar. 31, 1988]

Exportation

§ 179.114 Application and permit for exportation of firearms.

Any person desiring to export a firearm without payment of the transfer tax must file with the Director an application on Form 9 (Firearms), Application and Permit for Exportation of Firearms, in quadruplicate, for a permit providing for deferment of tax liability. Part 1 of the application shall show the name and address of the foreign consignee, number of firearms covered by the application, the intended port of exportation, a complete description of each firearm to be exported, the name, address, State Department license number (or date of application if not issued), and identification of the special (occupational) tax stamp of the transferor. Part 1 of the application shall be executed under the penalties of perjury by the transferor and shall be supported by a certified copy of a written order or contract of sale or other evidence showing that the firearm is to be shipped to a foreign designation Where it is desired to make a transfer free of tax to another person who in turn will export the firearm, the transferor shall likewise file an application supported by evidence that the transfer will start the firearm in course of exportation, except, however, that where such transferor and exporter are registered special-taxpayers the transferor will not be required to file an application on Form 9 (Firearms).

§ 179.115 Action by Director.

If the application is acceptable, the Director will execute the permit, Part 2 of Form 9 (Firearms), to export the firearm described on the form and return three copies thereof to the applicant. Issuance of the permit by the Director will suspend assertion of tax liability for a period of six (6) months from the date of issuance. If the application is disapproved, the Director will indicate thereon the reason for such action and return the forms to the applicant.

§ 179.116 Procedure by exporter.

Shipment may not be made until the permit, Form 9 (Firearms), is received from the Director. If exportation is to be made by means other than by parcel post, two copies of the form must be addressed to the District Director of Customs at the port of exportation, and must precede or accompany the shipment in order to permit appropriate inspection prior to lading. If exportation is to be made by parcel post, one copy of the form must be presented to the postmaster at the office receiving the parcel who will execute Part 4 of such form and return the form to the exporter for transmittal to the Director. In the event exportation is not effected, all copies of the form must be immediately returned to the Director for cancellation.

§ 179.117   Action by Customs.

Upon receipt of a permit Form 9 (Firearms), in duplicate, authorizing the exportation of firearms, the District Director of Customs may order such inspection as deemed necessary prior to lading of the merchandise If satisfied that the shipment is proper and the information contained in the permit to export is in agreement with information shown in the shipper's export declaration, the District Director of Customs will, after the merchandise has been duly exported, execute the certificate of exportation (Part 3 of Form 9 (Firearms)). One copy of the form will be retained with the shipper's export declaration and the remaining copy thereof will be transmitted to the Director.

§ 179.118   Proof of exportation.

Within a six-month's period from date of issuance of the permit to export firearms, the exporter shall furnish or cause to be furnished to the Director **(a)** the certificate of exportation (Part 3 of Form 9 (Firearms)) executed by the District Director of Customs as provided in § 179.117, ro **(b)** the certificate of mailing by parcel post (Part 4 of Form 9 (Firearms)) executed by the postmaster of the post office receiving the parcel containing the firearm, o **(c)** a certificate of landing executed by a Customs officer of the foreign country to which the firearm is exported, or **(d)** a sworn statement of the foreign consignee covering the receipt of the firearm; o **(e)** the return receipt, or a reproduced copy thereof, signed by the addressee or his agent, where the shipment of a firearm was made by insured or registered parcel post. Issuance of a permit to export a firearm and furnishing of evidence establishing such exportation under this section will relieve the actual exporter and the person selling to the exporter for exportation from transfer tax liability.   Where satisfactory evidence of exportation of a firearm is not furnished within the stated period, the transfer tax will be assessed.

§ 179.119   Transportation of firearms to effect exportation.

Notwithstanding any provision of § 178.28 of this chapter, it shall not be required that authorization be obtained from the Director for the transportation in interstate or foreign commerce of a firearm in order to effect the exportation of a firearm authorized under the provisions of this subpart.

[T.D. ATF-270, 53 FR 10511, Mar. 31, 1988]

§ 179.120   Refunds.

Where, after payment of tax by the manufacturer, a firearm is exported, and satisfactory proof of exportation (See § 179.118) is furnished, a claim for refund may be submitted on Form 843 (See § 179.172). If the manufacturer waives al claim for the amount to be refunded, the refund shall be made to the exporter A claim for refund by an exporter of tax paid by a manufacturer should be accompanied by waiver of the manufacturer and proof of tax payment by the latter.

§ 179.121   Insular possessions.

Transfers of firearms to persons in the insular possessions of the

United States are exempt from transfer tax, provided title in cases involving change of title (and custody or control, in cases not involving change of title), does not pass to the transferee or his agent in the United States However, such exempt transactions must be covered by approved permits and supporting documents corresponding to those required in the case of firearms exported to foreign countries (see §§ 179.114 and 179.115), except that the Director may vary the requirements herein set forth in accordance with the requirements of the governing authority of the insular possession. Shipments to the insular possessions will not be authorized without compliance with the requirements of the governing authorities thereof. In the case of a nontaxable transfer to a person in such insular possession, the exemption extends only to such transfer and not to prior transfers.

Arms Export Control Act

§ 179.122 Requirements.

(a) Person engaged in the business of importing firearms are required by the Arms Export Control Act (22 U.S.C. 2778) to register with the Director. (See Part 47 of this chapter.)

(b) Persons engaged in the business of exporting firearms caliber .22 or larger are subject to the requirements of a license issued by the Secretary of State. Application for such license should be made to the Office of Munitions Control, Department of State, Washington, DC 20502, prior to exporting firearms.

[36 FR 14256, Aug. 3,.1971 Redesignated at 40 FR 16835, Apr. 15, 1975, and amended by T.D. ATF-241, 51 FR 39634, Oct. 29, 1986; T.D. ATF-270, 53 FR 10511, Mar. 31, 1988]

Subpart I-Records and Returns

§ 179.131 Records.

For the purposes of this part, each manufacturer, importer, and dealer in firearms shall keep and maintain such records regarding the manufacture, importation, acquisition (whether by making, transfer, or otherwise), receipt, and disposition of firearms as are prescribed, and in the manner and place required, by Part 178 of this chapter. In addition, each manufacturer, importer, and dealer shall maintain, in chronological order, at his place of business a separate record consisting of the documents required by this part showing the registration of any firearm to him. If firearms owned or possessed by a manufacturer, importer, or dealer are stored or kept on premises other than the place of business shown on his special (occupational) tax stamp, the record establishing registration shall show where such firearms are stored or kept. The records required by this part shall be readily accessible for inspection at all reasonable times by ATF officers.

(Approved by the Office of Management and Budget under control number 1512-0387)

[36 FR 14256, Aug. 3,.1971 Redesignated at 40 FR 16835, Apr. 15, 1975, and amended by T.D. ATF-172, 49 FR 14942, Apr. 16, 1984]

Subpart J-Stolen or Lost Firearms or Documents

§ 179.141 Stolen or lost firearms.

Whenever any registered firearm is stolen or lost, the person losing possession thereof will, immediately upon discovery of such theft or loss, make a report to the Director showing the following:

(a) Name and address of the person in whose name the firearm is registered (b) kind of firearm (c) serial number (d) model, (e) caliber (f) manufacturer of the firearm (g) date and place of theft or loss, an (h) complete statement of facts and circumstances surrounding such theft or loss.

§ 179.142   Stolen or lost documents.

When any Forms 1, 2, 3, 4, 5, 6A, or 10 (Firearms) evidencing possession of a firearm is stolen, lost, or destroyed, the person losing possession will immediately upon discovery of the theft, loss, or destruction report the matter to the Director. The report will show in detail the circumstances of the theft, loss, or destruction and will include all know facts which may serve to identify the document. Upon receipt of the report, the Director will make such investigation as appears appropriate and may issue a duplicate document upon such conditions as the circumstances warrant.

Subpart K-Examination of Books and Records

§ 179.151   Failure to make returns: Substitute returns.

If any person required by this part to make returns shall fail or refuse to make any such return within the time prescribed by this part or designated by the Director, then the return shall be made by an ATF officer upon inspection of the books, but the making of such return by an ATF officer shall not relieve the person from any default or penalty incurred by reason of failure to make such return.

(53 Stat. 437; 26 U.S.C. 6020)

§ 179.152   Penalties (records and returns).

Any person failing to keep records or make returns, or making, or causing the making of, a false entry on any application, return or record, knowing such entry to be false, is liable to fine and imprisonment as provided in section 5871, I.R.C.

Subpart L-Distribution and Sale of Stamps

§ 179.161   National Firearms Act stamps.

"National Firearms Act" stamps evidencing payment of the transfer tax or tax on the making of a firearm are maintained by the Director. The remittance for purchase of the appropriate tax stamp shall be submitted with the application. Upon approval of the application, the Director will cause the appropriate tax to be paid by affixing the appropriate stamp to the application.

[T.D. ATF-270, 53 FR 10511, Mar. 31, 1988]

§ 179.162   Stamps authorized.

Adhesive stamps of the $5 and $200 denomination, bearing the words "National Firearms Act," have been prepared and only such stamps shall be used for the payment of the transfer tax and for the tax on the making of a firearm.
[T.D. ATF-270, 53 FR 10511, Mar. 31, 1988]

§ 179.163 Reuse of stamps prohibited.

A stamp once affixed to one document cannot lawfully be removed and affixed to another. Any person willfully hreusing suc a stamp shall be subject to the penalty prescribed by 26 U.S.C. 7208.

[36 FR 14256, Aug. 3,.1971 Redesignated at 40 FR 16835, Apr. 15, 1975, and amended by T.D. ATF-48, 44 FR 55843, Sept. 28, 1979]

Subpart M-Redemption of or Allowance for Stamps or Refunds

§ 179.171 Redemption of or allowance for stamps.

Where a National Firearms Act stamp is destroyed, mutilated or rendered useless after purchase, and before liability has been incurred, such stamp may be redeemed by giving another stamp in lieu thereof. Claim for redemption of the stamp should be filed on ATF Form 2635 (5620.8) with the Director. Such claim shall be accompanied by the stamp or by a satisfactory explanation of the reasons why the stamp cannot be returned, and shall be filed within 3 years after the purchase of the stamp.

(68A Stat. 830; 26 U.S.C. 6805)

[T.D. ATF-270, 53 FR 10511, Mar. 31, 1988]

§ 179.172 Refunds.

As indicated in this part, the transfer tax or tax on the making of a firearm is ordinarily paid by the purchase and affixing of stamps, while special tax stamps are issued in payment of special (occupational) taxes. However, in exceptional cases, transfer tax, tax on the making of firearms, and/or special (occupational) tax may be paid pursuant to assessment. Claims for refunds of such taxes, paid pursuant to assessment, shall be filed on ATF Form 2635 (5620.8) within 3 years next after payment of the taxes. Such claims shall be filed with the regional director (compliance) serving the region in which the tax was paid. (For provisions relating to hand-carried documents and manner of filing, see 26 CFR 301.6091-1(b) and 301.6402-2(a). When an applicant to make or transfer a firearm wishes a refund of the tax paid on an approved application where the firearm was not made pursuant to an approved Form 1 (Firearms) or transfer of the firearm did not take place pursuant to an approved Form 4 (Firearms), the applicant shall file a claim for refund of the tax on ATF Form 2635 (5620.8) with the Director. The claim shall be accompanied by the approved application bearing the stamp and an explanation why the tax liability was not incurred. Such claim shall be filed within 3 years next after payment of the tax.

(68A Stat. 808, 830; 26 U.S.C. 6511, 6805)

[T.D. ATF-270, 53 FR 10512, Mar. 31, 1988]

Subpart N-Penalties and Forfeitures

§ 179.181 Penalties.

Any person who violates or fails to comply with the requirements of 26 U.S.C. Chapter 53 shall, upon conviction, be subject to the penalties imposed under 26 U.S.C. 5871.

[T.D. ATF-48, 44 FR 55843, Sept. 28, 1979]

§ 179.182 Forfeitures.

Any firearm involved in any violation of the provisions of 26 U.S.C. Chapter 53, shall be subject to seizure, and forfeiture under the internal revenue law : **Provided, however,** That the disposition of forfeited firearms shall be in conformance with the requirements of 26 U.S.C. 5872. In addition, any vessel, vehicle or aircraft used to transport, carry, convey or conceal or possess any firearm with respect to which there has been committed any violation of any provision of 26 U.S.C. Chapter 53, or the regulations in this part issued pursuant thereto, shall be subject to seizure and forfeiture under the Customs laws, as provided by the act of August 9, 1939 (49 U.S.C. App., Chapter 11).

[T.D. ATF-270, 53 FR 10512, Mar. 31, 1988]

Subpart O-Other Laws Applicable

§ 179.191 Applicability of other provisions of internal revenue laws.

All of the provisions of the internal revenue laws not inconsistent with the provisions of 26 U.S.C. Chapter 53 shall be applicable with respect to the taxes imposed by 26 U.S.C. 5801, 5811, and 5821 (See 26 U.S.C. 5846).

[T.D. ATF-48, 44 FR 55843, Sept. 28, 1979]

§ 179.192   Commerce in firearms and ammunition.

For provisions relating to commerce in firearms and ammunition, including the movement of destructive devices, machine guns, short-barreled shotguns, or short-barreled rifles, see 18 U.S.C. Chapter 44, and Part 178 of this chapter issued pursuant thereto.

[36 FR 14256, Aug. 3,.1971 Redesignated at 40 FR 16835, Apr. 15, 1975, and amended by T.D. ATF-48, 44 FR 55843, Sept. 28, 1979]

§ 179.193   Arms Export Control Act.

Forp rovisions relating to the registration and licensing of persons engaged in the business of manufacturing, importing or exporting arms, ammunition, or implements of war, see the Arms Export Control Act (22 U.S.C. 2778), and the regulations issued pursuant thereto. (See also Part 47 of this chapter.)

[T.D. ATF-270, 53 FR 10512, Mar. 31, 1988]

PART 25 – DEPARTMENT OF JUSTICE INFORMATION SYSTEMS

Subpart A – The National Instant Criminal Background Check System

Authority: Pub. L. 103-159, 107 Stat. 1536.

Subpart A – The National Instant Criminal Background Check System

§ 25.1 Purpose and authority.

The purpose of this subpart is to establish policies and procedures implementing the Brady Handgun Violence Prevention Act (Brady Act), Public Law 103-159, 107 Stat. 1536. The Brady Act requires the Attorney General to establish a National Instant Criminal Background Check System (NICS) to be contacted by any licensed importer, licensed manufacturer, or licensed dealer of firearms for information as to whether the transfer of a firearm to any person who is not licensed under 18 U.S.C. 923 would be in violation of Federal or state law. The regulations in this subpart are issued pursuant to section 103(h) of the Brady Act, 107 Stat. 1542 (18 U.S.C. 922 note), and include requirements to ensure the privacy and security of the NICS and appeals procedures for persons who have been denied the right to obtain a firearm as a result of a NICS background check performed by the Federal Bureau of Investigation (FBI) or a state or local law enforcement agency.

§ 25.2 Definitions.

Appeal means a formal procedure to challenge the denial of a firearm transfer.

ARI means a unique Agency Record Identifier assigned by the agency submitting records for inclusion in the NICS Index.

ATF means the Bureau of Alcohol, Tobacco, and Firearms of the Department of Treasury.

Audit log means a chronological record of system (computer) activities that enables the reconstruction and examination of the sequence of events and/or changes in an event.

Business day means a 24-hour day (beginning at 12:01 a.m.) on which state offices are open in the state in which the proposed firearm transaction is to take place.

Control Terminal Agency means a state or territorial criminal justice agency recognized by the FBI as the agency responsible for providing state-or territory-wide service to criminal justice users of NCIC data.

Data source means an agency that provided specific information to the NICS.

Delayed means that more research is required prior to a NICS "Proceed" or "Denied" response. A "Delayed" response to the FFL indicates that it would be unlawful to transfer the firearm until receipt of a follow-up "Proceed" response from the NICS or the expiration of three business days, whichever occurs first.

Denied means denial of a firearm transfer based on a NICS response indicating one or more matching records were found providing information demonstrating that receipt of a firearm by a prospective transferee would violate 18 U.S.C. 922 or state law.

Denying agency means a POC or the NICS Operations Center, whichever determines that information in the NICS indicates that the transfer of a firearm to a person would violate Federal or state law, based on a background check.

Dial-up access means any routine access through commercial switched circuits on a continuous or temporary basis.

Federal agency means any authority of the United States that is an "Agency" under 44 U.S.C. 3502(1), other than those considered to be independent regulatory agencies, as defined in 44 U.S.C. 3502(10).

FFL (federal firearms licensee) means a person licensed by the ATF as a manufacturer, dealer, or importer of firearms.

Firearm has the same meaning as in 18 U.S.C. 921(a)(3).

Licensed dealer means any person defined in 27 CFR 178.11.

Licensed importer has the same meaning as in 27 CFR 178.11.

Licensed manufacturer has the same meaning as in 27 CFR 178.11.

NCIC (National Crime Information Center) means the nationwide computerized information system of criminal justice data established by the FBI as a service to local, state, and Federal criminal justice agencies.

NICS means the National Instant Criminal Background Check System, which an FFL must, with limited exceptions, contact for information on whether receipt of a firearm by a person who is not licensed under 18 U.S.C. 923 would violate Federal or state law.

NICS Index means the database, to be managed by the FBI, containing information provided by Federal and state agencies about persons prohibited under Federal law from receiving or possessing a firearm. The NICS Index is separate and apart from the NCIC and the Interstate Identification Index (III).

NICS Operations Center means the unit of the FBI that receives telephone or electronic inquires from FFLs to perform background checks, makes a determination based upon available information as to whether the receipt or transfer of a firearm would be in violation of Federal or state law, researches criminal history records, tracks and finalizes appeals, and conducts audits of system use.

NICS Operations Center's regular business hours means the hours of

9:00 a.m. to 2:00 a.m., Eastern Time, seven days a week.

NICS Representative means a person who receives telephone inquiries to the NICS Operations Center from FFLs requesting background checks and provides a response as to whether the receipt or transfer of a firearm may proceed or is delayed.

NRI (NICS Record Identifier) means the system-generated unique number associated with each record in the NICS Index.

NTN (NICS Transaction Number) means the unique number that will be assigned to each valid background check inquiry received by the NICS. Its primary purpose will be to provide a means of associating inquiries to the NICS with the responses provided by the NICS to the FFLs.

ORI (Originating Agency Identifier) means a nine-character identifier assigned by the FBI to an agency that has met the established qualifying criteria for ORI assignment to identify the agency in transactions on the NCIC System.

Originating Agency means an agency that provides a record to a database checked by the NICS.

POC (Point of Contact) means a state or local law enforcement agency serving as an intermediary between an FFL and the federal databases checked by the NICS. A POC will receive NICS background check requests from FFLs, check state or local record systems, perform NICS inquiries, determine whether matching records provide information demonstrating that an individual is disqualified from possessing a firearm under Federal or state law, and respond to FFLs with the results of a NICS background check. A POC will be an agency with express or implied authority to perform POC duties pursuant to state statute, regulation, or executive order.

Proceed means a NICS response indicating that the information available to the system at the time of the response did not demonstrate that transfer of the firearm would violate federal or state law. A "Proceed" response would not relieve an FFL from compliance with other provisions of Federal or state law that may be applicable to firearms transfers. For example, under 18 U.S.C. 922(d), an FFL may not lawfully transfer a firearm if he or she knows or has reasonable cause to believe that the prospective recipient is prohibited by law from receiving or possessing a firearm.

Record means any item, collection, or grouping of information about an individual that is maintained by an agency, including but not limited to information that disqualifies the individual from receiving a firearm, and that contains his or her name or other personal identifiers.

STN (State-Assigned Transaction Number) means a unique number that may be assigned by a POC to a valid background check inquiry.

System means the National Instant Criminal Background Check System (NICS).

§ 25.3 System information.

(a) There is established at the FBI a National Instant Criminal Background Check System.

(b) The system will be based at the Federal Bureau of Investigation, 1000 Custer Hollow Road, Clarksburg, West Virginia 26306-0147.

(c) The system manager and address are: Director, Federal Bureau of Investigation, J. Edgar Hoover F.B.I. Building, 935 Pennsylvania Avenue, NW, Washington, D.C. 20535.

§ 25.4 Record source categories.

It is anticipated that most records in the NICS Index will be obtained from Federal agencies. It is also anticipated that a limited number of authorized state and local law enforcement agencies will voluntarily contribute records to the NICS Index. Information in the NCIC and III systems that will be searched during a background check has been or will be contributed voluntarily by Federal, state, local, and international criminal justice agencies.

§ 25.5 Validation and data integrity of records in the system.

(a) The FBI will be responsible for maintaining data integrity during all NICS operations that are managed and carried out by the FBI. This responsibility includes:

(1) Ensuring the accurate adding, canceling, or modifying of NICS Index records supplied by Federal agencies;

(2) Automatically rejecting any attempted entry of records into the NICS Index that contain detectable invalid data elements;

(3) Automatic purging of records in the NICS Index after they are on file for a prescribed period of time; and

(4) Quality control checks in the form of periodic internal audits by FBI personnel to verify that the information provided to the NICS Index remains valid and correct.

(b) Each data source will be responsible for ensuring the accuracy and validity of the data it provides to the NICS Index and will immediately correct any record determined to be invalid or incorrect.

§ 25.6 Accessing records in the system.

(a) FFLs may initiate a NICS background check only in connection with a proposed firearm transfer as required by the Brady Act. FFLs are strictly prohibited from initiating a NICS background check for any other purpose. The process of accessing the NICS for the purpose of conducting a NICS background check is initiated by an FFL's contacting the FBI NICS Operations Center (by telephone or electronic dial-up access) or a POC. FFLs in each state will be advised by the ATF whether they are required to initiate NICS background checks with the NICS Operations Center or a POC and how they are to do so.

(b) Access to the NICS through the FBI NICS Operations Center. FFLs may contact the NICS Operations Center by use of a toll-free telephone number, only during its regular business hours. In addition to telephone access, toll-free electronic dial-up access to the NICS will be provided to FFLs after the beginning of the NICS operation. FFLs with electronic dial-up access will be able to contact the NICS 24 hours each day, excluding scheduled and unscheduled downtime.

(c)(1) The FBI NICS Operations Center, upon receiving an FFL telephone or electronic dial-up request for a background check, will:

(i) Verify the FFL Number and code word;

(ii) Assign a NICS Transaction Number (NTN) to a valid inquiry and provide the NTN to the FFL;

(iii) Search the relevant databases (i.e., NICS Index, NCIC, III) for any matching records; and

(iv) Provide the following NICS responses based upon the consolidated NICS search results to the FFL that requested the background check:

(A) "Proceed" response, if no disqualifying information was found in the NICS Index, NCIC, or III.

(B) "Delayed" response, if the NICS search finds a record that requires more research to determine whether the prospective transferee is disqualified from possessing a firearm by Federal or state law. A "Delayed" response to the FFL indicates that the firearm transfer should not proceed pending receipt of a follow-up "Proceed" response from the NICS or the expiration of three business days (exclusive of the day on which the query is made), whichever occurs first. (Example: An FFL requests a NICS check on a prospective firearm transferee at 9:00 a.m. on Friday and shortly thereafter receives a "Delayed" response from the NICS. If state offices in the state in which the FFL is located are closed on Saturday and Sunday and open the following Monday, Tuesday, and Wednesday, and the NICS has not yet responded with a "Proceed" or "Denied" response, the FFL may transfer the firearm at 12:01 a.m. Thursday.)

(C) "Denied" response, when at least one matching record is found in either the NICS Index, NCIC, or III that provides information demonstrating that receipt of a firearm by the prospective transferee would violate 18 U.S.C. 922 or state law. The "Denied" response will be provided to the requesting FFL by the NICS Operations Center during its regular business hours.

(2) None of the responses provided to the FFL under paragraph (c)(1) of this section will contain any of the underlying information in the records checked by the system.

(d) Access to the NICS through POCs. In states where a POC is designated to process background checks for the NICS, FFLs will contact the POC to initiate a NICS background check. Both ATF and the POC will notify FFLs in the POC's state of the means by which FFLs can contact the POC. The NICS will provide POCs with electronic access to the system virtually 24 hours each day through the NCIC communication network. Upon receiving a request for a background check from an FFL, a POC will:

(1) Verify the eligibility of the FFL either by verification of the FFL number or an alternative POC-verification system;

(2) Enter a purpose code indicating that the query of the system is for the purpose of performing a NICS background check in connection with the transfer of a firearm; and

(3) Transmit the request for a background check via the NCIC interface to the NICS.

(e) Upon receiving a request for a NICS background check, POCs may also conduct a search of available files in state and local law enforcement and other relevant record systems, and may provide a unique State-Assigned Transaction Number (STN) to a valid inquiry for a background check.

(f) When the NICS receives an inquiry from a POC, it will search the relevant databases (i.e., NICS Index, NCIC, III) for any matching record(s) and will provide an electronic response to the POC. This response will consolidate the search results of the relevant databases and will include the NTN. The following types of responses may be provided by the NICS to a state or local agency conducting a background check:

(1) No record response, if the NICS determines, through a complete search, that no matching record exists.

(2) Partial response, if the NICS has not completed the search of all of its records. This response will indicate the databases that have been searched (i.e., III, NCIC, and/or NICS Index) and the databases that have not been searched. It will also provide any potentially disqualifying information found in any of the databases searched. A follow-up response will be sent as soon as all the relevant databases have been searched. The follow-up response will provide the complete search results.

(3) Single matching record response, if all records in the relevant databases have been searched and one matching record was found.

(4) Multiple matching record response, if all records in the relevant databases have been searched and more than one matching record was found.

(g) Generally, based on the response(s) provided by the NICS, and other information available in the state and local record systems, a POC will:

(1) Confirm any matching records; and

(2) Notify the FFL that the transfer may proceed, is delayed pending further record analysis, or is denied. "Proceed" notifications made within three business days will be accompanied by the NTN or STN traceable to the NTN. The POC may or may not provide a transaction number (NTN or STN) when notifying the FFL of a "Denied" response.

(h) In cases where a transfer is denied by a POC, the POC should provide a denial notification to the NICS. This denial notification will include the name of the person who was denied a firearm and the NTN. The information provided in the denial notification will be maintained in the NICS Audit Log described in § 25.9(b). This notification may be provided immediately by electronic message to the NICS (i.e., at the time the transfer is denied) or as soon thereafter as possible. If a denial notification is not provided by a POC, the NICS will assume that the transfer was allowed and will destroy its records regarding the transfer in accordance with the procedures detailed in § 25.9.

(i) **Response recording.** FFLs are required to record the system response, whether provided by the FBI NICS Operations Center or a POC, on the appropriate ATF form for audit and inspection purposes, under 27 CFR part 178 recordkeeping requirements. The FBI NICS Operations Center response will always include an NTN and associated "Proceed," "Delayed," or "Denied" determination. POC responses may vary as discussed in Paragraph (g) of this section. In these instances, FFLs will record the POC response, including any transaction number and/or determination.

(j) **Access to the NICS Index for purposes unrelated to NICS background checks required by the Brady Act.** Access to the NICS Index for purposes unrelated to NICS background checks pursuant to 18 U.S.C. 922(t) shall be limited to uses for the purpose of:

(1) Providing information to Federal, state, or local criminal justice agencies in connection with the issuance of a firearm-related or explosives-related permit or license, including permits or licenses to possess, acquire, or transfer a firearm, or to carry a concealed firearm, or to import, manufacture, deal in, or purchase explosives; or

(2) Responding to an inquiry from the ATF in connection with a civil or criminal law enforcement activity relating to the Gun Control Act (18

U.S.C. Chapter 44) or the National Firearms Act (26 U.S.C. Chapter 53).

§ 25.7   Querying records in the system.

(a) The following search descriptors will be required in all queries of the system for purposes of a background check:

(1) Name;
(2) Sex;
(3) Race;
(4) Complete date of birth; and
(5) State of residence.

(b) A unique numeric identifier may also be provided to search for additional records based on exact matches by the numeric identifier. Examples of unique numeric identifiers for purposes of this system are: Social Security number (to comply with Privacy Act requirements, a Social Security number will not be required by the NICS to perform any background check) and miscellaneous identifying numbers (e.g., military number or number assigned by Federal, state, or local authorities to an individual's record). Additional identifiers that may be requested by the system after an initial query include height, weight, eye and hair color, and place of birth. At the option of the querying agency, these additional identifiers may also be included in the initial query of the system.

§ 25.8   System safeguards.

(a) Information maintained in the NICS Index is stored electronically for use in an FBI computer environment. The NICS central computer will reside inside a locked room within a secure facility. Access to the facility will be restricted to authorized personnel who have identified themselves and their need for access to a system security officer.

(b) Access to data stored in the NICS is restricted to duly authorized agencies. The security measures listed in paragraphs (c) through (f) of this section are the minimum to be adopted by all POCs and data sources having access to the NICS.

(c) State or local law enforcement agency computer centers designated by a Control Terminal Agency as POCs shall be authorized NCIC users and shall observe all procedures set forth in the NCIC Security Policy of 1992 when processing NICS background checks. The responsibilities of the Control Terminal Agencies and the computer centers include the following:

(1) The criminal justice agency computer site must have adequate physical security to protect against any unauthorized personnel gaining access to the computer equipment or to any of the stored data.

(2) Since personnel at these computer centers can have access to data stored in the NICS, they must be screened thoroughly under the authority and supervision of a state Control Terminal Agency. This authority and supervision may be delegated to responsible criminal justice agency personnel in the case of a satellite computer center being serviced through a state Control Terminal Agency. This screening will also apply to non-criminal justice maintenance or technical personnel.

(3) All visitors to these computer centers must be accompanied by staff personnel at all times.

(4) POCs utilizing a state/NCIC terminal to access the NICS must have the proper computer instructions written and other built-in controls to prevent data from being accessible to any terminals other than authorized terminals.

(5) Each state Control Terminal Agency shall build its data system around a central computer, through which each inquiry must pass for screening and verification.

(d) Authorized state agency remote terminal devices operated by POCs and having access to the NICS must meet the following requirements:

(1) POCs and data sources having terminals with access to the NICS must physically place these terminals in secure locations within the authorized agency;

(2) The agencies having terminals with access to the NICS must screen terminal operators and must restrict access to the terminals to a minimum number of authorized employees; and

(3) Copies of NICS data obtained from terminal devices must be afforded appropriate security to prevent any unauthorized access or use.

(e) FFL remote terminal devices may be used to transmit queries to the NICS via electronic dial-up access. The following procedures will apply to such queries:

(1) The NICS will incorporate a security authentication mechanism that performs FFL dial-up user authentication before network access takes place;

(2) The proper use of dial-up circuits by FFLs will be included as part of the periodic audits by the FBI; and

(3) All failed authentications will be logged by the NICS and provided to the NICS security administrator.

(f) FFLs may use the telephone to transmit queries to the NICS, in accordance with the following procedures:

(1) FFLs may contact the NICS Operations Center during its regular business hours by a telephone number provided by the FBI;

(2) FFLs will provide the NICS Representative with their FFL Number and code word, the type of sale, and the name, sex, race, date of birth, and state of residence of the prospective buyer; and

(3) The NICS will verify the FFL Number and code word before processing the request.

(g) The following precautions will be taken to help ensure the security and privacy of NICS information when FFLs contact the NICS Operations Center:

(1) Access will be restricted to the initiation of a NICS background check in connection with the proposed transfer of a firearm.

(2) The NICS Representative will only provide a response of "Proceed" or "Delayed" (with regard to the prospective firearms transfer), and will not provide the details of any record information about the transferee. In cases where potentially disqualifying information is found in response to an FFL query, the NICS Representative will provide a "Delayed" response to the FFL. Follow-up "Proceed" or "Denied" responses will be provided by the NICS Operations Center during its regular business hours.

(3) The FBI will periodically monitor telephone inquiries to ensure proper use of the system.

(h) All transactions and messages sent and received through electronic access by POCs and FFLs will be automatically logged in the NICS Audit Log described in § 25.9(b). Information in the NICS Audit Log will include initiation and termination messages, failed authentications, and matching records located by each search transaction.

(i) The FBI will monitor and enforce compliance by NICS users with the applicable system security requirements outlined in the NICS POC Guidelines and the NICS FFL Manual (available from the NICS Operations Center, Federal Bureau of Investigation, 1000 Custer Hollow Road, Clarksburg, West Virginia 26306-0147).

§ 25.9   Retention and destruction of records in the system.

(a) The NICS will retain NICS Index records that indicate that receipt of a firearm by the individuals to whom the records pertain would violate Federal or state law. The NICS will retain such records indefinitely, unless they are canceled by the originating agency. In cases where a firearms disability is not permanent, e.g., a disqualifying restraining order, the NICS will automatically purge the pertinent record when it is no longer disqualifying. Unless otherwise removed, records contained in the NCIC and III files that are accessed during a background check will remain in those files in accordance with established policy.

(b) The FBI will maintain an automated NICS Audit Log of all incoming and outgoing transactions that pass through the system.

(1) The Audit Log will record the following information: type of transaction (inquiry or response), line number, time, date of inquiry, header, message key, ORI, and inquiry/response data (including the name and other identifying information about the prospective transferee and the NTN). In cases of allowed transfers, all information in the Audit Log related to the person or the transfer, other than the NTN assigned to the transfer and the date the number was assigned, will be destroyed after not more than six months after the transfer is allowed. Audit Log records relating to denials will be retained for 10 years, after which time they will be transferred to a Federal Records Center for storage. The NICS will not be used to establish any system for the registration of firearms, firearm owners, or firearm transactions or dispositions, except with respect to persons prohibited from receiving a firearm by 18 U.S.C. 922(g) or (n) or by state law.

(2) The Audit Log will be used to analyze system performance, assist users in resolving operational problems, support the appeals process, or support audits of the use of the system. Searches may be conducted on the Audit Log by time frame, i.e., by day or month, or by a particular state or agency. Information in the Audit Log pertaining to allowed transfers may only be used by the FBI for the purpose of conducting audits of the use and performance of the NICS. Such information, however, may be retained and used as long as needed to pursue cases of identified misuse of the system. The NICS, including the NICS Audit Log, may not be used by any department, agency, officer, or employee of the United States to establish any system for the registration of firearms, firearm owners, or firearm transactions or dispositions. The Audit Log will be monitored and reviewed on a regular basis to detect any possible misuse of the NICS data.

(c) The following records in the FBI-operated terminals of the NICS will be subject to the Brady Act's requirements for destruction:

(1) All inquiry and response messages (regardless of media) relating to a background check that results in an allowed transfer; and

(2) All information (regardless of media) contained in the NICS Audit Log relating to a background check that results in an allowed transfer.

(d) The following records of state and local law enforcement units serving as POCs will be subject to the Brady Act's requirements for destruction:

(1) All inquiry and response messages (regardless of media) relating to the initiation and result of a check of the NICS that allows a transfer that are not part of a record system created and maintained pursuant to independent state law regarding firearms transactions; and

(2) All other records relating to the person or the transfer created as a result of a NICS check that are not part of a record system created and maintained pursuant to independent state law regarding firearms transactions.

§ 25.10 Correction of erroneous system information.

(a) An individual may request the reason for the denial from the agency that conducted the check of the NICS (the "denying agency," which will be either the FBI or the state or local law enforcement agency serving as a POC). The FFL will provide to the denied individual the name and address of the denying agency and the unique transaction number (NTN or STN) associated with the NICS background check. The request for the reason for the denial must be made in writing to the denying agency. (POCs at their discretion may waive the requirement for a written request.)

(b) The denying agency will respond to the individual with the reasons for the denial within five business days of its receipt of the individual's request. The response should indicate whether additional information or documents are required to support an appeal, such as fingerprints in appeals involving questions of identity (i.e., a claim that the record in question does not pertain to the individual who was denied).

(c) If the individual wishes to challenge the accuracy of the record upon which the denial is based, or if the individual wishes to assert that his or her rights to possess a firearm have been restored, he or she may make application first to the denying agency, i.e., either the FBI or the POC. If the denying agency is unable to resolve the appeal, the denying agency will so notify the individual and shall provide the name and address of the agency that originated the document containing the information upon which the denial was based. The individual may then apply for correction of the record directly to the agency from which it originated. If the record is corrected as a result of the appeal to the originating agency, the individual may so notify the denying agency, which will, in turn, verify the record correction with the originating agency (assuming the originating agency has not already notified the denying agency of the correction) and take all necessary steps to correct the record in the NICS.

(d) As an alternative to the above procedure where a POC was the denying agency, the individual may elect to direct his or her challenge to the accuracy of the record, in writing, to the FBI, NICS Operations Center, Criminal Justice Information Services Division, 1000 Custer Hollow Road, Module C-3, Clarksburg, West Virginia 26306-0147. Upon receipt of the information, the FBI will investigate the matter by contacting the POC that denied the transaction or the data source. The FBI will request the POC or the data source to verify that the record in question pertains to the individual who was denied, or to verify or correct the challenged record. The FBI will consider the information it receives from the individual and the response it receives from the POC or the data source.

If the record is corrected as a result of the challenge, the FBI shall so notify the individual, correct the erroneous information in the NICS, and give notice of the error to any Federal department or agency or any state that was the source of such erroneous records.

(e) Upon receipt of notice of the correction of a contested record from the originating agency, the FBI or the agency that contributed the record shall correct the data in the NICS and the denying agency shall provide a written confirmation of the correction of the erroneous data to the individual for presentation to the FFL. If the appeal of a contested record is successful and thirty (30) days or less have transpired since the initial check, and there are no other disqualifying records upon which the denial was based, the NICS will communicate a "Proceed" response to the FFL. If the appeal is successful and more than thirty (30) days have transpired since the initial check, the FFL must recheck the NICS before allowing the sale to continue. In cases where multiple disqualifying records are the basis for the denial, the individual must pursue a correction for each record.

(f) An individual may also contest the accuracy or validity of a disqualifying record by bringing an action against the state or political subdivision responsible for providing the contested information, or responsible for denying the transfer, or against the United States, as the case may be, for an order directing that the contested information be corrected or that the firearm transfer be approved.

§ 25.11 Prohibited activities and penalties.

(a) State or local agencies, FFLs, or individuals violating this subpart A shall be subject to a fine not to exceed $10,000 and subject to cancellation of NICS inquiry privileges.

(b) Misuse or unauthorized access includes, but is not limited to, the following:

(1) State or local agencies', FFLs', or individuals' purposefully furnishing incorrect information to the system to obtain a "Proceed" response, thereby allowing a firearm transfer;

(2) State or local agencies', FFLs', or individuals' purposefully using the system to perform a check for unauthorized purposes; and

(3) Any unauthorized person's accessing the NICS.

[AG Order No. 2186-98, 63 FR 58303, Oct. 30, 1998]

PART 25 - DEPARTMENT OF JUSTICE INFORMATION SYSTEMS

Subpart A - The National Instant Criminal
Background Check System

Authority: Pub. L. 103-159, 107 Stat. 1536.

Subpart A - The National Instant Criminal
Background Check System

§ 25.1 Purpose and authority.

The purpose of this subpart is to establish policies and procedures implementing the Brady Handgun Violence Prevention Act (Brady Act), Public Law 103-159, 107 Stat. 1536. The Brady Act requires the Attorney General to establish a National Instant Criminal Background Check System (NICS) to be contacted by any licensed importer, licensed manufacturer, or licensed dealer of firearms for information as to whether the transfer of a firearm to any person who is not licensed under 18 U.S.C. 923 would be in violation of Federal or state law. The regulations in this subpart are issued pursuant to section 103(h) of the Brady Act, 107 Stat. 1542 (18 U.S.C. 922 note), and include requirements to ensure the privacy and security of the NICS and appeals procedures for persons who have been denied the right to obtain a firearm as a result of a NICS background check performed by the Federal Bureau of Investigation (FBI) or a state or local law enforcement agency.

§ 25.2 Definitions.

Appeal means a formal procedure to challenge the denial of a firearm transfer.

ARI means a unique Agency Record Identifier assigned by the agency submitting records for inclusion in the NICS Index.

ATF means the Bureau of Alcohol, Tobacco, and Firearms of the Department of Treasury.

Audit log means a chronological record of system (computer) activities that enables the reconstruction and examination of the sequence of events and/or changes in an event.

Business day means a 24-hour day (beginning at 12:01 a.m.) on which state offices are open in the state in which the proposed firearm transaction is to take place.

Control Terminal Agency means a state or territorial criminal justice agency recognized by the FBI as the agency responsible for providing state-or territory-wide service to criminal justice users of NCIC data.

Data source means an agency that provided specific information to the NICS.

Delayed means that more research is required prior to a NICS "Proceed" or "Denied" response. A "Delayed" response to the FFL indicates that it would be unlawful to transfer the firearm until receipt of a follow-up "Proceed" response from the NICS or the expiration of three business days, whichever occurs first.

Denied means denial of a firearm transfer based on a NICS response indicating one or more matching records were found providing information demonstrating that receipt of a firearm by a prospective transferee would violate 18 U.S.C. 922 or state law.

Denying agency means a POC or the NICS Operations Center, whichever determines that information in the NICS indicates that the transfer of a firearm to a person would violate Federal or state law, based on a background check.

Dial-up access means any routine access through commercial switched circuits on a continuous or temporary basis.

Federal agency means any authority of the United States that is an "Agency" under 44 U.S.C. 3502(1), other than those considered to be independent regulatory agencies, as defined in 44 U.S.C. 3502(10).

FFL (federal firearms licensee) means a person licensed by the ATF as a manufacturer, dealer, or importer of firearms.

Firearm has the same meaning as in 18 U.S.C. 921(a)(3).

Licensed dealer means any person defined in 27 CFR 178.11.

Licensed importer has the same meaning as in 27 CFR 178.11.

Licensed manufacturer has the same meaning as in 27 CFR 178.11.

NCIC (National Crime Information Center) means the nationwide computerized information system of criminal justice data established by the FBI as a service to local, state, and Federal criminal justice agencies.

NICS means the National Instant Criminal Background Check System, which an FFL must, with limited exceptions, contact for information on whether receipt of a firearm by a person who is not licensed under 18 U.S.C. 923 would violate Federal or state law.

NICS Index means the database, to be managed by the FBI, containing information provided by Federal and state agencies about persons prohibited under Federal law from receiving or possessing a firearm. The NICS Index is separate and apart from the NCIC and the Interstate Identification Index (III).

NICS Operations Center means the unit of the FBI that receives telephone or electronic inquires from FFLs to perform background checks, makes a determination based upon available information as to whether the receipt or transfer of a firearm would be in violation of Federal or state law, researches criminal history records, tracks and finalizes appeals, and conducts audits of system use.

NICS Operations Center's regular business hours means the hours of

9:00 a.m. to 2:00 a.m., Eastern Time, seven days a week.

NICS Representative means a person who receives telephone inquiries to the NICS Operations Center from FFLs requesting background checks and provides a response as to whether the receipt or transfer of a firearm may proceed or is delayed.

NRI (NICS Record Identifier) means the system-generated unique number associated with each record in the NICS Index.

NTN (NICS Transaction Number) means the unique number that will be assigned to each valid background check inquiry received by the NICS. Its primary purpose will be to provide a means of associating inquiries to the NICS with the responses provided by the NICS to the FFLs.

ORI (Originating Agency Identifier) means a nine-character identifier assigned by the FBI to an agency that has met the established qualifying criteria for ORI assignment to identify the agency in transactions on the NCIC System.

Originating Agency means an agency that provides a record to a database checked by the NICS.

POC (Point of Contact) means a state or local law enforcement agency serving as an intermediary between an FFL and the federal databases checked by the NICS. A POC will receive NICS background check requests from FFLs, check state or local record systems, perform NICS inquiries, determine whether matching records provide information demonstrating that an individual is disqualified from possessing a firearm under Federal or state law, and respond to FFLs with the results of a NICS background check. A POC will be an agency with express or implied authority to perform POC duties pursuant to state statute, regulation, or executive order.

Proceed means a NICS response indicating that the information available to the system at the time of the response did not demonstrate that transfer of the firearm would violate federal or state law. A "Proceed" response would not relieve an FFL from compliance with other provisions of Federal or state law that may be applicable to firearms transfers. For example, under 18 U.S.C. 922(d), an FFL may not lawfully transfer a firearm if he or she knows or has reasonable cause to believe that the prospective recipient is prohibited by law from receiving or possessing a firearm.

Record means any item, collection, or grouping of information about an individual that is maintained by an agency, including but not limited to information that disqualifies the individual from receiving a firearm, and that contains his or her name or other personal identifiers.

STN (State-Assigned Transaction Number) means a unique number that may be assigned by a POC to a valid background check inquiry.

System means the National Instant Criminal Background Check System (NICS).

§ 25.3 System information.

(a) There is established at the FBI a National Instant Criminal Background Check System.

(b) The system will be based at the Federal Bureau of Investigation, 1000 Custer Hollow Road, Clarksburg, West Virginia 26306-0147.

(c) The system manager and address are: Director, Federal Bureau of Investigation, J. Edgar Hoover F.B.I. Building, 935 Pennsylvania Avenue, NW, Washington, D.C. 20535.

§ 25.4 Record source categories.

It is anticipated that most records in the NICS Index will be obtained from Federal agencies. It is also anticipated that a limited number of authorized state and local law enforcement agencies will voluntarily contribute records to the NICS Index. Information in the NCIC and III systems that will be searched during a background check has been or will be contributed voluntarily by Federal, state, local, and international criminal justice agencies.

§ 25.5 Validation and data integrity of records in the system.

(a) The FBI will be responsible for maintaining data integrity during all NICS operations that are managed and carried out by the FBI. This responsibility includes:

(1) Ensuring the accurate adding, canceling, or modifying of NICS Index records supplied by Federal agencies;

(2) Automatically rejecting any attempted entry of records into the NICS Index that contain detectable invalid data elements;

(3) Automatic purging of records in the NICS Index after they are on file for a prescribed period of time; and

(4) Quality control checks in the form of periodic internal audits by FBI personnel to verify that the information provided to the NICS Index remains valid and correct.

(b) Each data source will be responsible for ensuring the accuracy and validity of the data it provides to the NICS Index and will immediately correct any record determined to be invalid or incorrect.

§ 25.6 Accessing records in the system.

(a) FFLs may initiate a NICS background check only in connection with a proposed firearm transfer as required by the Brady Act. FFLs are strictly prohibited from initiating a NICS background check for any other purpose. The process of accessing the NICS for the purpose of conducting a NICS background check is initiated by an FFL's contacting the FBI NICS Operations Center (by telephone or electronic dial-up access) or a POC. FFLs in each state will be advised by the ATF whether they are required to initiate NICS background checks with the NICS Operations Center or a POC and how they are to do so.

(b) Access to the NICS through the FBI NICS Operations Center. FFLs may contact the NICS Operations Center by use of a toll-free telephone number, only during its regular business hours. In addition to telephone access, toll-free electronic dial-up access to the NICS will be provided to FFLs after the beginning of the NICS operation. FFLs with electronic dial-up access will be able to contact the NICS 24 hours each day, excluding scheduled and unscheduled downtime.

(c)(1) The FBI NICS Operations Center, upon receiving an FFL telephone or electronic dial-up request for a background check, will:

(i) Verify the FFL Number and code word;

(ii) Assign a NICS Transaction Number (NTN) to a valid inquiry and provide the NTN to the FFL;

(iii) Search the relevant databases (i.e., NICS Index, NCIC, III) for any matching records; and

(iv) Provide the following NICS responses based upon the consolidated NICS search results to the FFL that requested the background check:

(A) "Proceed" response, if no disqualifying information was found in the NICS Index, NCIC, or III.

(B) "Delayed" response, if the NICS search finds a record that requires more research to determine whether the prospective transferee is disqualified from possessing a firearm by Federal or state law. A "Delayed" response to the FFL indicates that the firearm transfer should not proceed pending receipt of a follow-up "Proceed" response from the NICS or the expiration of three business days (exclusive of the day on which the query is made), whichever occurs first. (Example: An FFL requests a NICS check on a prospective firearm transferee at 9:00 a.m. on Friday and shortly thereafter receives a "Delayed" response from the NICS. If state offices in the state in which the FFL is located are closed on Saturday and Sunday and open the following Monday, Tuesday, and Wednesday, and the NICS has not yet responded with a "Proceed" or "Denied" response, the FFL may transfer the firearm at 12:01 a.m. Thursday.)

(C) "Denied" response, when at least one matching record is found in either the NICS Index, NCIC, or III that provides information demonstrating that receipt of a firearm by the prospective transferee would violate 18 U.S.C. 922 or state law. The "Denied" response will be provided to the requesting FFL by the NICS Operations Center during its regular business hours.

(2) None of the responses provided to the FFL under paragraph (c)(1) of this section will contain any of the underlying information in the records checked by the system.

(d) Access to the NICS through POCs. In states where a POC is designated to process background checks for the NICS, FFLs will contact the POC to initiate a NICS background check. Both ATF and the POC will notify FFLs in the POC's state of the means by which FFLs can contact the POC. The NICS will provide POCs with electronic access to the system virtually 24 hours each day through the NCIC communication network. Upon receiving a request for a background check from an FFL, a POC will:

(1) Verify the eligibility of the FFL either by verification of the FFL number or an alternative POC-verification system;

(2) Enter a purpose code indicating that the query of the system is for the purpose of performing a NICS background check in connection with the transfer of a firearm; and

(3) Transmit the request for a background check via the NCIC interface to the NICS.

(e) Upon receiving a request for a NICS background check, POCs may also conduct a search of available files in state and local law enforcement and other relevant record systems, and may provide a unique State-Assigned Transaction Number (STN) to a valid inquiry for a background check.

(f) When the NICS receives an inquiry from a POC, it will search the relevant databases (i.e., NICS Index, NCIC, III) for any matching record(s) and will provide an electronic response to the POC. This response will consolidate the search results of the relevant databases and will include the NTN. The following types of responses may be provided by the NICS to a state or local agency conducting a background check:

(1) No record response, if the NICS determines, through a complete search, that no matching record exists.

(2) Partial response, if the NICS has not completed the search of all of its records. This response will indicate the databases that have been searched (i.e., III, NCIC, and/or NICS Index) and the databases that have not been searched. It will also provide any potentially disqualifying information found in any of the databases searched. A follow-up response will be sent as soon as all the relevant databases have been searched. The follow-up response will provide the complete search results.

(3) Single matching record response, if all records in the relevant databases have been searched and one matching record was found.

(4) Multiple matching record response, if all records in the relevant databases have been searched and more than one matching record was found.

(g) Generally, based on the response(s) provided by the NICS, and other information available in the state and local record systems, a POC will:

(1) Confirm any matching records; and

(2) Notify the FFL that the transfer may proceed, is delayed pending further record analysis, or is denied. "Proceed" notifications made within three business days will be accompanied by the NTN or STN traceable to the NTN. The POC may or may not provide a transaction number (NTN or STN) when notifying the FFL of a "Denied" response.

(h) In cases where a transfer is denied by a POC, the POC should provide a denial notification to the NICS. This denial notification will include the name of the person who was denied a firearm and the NTN. The information provided in the denial notification will be maintained in the NICS Audit Log described in § 25.9(b). This notification may be provided immediately by electronic message to the NICS (i.e., at the time the transfer is denied) or as soon thereafter as possible. If a denial notification is not provided by a POC, the NICS will assume that the transfer was allowed and will destroy its records regarding the transfer in accordance with the procedures detailed in § 25.9.

(i) **Response recording.** FFLs are required to record the system response, whether provided by the FBI NICS Operations Center or a POC, on the appropriate ATF form for audit and inspection purposes, under 27 CFR part 178 recordkeeping requirements. The FBI NICS Operations Center response will always include an NTN and associated "Proceed," "Delayed," or "Denied" determination. POC responses may vary as discussed in Paragraph (g) of this section. In these instances, FFLs will record the POC response, including any transaction number and/or determination.

(j) **Access to the NICS Index for purposes unrelated to NICS background checks required by the Brady Act.** Access to the NICS Index for purposes unrelated to NICS background checks pursuant to 18 U.S.C. 922(t) shall be limited to uses for the purpose of:

(1) Providing information to Federal, state, or local criminal justice agencies in connection with the issuance of a firearm-related or explosives-related permit or license, including permits or licenses to possess, acquire, or transfer a firearm, or to carry a concealed firearm, or to import, manufacture, deal in, or purchase explosives; or

(2) Responding to an inquiry from the ATF in connection with a civil or criminal law enforcement activity relating to the Gun Control Act (18

U.S.C. Chapter 44) or the National Firearms Act (26 U.S.C. Chapter 53).

§ 25.7   Querying records in the system.

(a) The following search descriptors will be required in all queries of the system for purposes of a background check:

(1) Name;
(2) Sex;
(3) Race;
(4) Complete date of birth; and
(5) State of residence.

(b) A unique numeric identifier may also be provided to search for additional records based on exact matches by the numeric identifier. Examples of unique numeric identifiers for purposes of this system are: Social Security number (to comply with Privacy Act requirements, a Social Security number will not be required by the NICS to perform any background check) and miscellaneous identifying numbers (e.g., military number or number assigned by Federal, state, or local authorities to an individual's record). Additional identifiers that may be requested by the system after an initial query include height, weight, eye and hair color, and place of birth. At the option of the querying agency, these additional identifiers may also be included in the initial query of the system.

§ 25.8   System safeguards.

(a) Information maintained in the NICS Index is stored electronically for use in an FBI computer environment. The NICS central computer will reside inside a locked room within a secure facility. Access to the facility will be restricted to authorized personnel who have identified themselves and their need for access to a system security officer.

(b) Access to data stored in the NICS is restricted to duly authorized agencies. The security measures listed in paragraphs (c) through (f) of this section are the minimum to be adopted by all POCs and data sources having access to the NICS.

(c) State or local law enforcement agency computer centers designated by a Control Terminal Agency as POCs shall be authorized NCIC users and shall observe all procedures set forth in the NCIC Security Policy of 1992 when processing NICS background checks. The responsibilities of the Control Terminal Agencies and the computer centers include the following:

(1) The criminal justice agency computer site must have adequate physical security to protect against any unauthorized personnel gaining access to the computer equipment or to any of the stored data.

(2) Since personnel at these computer centers can have access to data stored in the NICS, they must be screened thoroughly under the authority and supervision of a state Control Terminal Agency. This authority and supervision may be delegated to responsible criminal justice agency personnel in the case of a satellite computer center being serviced through a state Control Terminal Agency. This screening will also apply to non-criminal justice maintenance or technical personnel.

(3) All visitors to these computer centers must be accompanied by staff personnel at all times.

(4) POCs utilizing a state/NCIC terminal to access the NICS must have the proper computer instructions written and other built-in controls to prevent data from being accessible to any terminals other than authorized terminals.

(5) Each state Control Terminal Agency shall build its data system around a central computer, through which each inquiry must pass for screening and verification.

(d) Authorized state agency remote terminal devices operated by POCs and having access to the NICS must meet the following requirements:

(1) POCs and data sources having terminals with access to the NICS must physically place these terminals in secure locations within the authorized agency;

(2) The agencies having terminals with access to the NICS must screen terminal operators and must restrict access to the terminals to a minimum number of authorized employees; and

(3) Copies of NICS data obtained from terminal devices must be afforded appropriate security to prevent any unauthorized access or use.

(e) FFL remote terminal devices may be used to transmit queries to the NICS via electronic dial-up access. The following procedures will apply to such queries:

(1) The NICS will incorporate a security authentication mechanism that performs FFL dial-up user authentication before network access takes place;

(2) The proper use of dial-up circuits by FFLs will be included as part of the periodic audits by the FBI; and

(3) All failed authentications will be logged by the NICS and provided to the NICS security administrator.

(f) FFLs may use the telephone to transmit queries to the NICS, in accordance with the following procedures:

(1) FFLs may contact the NICS Operations Center during its regular business hours by a telephone number provided by the FBI;

(2) FFLs will provide the NICS Representative with their FFL Number and code word, the type of sale, and the name, sex, race, date of birth, and state of residence of the prospective buyer; and

(3) The NICS will verify the FFL Number and code word before processing the request.

(g) The following precautions will be taken to help ensure the security and privacy of NICS information when FFLs contact the NICS Operations Center:

(1) Access will be restricted to the initiation of a NICS background check in connection with the proposed transfer of a firearm.

(2) The NICS Representative will only provide a response of "Proceed" or "Delayed" (with regard to the prospective firearms transfer), and will not provide the details of any record information about the transferee. In cases where potentially disqualifying information is found in response to an FFL query, the NICS Representative will provide a "Delayed" response to the FFL. Follow-up "Proceed" or "Denied" responses will be provided by the NICS Operations Center during its regular business hours.

(3) The FBI will periodically monitor telephone inquiries to ensure proper use of the system.

(h) All transactions and messages sent and received through electronic access by POCs and FFLs will be automatically logged in the NICS Audit Log described in § 25.9(b). Information in the NICS Audit Log will include initiation and termination messages, failed authentications, and matching records located by each search transaction.

(i) The FBI will monitor and enforce compliance by NICS users with the applicable system security requirements outlined in the NICS POC Guidelines and the NICS FFL Manual (available from the NICS Operations Center, Federal Bureau of Investigation, 1000 Custer Hollow Road, Clarksburg, West Virginia 26306-0147).

§ 25.9 Retention and destruction of records in the system.

(a) The NICS will retain NICS Index records that indicate that receipt of a firearm by the individuals to whom the records pertain would violate Federal or state law. The NICS will retain such records indefinitely, unless they are canceled by the originating agency. In cases where a firearms disability is not permanent, e.g., a disqualifying restraining order, the NICS will automatically purge the pertinent record when it is no longer disqualifying. Unless otherwise removed, records contained in the NCIC and III files that are accessed during a background check will remain in those files in accordance with established policy.

(b) The FBI will maintain an automated NICS Audit Log of all incoming and outgoing transactions that pass through the system.

(1) The Audit Log will record the following information: type of transaction (inquiry or response), line number, time, date of inquiry, header, message key, ORI, and inquiry/response data (including the name and other identifying information about the prospective transferee and the NTN). In cases of allowed transfers, all information in the Audit Log related to the person or the transfer, other than the NTN assigned to the transfer and the date the number was assigned, will be destroyed after not more than six months after the transfer is allowed. Audit Log records relating to denials will be retained for 10 years, after which time they will be transferred to a Federal Records Center for storage. The NICS will not be used to establish any system for the registration of firearms, firearm owners, or firearm transactions or dispositions, except with respect to persons prohibited from receiving a firearm by 18 U.S.C. 922(g) or (n) or by state law.

(2) The Audit Log will be used to analyze system performance, assist users in resolving operational problems, support the appeals process, or support audits of the use of the system. Searches may be conducted on the Audit Log by time frame, i.e., by day or month, or by a particular state or agency. Information in the Audit Log pertaining to allowed transfers may only be used by the FBI for the purpose of conducting audits of the use and performance of the NICS. Such information, however, may be retained and used as long as needed to pursue cases of identified misuse of the system. The NICS, including the NICS Audit Log, may not be used by any department, agency, officer, or employee of the United States to establish any system for the registration of firearms, firearm owners, or firearm transactions or dispositions. The Audit Log will be monitored and reviewed on a regular basis to detect any possible misuse of the NICS data.

(c) The following records in the FBI-operated terminals of the NICS will be subject to the Brady Act's requirements for destruction:

(1) All inquiry and response messages (regardless of media) relating to a background check that results in an allowed transfer; and

(2) All information (regardless of media) contained in the NICS Audit Log relating to a background check that results in an allowed transfer.

(d) The following records of state and local law enforcement units serving as POCs will be subject to the Brady Act's requirements for destruction:

(1) All inquiry and response messages (regardless of media) relating to the initiation and result of a check of the NICS that allows a transfer that are not part of a record system created and maintained pursuant to independent state law regarding firearms transactions; and

(2) All other records relating to the person or the transfer created as a result of a NICS check that are not part of a record system created and maintained pursuant to independent state law regarding firearms transactions.

§ 25.10 Correction of erroneous system information.

(a) An individual may request the reason for the denial from the agency that conducted the check of the NICS (the "denying agency," which will be either the FBI or the state or local law enforcement agency serving as a POC). The FFL will provide to the denied individual the name and address of the denying agency and the unique transaction number (NTN or STN) associated with the NICS background check. The request for the reason for the denial must be made in writing to the denying agency. (POCs at their discretion may waive the requirement for a written request.)

(b) The denying agency will respond to the individual with the reasons for the denial within five business days of its receipt of the individual's request. The response should indicate whether additional information or documents are required to support an appeal, such as fingerprints in appeals involving questions of identity (i.e., a claim that the record in question does not pertain to the individual who was denied).

(c) If the individual wishes to challenge the accuracy of the record upon which the denial is based, or if the individual wishes to assert that his or her rights to possess a firearm have been restored, he or she may make application first to the denying agency, i.e., either the FBI or the POC. If the denying agency is unable to resolve the appeal, the denying agency will so notify the individual and shall provide the name and address of the agency that originated the document containing the information upon which the denial was based. The individual may then apply for correction of the record directly to the agency from which it originated. If the record is corrected as a result of the appeal to the originating agency, the individual may so notify the denying agency, which will, in turn, verify the record correction with the originating agency (assuming the originating agency has not already notified the denying agency of the correction) and take all necessary steps to correct the record in the NICS.

(d) As an alternative to the above procedure where a POC was the denying agency, the individual may elect to direct his or her challenge to the accuracy of the record, in writing, to the FBI, NICS Operations Center, Criminal Justice Information Services Division, 1000 Custer Hollow Road, Module C-3, Clarksburg, West Virginia 26306-0147. Upon receipt of the information, the FBI will investigate the matter by contacting the POC that denied the transaction or the data source. The FBI will request the POC or the data source to verify that the record in question pertains to the individual who was denied, or to verify or correct the challenged record. The FBI will consider the information it receives from the individual and the response it receives from the POC or the data source.

If the record is corrected as a result of the challenge, the FBI shall so notify the individual, correct the erroneous information in the NICS, and give notice of the error to any Federal department or agency or any state that was the source of such erroneous records.

(e) Upon receipt of notice of the correction of a contested record from the originating agency, the FBI or the agency that contributed the record shall correct the data in the NICS and the denying agency shall provide a written confirmation of the correction of the erroneous data to the individual for presentation to the FFL. If the appeal of a contested record is successful and thirty (30) days or less have transpired since the initial check, and there are no other disqualifying records upon which the denial was based, the NICS will communicate a "Proceed" response to the FFL. If the appeal is successful and more than thirty (30) days have transpired since the initial check, the FFL must recheck the NICS before allowing the sale to continue. In cases where multiple disqualifying records are the basis for the denial, the individual must pursue a correction for each record.

(f) An individual may also contest the accuracy or validity of a disqualifying record by bringing an action against the state or political subdivision responsible for providing the contested information, or responsible for denying the transfer, or against the United States, as the case may be, for an order directing that the contested information be corrected or that the firearm transfer be approved.

§ 25.11 Prohibited activities and penalties.

(a) State or local agencies, FFLs, or individuals violating this subpart A shall be subject to a fine not to exceed $10,000 and subject to cancellation of NICS inquiry privileges.

(b) Misuse or unauthorized access includes, but is not limited to, the following:

(1) State or local agencies', FFLs', or individuals' purposefully furnishing incorrect information to the system to obtain a "Proceed" response, thereby allowing a firearm transfer;

(2) State or local agencies', FFLs', or individuals' purposefully using the system to perform a check for unauthorized purposes; and

(3) Any unauthorized person's accessing the NICS.

[AG Order No. 2186-98, 63 FR 58303, Oct. 30, 1998]

RULINGS, PROCEDURES, AND INDUSTRY CIRCULARS

TABLE OF CONTENTS

ATF Rul. 28 -2   KG-9 pistol classified.

ATF Rul. 28 -8   SM10 and SM11A1 pistols; SAC carbines classified.

ATF Rul. 38 -5   YAC STEN MK II carbine classified.

ATF Rul. 58 -2   NFA curio or relic importation as sales samples:
                     Limitation.

ATF Rul. 58 -10   Surplus military frames or receivers, alone,
                     denied importation.

ATF Rul. 94-1   USAS-12 shotgun classified.

ATF Rul. 94-2   Striker-12/Streetsweeper classified.

ATF Rul. 95-3   37/38 mm gas/flare guns possessed with anti-personnel
                     ammunition classified.

ATF Rul. 96-2   Activities of auctioneers requiring a dealer's license.

PROCEDURES:

ATF Proc. 57 -3   Drop shipments of firearms: Recordkeeping.

ATF Proc. 90-1   Identification number for special (occupational)
                     taxpayers

INDUSTRY CIRCULARS:

I.C. 72 -23   Shipment to employees, agents, representatives,
                     writers & evaluators.

I.C. 72-30   Identification: Personal firearms on premises not
                     offered for sale.

I.C. 74-13   Verifying identity and licensed status of transferee.

I.C. 77 -20   Duplication of serial numbers on imported firearms.

RULINGS

27 CFR 178.41: General
(Also 178.42, 178.50)

Firearms or ammunition may not be sold at gun shows by a licensed dealer, but orders may be taken under specified conditions; Revenue Ruling 66-265 superseded.

Rev. Rul. 69-59

Advice has been requested whether a person who is licensed under 18 U.S.C. Chapter 44 (which superseded the Federal Firearms Act (15 U.S.C. Chapter 18)) or who is continuing operations under a license issued to him under the Federal Firearms Act, as a manufacturer, importer or dealer in firearms or ammunition may sell firearms or ammunition at a gun show held on premises other than those covered by his outstanding license.

Under 18 U.S.C. 923(a), "a separate fee" is required to be paid for each place at which business as a licensee is to be conducted. Further, each applicant for a license is required to have in a State "premises from which he conducts business" (18 U.S.C. 923(d)(1)(E)) and to specify such premises in the license application. In addition, record are required to be maintained at the business premises covered by the license (18 U.S.C. 923(g)).

Therefore, a person holding a valid license may engage in the business covered by the license only at the specific business premises for which his license has been obtained. Thus, a licensee may not sell firearms or ammunition at a gun show held on premises other than those covered by his license. He may, however, have a booth or table at such a gun show at which he displays his wares and takes orders for them, provided that the sale and delivery of the firearms or ammunition are to be lawfully effected from his licensed business premises only and his records properly reflect such transactions.

There are no provisions in the law for the issuance of temporary licenses to cover sales at gun shows and licenses will be issued only for premises where the applicant regularly intends to engage in the business to be covered by the license.

This ruling does not apply to the activities of licensed collectors with respect to the receipt or disposition of curios and relics by such collectors. For provisions relating to transactions by licensed collectors of curios and relics, see 26 CFR 178.50 of the regulations.

Rev. Rul. 66-265, 1966-2 C.B. 559, is hereby superseded.

Editor's Note

In 1986, the GCA was amended to allow licensees to sell firearms at gunshows in the State in which their licensed premises are located, and was further amended in 1997 to allow licensees to sell curio or relic firearms to other licensees at any location. The ruling is still applicable to licensees' off-premises sales not addressed by these amendments.

27 CFR 178.99: CERTAIN PROHIBITED SALES OR DELIVERIES

A licensed importer, manufacturer or dealer may, without transferring title, ship firearms interstate in care of his nonlicensed employees, agents and representatives for the use and benefit of the licensee's business.

Rev. Rul. 69-248

Under 18 U.S.C. 922(a)(2), it is unlawful (with certain exceptions not here pertinent) for any licensed importer, manufacturer, or dealer to ship, transport or deliver in interstate commerce any firearm to any person other than another licensee. However, there is no provision in 18 U.S.C. Chapter 44 which would prohibit licensees from shipping, transporting or delivering firearms in interstate commerce to themselves for business purposes (exclusive of sale or disposition) in care of their employees, agents or representatives.

Licensees engaged in the firearms or ammunition business typically are corporations which can only conduct their operations through employees, agents and representatives. In the course of such operations, it is frequently necessary to ship, transport or deliver firearms in interstate commerce for bona fide business purposes such as display, advertising research, testing, comparative evaluations and marketing promotions. In such cases, title to, and ultimate control of, the firearms remain in the licensee even though the firearms or ammunition are placed in the temporary custody of an employee, agent or representative for limited lengths of time.

Therefore, under the circumstances described herein, licensed manufacturers, importers and dealers may ship, transport or deliver firearms in interstate commerce to themselves in care of employees, agents and representatives without being in violation of 18 U.S.C. 922(a)(2).

However, such shipment, transportation or delivery should not be made in care of persons who are ineligible "to ship or transport in interstate or foreign commerce, or possess in or affecting commerce, any firearm or ammunition; or to receive any firearm or ammunition which has been shipped or transported in interstate or foreign commerce" under 922(g).

This ruling is specifically limited to shipments made by a licensed importer, manufacturer or dealer in care of hi nonlicensed employees, agents or representatives for bona fide business purposes (exclusive of sale or disposition), where the actual custody of the firearms is transferred for a limited period of time and where title and ultimate control of the firearms remain in such licensee. When no longer needed by the employee, agent or representative for the business purpose fo which received, disposition of all such firearms must be by return to the licensee or in a manner consistent with the provisions of 18 U.S.C. Chapter 44., (See also I.C. 72-23.)

[This Revenue Ruling does not apply to firearms and ammunition within the purview of the National Firearms Act (26 U.S.C. Chapter 53), pe I.C. 72-23]

[190 -1 C.B. 360] [Amended]

27 CFR 178.114: IMPORTATION BY
MEMBERS OF THE U.S. ARMED FORCES

Members of U.S. Armed Forces may, under specified conditions, import up to 3 rifles or shotguns and up to 1,000 rounds of ammunition without obtaining the permit required by 27 CFR 178.114.

Rev. Rul. 69-309

Under the authority contained in 27 CFR 178.22, the requirement of a permit (Form 6 (Firearms)) is hereby waived as to any member of the U.S. Armed Forces desiring to transport, ship, receive or import firearms or ammunition into the United States under the following circumstances:

(1) The member of the U.S. Armed Forces is on active duty outside the United States or has been on active duty outside the United States within the 60-day period immediately preceding the transportation, shipment, receipt or importation;

(2) The importation consists of rifles or shotguns or any combination thereof (excluding any firearm coming within the purview of the National Firearms Act and any firearms of military surplus origin) not to exceed a total of three, and not more than 1,000 rounds of ammunition for rifles and shotguns (excluding tracer or incendiary ammunition) when those firearms or ammunition are on the person of a returning member of the U.S. Armed Forces or with his baggage or effects, whether accompanied or unaccompanied (but not mailed unless they are included in unaccompanied baggage or effects which are officially shipped through the mails by a Transportation Officer of the U.S. Armed Forces incident to a permanent change of duty station);

(3) The rifles and shotguns and ammunition are being transported, shipped, received and imported into the United States to the place of residence, and are intended for the personal use of the member of the U.S. Armed Forces importing them;

(4) The importation is incident to the return of the importer to a permanent duty station in the United States from a permanent duty station abroad, or his release from active duty;

(5) The importer of the firearms and ammunition completes, and he or his authorized agent furnishes to the Customs officer releasing the firearms and ammunition, a Form 6A (Firearms) pursuant to 27 CFR 178.114(b) and a certification that the importation of and the transportation to and the receipt and possession by the importer at his place of residence would not constitute any violation of Title I of the Gun Control Act (Title 18, U.S.C., Chapter 44), or Section 38 of the Arms Export Control Act of 1976 (Title 22, U.S.C., Sec. 2778), or any applicable State law or published ordinance.

[190 -1 C.B. 361] [Amended]

27 CFR 178.41: GENERAL
(Also 178.42, 178.45, 178.49)

Licensed firearms dealers operating at multiple locations may establish a common expiration date for all licenses.

ATF Rul. 73-9

Licensed firearms dealers operating more than one location for which a license is required may establish a common expiration date for all licenses issued to their several locations. Dealers wishing to establish such a date for all licenses issued to them may make application in writing to the Regional Director (Compliance) of the region in which the businesses or activities are operated. The application should set out the requested common expiration date and should list all licensed premises in the region covered by the application. The Regional Director (Compliance) will advise the dealer whether the request may be approved and, if approved, will provide the necessary instructions and renewal application. It is pointed out that approval of a request will entail a one time loss associated with the existing license, as it will be cancelled on and after the date of issuance of the license bearing the requested common expiration date, and the regulations do not provide for prorated refunds.

[73 AF C.B. 102] [Amended]

27 CFR 178.11: MEANING OF TERMS
(Also 178.23, 178.44)

Because of the nature of operations conducted by a gunsmith, he shall not be required to have business premises open to the general public or to have regular business hours.

ATF Rul. 73-13

Because of the nature of operations conducted by a gunsmith, any applicant for a license who intends to engage solely in this type of business and so specifies on his application will not be required to maintain regular business hours. Further, if the business is conducted from a private dwelling, a separate portion should be designated as the business premises, which need not be open to all segments of the public but only accessible to the clientele that the business is set up to serve. However, the licensed premises of the gunsmith are subject to the inspection requirements of 18 U.S.C. 923(g) and 27 CFR 178.23, and the gunsmith must maintain the required records as specified in 27 CFR 178.121 et seq.

Further, since a gunsmith is a licensed firearms dealer, if he engages in the business of buying and selling firearms, he must record his transactions on Form 4473 (Firearms Transaction Record) for each sale, and maintain the firearms acquisition and disposition records required of all licensed dealers. However, if a gunsmith engages in the business of buying and selling firearms during the term of his current license, he may be required to submit a new Form 7 (Firearms) at the time of renewal in accordance with 27 CFR 178.45 and meet the requirements of an applicant engaging in the business of buying and selling firearms, such as having business premises open to the general public and having regular business hours.

(Amplified by ATFR 77-1)

[73 AF C.B. 92]

27 CFR 178.11: MEANING OF TERMS
(Also 178.23, 178.44, 178.99, 178.124)

Because of the nature of operations conducted by a consultant or expert, he shall not be required to have business premises open to the general public or to have regular business hours.

ATF Rul. 73-19

Revenue Ruling 69-248, C.B. 1996 -1, 360 (Internal Revenue) permits firearms licensees to ship, transport, or deliver firearms in interstate commerce to thei nonlicensed employees, agents, or representatives for business purposes. As was clarified in Industry Circular 72-23 the ruling also permits firearms licensees to similarly transfero firearms t nonlicensed professional writers, consultants, and evaluators for research or evaluation.

Title 18 U.S.C., Section 922(a)(2)(A), permits an individual to ship (and have returned to him) in interstate commerce a firearm to a firearms licensee for repair or customizing. Furthermore, the definition of a firearms dealer in 18 U.S.C. 921 and 27 CFR 178.11 is sufficiently broad that it can be interpreted to include a qualified firearms consultant or expert who is engaged in the business of testing or examining firearms. In view of these provisions, the Bureau has determined tha firearms consultants or experts may be licensed as firearms dealers in order that they may receive firearms from nonlicensed individuals for testing and examination.

Because of the nature of operations conducted by a firearms consultant or expert, any licensed dealer who engages solely in this type of business will not be required to maintain regular business hours. If the business is conducted from a private residence, a separate portion of the dwelling should be designated as "business premises." Such premises need not be open to all segments of the public but only accessible to the clientele that the business is set up to serve. However, the licensed premises of th firearms consultah -expert shall be subject to inspection under the authority of 18 U.S.C. 923(g) and 27 CFR 178.23.

A licensed firearms consultant or expert shall maintain records of receipt and delivery of firearms, as is required by 27 CFR 178, Subpart H, except that the licensee need not prepare Forms 4473, Firearms Transaction Record, reflecting the firearms examined.

However, shipments and deliveries of firearms shall not be made in care of persons who are ineligible "to ship or transport in interstate or foreign commerce, or possess in or affecting commerce, any firearm or ammunition; or to receive any firearm or ammunition which has been shipped or transported in interstate or foreign commerce" under 922(g).

A firearms consultant or expert who desires to obtain a license as a dealer in firearms shall file Form 7 (Firearms), Application for License Under 18 U.S.C. Chapter 44, Firearms, in the manner prescribed by 27 CFR 178.44. The application shall include a statement that the applicant is engaged in business as a bona fide firearms consultant or expert and, where the applicant intends to perform testing or examination services for one or more persons on a continuing basis, the statement shall include the name, address, and nature of business of such persons. A license as a dealer in firearms will be issued only after the Regional Director (Compliance) is satisfied that the applicant is a bona fide consultant or expert and is otherwise qualified under the law.

Since a licensed firearms consultant or expert is a firearms dealer, if he engages in the business of buying and selling firearms, he must record his transactions on Form 4473, Firearms Transaction Record, for each sale, and maintain the firearms acquisition and disposition records required of all licensed dealers If a firearms consultant or expert engages in the business of buying and selling firearms during the term of his current license, he may be required to submit a new Form 7 (Firearms) at the time of renewal in accordance with 27 CFR 178.45 and meet the requirements of an applicant engaging in the business of buying and selling firearms, such as having business premises open to the general public and having regular business hours.

[This ATF ruling does not apply to firearms within the purview of the National Firearms Act (26 U.S.C. Chapter 53)]

[73 AT C.B. 93] [Amended]

27 CFR 179.104: REGISTRATION OF
FIREARMS BY CERTAIN
GOVERNMENTAL ENTITIES

When NFA firearms are registered on Form 10 by governmental entities, subsequent transfers of such firearms shall be made only to other governmental entities.

ATF Rul. 74-8

Advice has been requested whether the Bureau will approve transfer of National Firearms Act weapons by a State or political subdivision (police department) to a special occupational taxpayer where such firearms were registered in the National Firearms Registration and Transfer Record pursuant to 27 CFR 179.104.

27 CFR 179.104 provides that any State, any political subdivision thereof, or any official police organization of such a government entity engaged in criminal investigations, which acquires for official use a firearm not registered to it, such as by abandonment or by forfeiture, will register such firearm with the Director by filing Form 10 (Firearms), Application for Registration of Firearms Acquired by Certain Governmental Entities, and that such registration shall become a part of the National Firearms Registration and Transfer Record.

The purp se of the above regulation was to permit the limited registration of firearms by certain governmental entities for official use only. 27 CFR 179.104 may not be used as a vehicle to register otherwise unregisterable firearms for the purpose of introducing such firearms into ordinary commercial channels. Accordingly, when registration of firearms by governmental entities is approved on Form 10, the form will be marked "official use only." The Bureau will approve subsequent transfers of such firearms only to other governmental entities for official use. Otherwise, such firearms must be destroyed or abandoned to the Bureau.

[74 AF C.B. 67]

27 CFR 178.114: IMPORTATION BY
MEMBERS OF THE U.S. ARMED FORCES

A member of the U.S. Armed Forces who is a resident of any State or territory which requires that a permit or other authorization be issued prior to possessing or owning a handgun shall submit evidence of compliance with State law before an application to import a handgun may be approved.

ATF Rul. 74-13

Handguns have been transported, shipped, received, or imported into the United States by members of the United States Armed Forces to their place of residence without such members having obtained the required permit or other authorization required by their State of residence which would permit them to possess or own (as opposed to a license to purchase) handguns in that State.

18 U.S.C. 925(a)(4) provides that when established to the satisfaction of the Secretary to be consistent with the provisions of 18 U.S.C. Chapter 44 and other applicable Federal and State laws and published ordinances, the Secretary may authorize the transportation, shipment, receipt, or importation into the United States to the place of residence of any member of the United States Armed Forces who is on active duty outside the United States (or has been on active duty outside the United States within the 60-day period

immediately preceding the transportation, shipment, receipt, or importation), of any firearm or ammunition which is:

(a) Determined by the Secretary to be generally recognized as particularly suitable for sporting purposes, or determined by the Department of Defense to be a type of firearm normally classified as a war souvenir; and

(b) Intended for the personal use of such member.

27 CFR 178.114(a) provides that an application for a permit to import a firearm or ammunition into the United States to the place of residence of any military member of the United States Armed Forces on active duty outside the United States shall include a certification by the applicant that the transportation, receipt, or possession of the firearm or ammunition to be imported, would not constitute a violation of any State law or local ordinance at the place of the applicant's residence.

In order to assure that the transportation, shipment, receipt, or importation of handguns under 27 CFR 178.114 is not in violation of applicable State laws, it is held that, any member of the United States Armed Forces who is a resident of any State or territory which requires that a permit or authorization be obtained prior to possessing or owning a handgun shall, in addition to making the required certification in the application, submit with his application to the Director a copy of the license, permit, certificate of registration, or firearm identification card, as applicable and as required by his State, in order to obtain a permit to import a handgun into the United States.

[74 AF C.B. 60]

27 CFR 178.124: FIREARMS TRANSACTION RECORDS
(Also 178.123, 178.125, 178.147)

Form 4473 shall not be required to record disposition of a like replacement firearm when such firearm is delivered by a licensee to the person from whom the malfunctioning or damaged firearm was received, provided such disposition is recorded in the licensee's permanent records.

ATF Rul. 74-20

It is held that a firearms transaction record, Form 4473, shall not be required to record the disposition of a replacement firearm of the same kind and type where such a firearm is delivered by a licensee to the person from whom the malfunctioning or damaged firearm was received.

It should be noted, however, that the licensee is required by 27 CFR 178.125 to maintain in his permanent records the disposition of such a replacement firearm. (See also ATFR 76-25)

[74 AF C.B. 61]

27 CFR 178.11: MEANING OF TERMS
(Also 179.11)

A small caliber weapon ostensibly designed to expel only tear gas, similar substances, or pyrotechnic signals, which may readily be converted to expel a projectile by means of an explosive, classified as a firearm.

ATF Rul. 75-7

The term "firearm" as used in 18 U.S.C. 921(a)(3) includes "any weapon (including a starter gun) which will or is designed to or may readily be converted to expel a projectile by the action of an explosive."

A small caliber weapon ostensibly designed to expel only tear gas, similar substances or pyrotechnic signals by the action of an explosive, which may readily be converted to expel a projectile by means of an explosive, constitutes, a "firearm" within the purview of 18 U.S.C. 921(a)(3)(A).

Tests performed on these weapons have established that they may readily be converted to expel a projectile by the action of an explosive, normally by means of a minor alteration of the expended Helix cartridge and/or the simple attachment of a barrel/chamber to the firing mechanism.

Such weapons manufactured within the United States on or after June 1, 1975, will be subject to all of the provisions of Chapter 44 and 27 CFR Part 178. Such weapons manufactured before June 1 1975, will not be treated as subject to the provisions of Chapter 44 and 27 CFR Part 178 in order to allow persons manufacturing and dealing in such weapons to comply with the provisions of Chapter 44 and 27 CFR Part 178.

Since such weapons are not generally recognized as particularly suitable for or readily adaptable to sporting purposes (18 U.S.C. 925(d)(3)), the importation of such weapons is prohibited unless such importation comes within one of the statutory exceptions provided in 18 U.S.C. 925.

[75 AF C.B. 55]

27 CFR 178.94: SALES OR DELIVERIES
BETWEEN LICENSEES

A firearms licensee may continue operations until his renewal application for a license is finally acted upon.

ATF Rul. 75-27

Under 5 U.S.C. 558, when a licensee has made timely and sufficient application for a renewal in accordance with agency rules, a license with reference to an activity of a continuing nature does not expire until the application has been finally determined by the agency. In accordance with section 558, a firearms licensee who timely applies for renewal of his license is authorized to continue his firearms operations as authorized by his license until his renewal application is finally acted upon. As provided by 27 CFR 178.94, a transferor licensee is authorized to continue to make shipments to a licensee for not more than 45 days following the expiration date of the transferee's license.

Held, a transferor licensee may continue to make firearms and ammunition shipments to a licensee who has timely applied for renewal of his license but has not had his application acted upon within 45 days after the expiration of his license. The transferor licensee shall, however, in cases where the 45-day period has passed, obtai appropriate evidence that the transferee's license renewal application is still pending in the office of the Regional Director (Compliance), Bureau of Alcohol, Tobacco and Firearms. Such evidence should cn sist of a letter from the Regional Director (Compliance), to the transferee licensee stating that his renewal application has been timely received and that action thereon is currently pending.

[75 AF C.B. 60]

27 CFR 178.92: IDENTIFICATION OF FIREARMS

Importers may adopt serial numbers placed on certain firearms by foreign manufacturers.

ATF Rul. 75-28

The Bureau has determined that in some cases the serial number placed on a firearm by a foreign manufacturer is adequate to provide the identification number required by section 178.92. See, also, section 178.22(a).

Held, where a serial number has been placed on the frame or receiver of a firearm by a foreign manufacturer in the manner contemplated by 27 CFR 178.92, and such serial number does not duplicate a number previously adopted or assigned by the importer to any other firearm, the importer may adopt the serial number of the foreign manufacturer:

Provided, the importer shall in all cases place his name and address (city and State, or recognized abbreviation thereof), and any other marks necessary to comply with the identification requirements of 27 CFR 178.92, on such imported firearms.

[75 AF C.B. 59] [Amended]

27 CFR 178.11: MEANING OF TERMS
(Also 179.11)

A hand-held device designed to expel by means of an explosive two electrical contacts (barbs) connected by two wires attached to a high voltage source in the device classified as a firearm.

ATF Rul. 76-6

Taser Model TF-1, a hand-held device designed to expel by means of an explosive two electrical contacts (barbs) connected by two wires attached to a high voltage source in the device, is a "firearm" within the purview of 18 U.S.C. 921(a)(3)(A). It is also an "any other weapon" under the National Firearms Act (26 U.S.C. 5845(e)).

In order to allow persons manufacturing and dealing in such weapons to comply with the provisions of Chapter 44 and 27 CFR Part 178, this ruling will be applicable to such weapons manufactured within the United States on or after May 1, 1976. Such weapons manufactured before May 1 1976, will not be treated as subject to the provisions of Chapter 44 and 27 CFR Part 178. With respect to the "any other weapon" classification under the National Firearms Act, pursuant to the 26 U.S.C. 7805(b), this ruling will not be applied to such weapons manufactured before May 1, 1976. Accordingly, such weapons manufactured on or after May 1, 1976, will be subject to all the provisions of the National Firearms Act and 27 CFR Part 179.

(Amplified by ATFR 80-20)

[76 AF C.B. 96]

27 CFR 178.124: FIREARMS
TRANSACTION RECORD
(Also 178.125, 178.126a)

Certain reporting and recordkeeping requirements of pawnbrokers are explained.

ATF Rul. 76-15

The regulations do not require that a pawnbroker execute Form 4473 when a firearm is pledged for a loan. However, he must record the receipt thereof in his permanent acquisition and disposition record as required by 27 CFR 178.125(e). At the time a firearm is redeemed by a nonlicense pledgor, Form 4473 must be executed and the appropriate entry made in the permanent acquisition and disposition record. Although a redemption is not considered a sale, it is a disposition for purposes of 27 CFR 178.124(a), 178.125(e), and 178.126a. See Huddleston v. United States, 415 U.S. 814 (1974).

However, no report of multiple sales and other dispositions is required to be filed with ATF when the handguns are returned to the person fro whom received.

Held, Form 4473, Firearms Transaction Record, need not be executed when a pawnbroker accepts a firearm as a pledge for a loan. However, if a nonlicense pledgor redeems the firearm or if disposition of the firearm is made to any other nonlicensee, Form 4473 must be executed. Held further, pawnbrokers must enter into their permanent acquisition and disposition record the receipt of a firearm as a pledge for a loan and any disposition, including redemption, of such firearm. Held further, pawnbrokers must submit reports of multiple sales and other dispositions of pistols and revolvers as required by 27 CFR 178.126a when the person receiving them is not the person who pawned the firearms.

[76 ATF C.B. 100] [Amended]

27 CFR 179.11: MEANING OF TERMS

Mere possession of a license and a special tax stamp as a dealer in firearms does not qualify a person to receive firearms transfer-tax-free.

ATF Rul. 76-22

The mere possession of a license and a special (occupational) tax stamp as a dealer in firearms does not qualify a person to receive firearms transfer-ta -free. Any person holding a license and a special tax stamp as a dealer in firearms and not actually engaged within the United States in the business of selling NFA firearms may not lawfully receive NFA firearms without the transfer tax having been paid by the transferor. Where it is, therefore, determined that the proposed transferee on a Form 3, Application for Tax-Exempt Transfer of Firearm and Registration to Special (Occupational) Taxpayer, is not actually engaged in the business of dealing in NFA firearms, such application will be denied. In addition, if such person receives NFA firearms without the transfer tax having been paid, such firearms may be subject to seizure for forfeiture as having been unlawfully transferred without payment of the transfer tax.

[76 ATF C.B. 103]

27 CFR 178.121: GENERAL
(Also 178.147)

Recordkeeping requirements for firearms from which parts are salvaged for use in repairing firearms are clarified.

ATF Rul. 76-25

Section 921(a)(3) of Title 18, United States Code, and the regulations at 27 CFR 178.11 define the term "firearm" to include any weapon which will, or is designed to, or may readily be converted to, expel a projectile by the action of an explosive, and the frame or receiver of any such weapon.

The regulations in 27 CFR 178.122, 178.123 and 178.125 require each licensed importer, licensed manufacturer, and licensed dealer, respectively, to maintain such records of acquisition (including by manufacture) or disposition, whether temporary or permanent, of firearms as therein prescribed.

Held, a licensee who purchases a damaged firearm for the purpose of salvaging parts therefrom shall enter receipt of the firearm in his firearms acquisition and disposition record. If the frame or receiver of the firearm is damaged to the extent that it cannot be repaired, or if the licensee does not desire to repair the frame or receiver, he may destroy it and show the disposition of the firearm in his records as having been destroyed. Before a firearm may be considered destroyed, it must be cut, severed or mangled in such a manner as to render the firearm completely inoperative and such that it cannot be restored to an operative condition.

Where the repair of a customer's firearm results in an exchange of a frame or receiver, an entry shall be made in the licensee's records to show the transfer of such replacement part, as it is a "firearm" as defined in 18 U.S.C. 921(a)(3) Further, as held in ATF Ruling 74-20, 1974 AT C.B. 61, a Form 4473, Firearms Transaction Record, shall not be required to record the disposition of a replacement firearm of the same kind and type where the firearm is delivered by the licensee to the person from whom the malfunctioning or damaged firearm was received. The frame or receiver received from the customer shall be entered as an acquisition, and if destroyed, it shall be entered in the disposition record as destroyed.

With regard to National Firearm Act firearms as defined in 26 U.S.C. 5845(a), in addition to the above recordkeeping requirements, the registration and transfer procedures of 27 CFR Part 179 must be complied with.

[76 AT C.B. 99]

27 CFR 178.121: GENERAL (RECORDS)

The recordkeeping requirements for licensed gunsmiths are clarified. ATF Rul. 73-13 amplified.

ATF Rul. 77-1

ATF Ruling 73-13, 1973 AT C.B. 92, held that a licensed gunsmith must maintain the required records as specified in 27 CFR 178.121 et seq., and if a gunsmith engages in the business of buying and selling firearms, he must record these transactions on a Form 4473 (Firearms Transaction Record) for each sale. However, as provided in Section 178.124(a), a Form 4473 is not required to record the disposition made of a firearm delivered to a gunsmith for repair or customizing when the firearm is returned to the person from whom received.

The Bureau recognizes the necessity for having on -the -spot repairs made to firearms at skeet, trap, target, and similar organized events. It is, therefore, held that licensed gunsmiths may take immediate on-the -spot repairs to firearms at skeet, trap, target, and similar organized shooting events.

Held further, a licensed gunsmith must enter into his bound acquisition and disposition record, required to be maintained by 27 CFR 178.125(e), each receipt and disposition of firearms, except that a firearm need not be entered in the bound acquisition and disposition record if the firearm is brought in for adjustment or repair and the owner waits while it is being adjusted or repaired or if the gunsmith returns the firearm to the owner during the same business day it is brought in. If the firearm is retained from one business day to another or longer, it must be recorded in the bound acquisition and disposition record.

Held further, a licensed gunsmith is not required to prepare a Form 4473 (Firearms Transaction Record) where a firearm is delivered to him for the sole purpose of customizing, adjustment, or repair and the firearm is returned to the person from whom received. However, if a licensed gunsmith engages in the business of selling firearms, he must record these transactions on a Form 4473 for each sale in addition to maintaining the bound firearms acquisition and disposition record required by 27 CFR 178.125(e).

ATF Rul. 37 -13, 1973 AT C.B. 92, is hereby amplified.

[77 AF C.B. 185]

27 CFR 178.124: FIREARMS
TRANSACTION RECORD

Means of identification furnished by a nonlicensee purchasing a firearm.

ATF Rul. 79-7

The law places upon the licensee the responsibility for establishing the identity, place of residence, and age of an unlicensed person before selling or delivering a firearm to such person.

Under 27 CFR 178.124, nonlicensee's eligibility to purchase a firearm is established through the use of Form 4473 (Firearms Transaction Record). The regulation provides that before a licensee may sell or deliver a firearm to a nonlicensee, the form must be completed showing the purchaser's name, address, date of birth, and other pertinent information. Further, the regulation and form require the purchaser to identify himself in any manner customarily used in commercial transactions (e.g., a driver's license) and the licensee must indicate the manner in which the purchaser was identified. (See, also, 178.125a, personal firearms collection sale by a licensee.)

Satisfactory identification of firearms purchaser must identify the purchaser's name, age or date of birth, place of residence, and signature. A driver's license or identification card issued by a State in lieu of a driver's license is particularly appropriate. Social Security cards, alien registration receipt cards, and military identification cards are not, in and of themselves, acceptable to identify potentia firearms purchasers. The Social Security card is unsatisfactory because no address or date of birth is shown. While the alien registration card and military identification show name, age or date of birth, as well as other identifying information, the State of residence is not shown. While a particular document may not be sufficient to meet the statutory requirement for identifying the purchaser, any combinationfo documents which together disclose the required information concerning the purchaser is acceptable.

[ATFQB 1979-1 26] [Amended]

Editor's Note

ATF Rul. 79-7 was modified by T.D. ATF-389, 62 FR 19442, April 21, 1997, to require additional information on Form 4473. Under the new regulations, the form also requires the purchaser to state whether he or she is a citizen of the United States and to affirmatively provide his or her State of residence. In the case of a purchaser who is an alien legally in the United States, the regulations also require the purchaser to provide proof of residency through the use of substantiating documentation showing that the purchaser has resided in a State continuously for at least 90 days prior to the transfer of the firearm. In addition, the licensee must verify the identity of the alien by examining a government issued photo identification document. It should also be noted that Public Law 105-277, enacted on October 21, 1998, makes it unlawful, with certain exceptions, to transfer firearms to aliens admitted to the United States under a nonimmigrant visa and for such aliens to possess firearms.

27 CFR 178.112: IMPORTATION BY A LICENSED IMPORTER

Applications to import surplus military firearms or nonsporting firearms or ammunition for individual law enforcement officers for official use must be accompanied by the agency's purchase order.

ATF Rul. 80-8

The Bureau of Alcohol, Tobacco and Firearms has received several inquiries from firearms importers and dealers, law enforcement agencies, and the public requesting clarification of the statutes, regulations and procedures regarding the importation of firearms for law enforcement agencies.

Importation of surplus military firearms or firearms not particularly suitable for or readily adaptable to sporting purposes is generally prohibited by section 925(d)(3) of Title 18, United States Code. However, section 925(a)(1) provides that this prohibition does not apply to the importation of firearms or ammunition sold or shipped to, or issued for the use of the United States or any department or agency thereof or any State or any department, agency, or political subdivision thereof.

Pursuant to section 925(a)(1), the Bureau has previously allowed the importation of surplus military firearms an nonsporting firearms for individual law enforcement officers for official use. In approving such importation applications, the Bureau required federa firearms licensees to obtain from the agency employing the officer a certificate of the chief law enforcement officer stating that the firearm ammunition is for use in the performance of official duties.

However, once these firearms are imported for the individual officer for "official use," there is no prohibition in the law against the officer's resale or retention of the firearms for personal use. The purpose of section 925(a)(1) is to permit importation of firearms for the exclusive use of government agencies. The statute was not intended and may not be used as a vehicle by whic unimportable firearms can be introduced into ordinary commercial channels in the United States.

Held, a licensee's application to import surplus military firearms or nonsporting firearms or ammunition for law enforcement officers will not be approved unless accompanied by a purchase order from a department or agency of the United States or any department, agency or political subdivision of any State. The term "State" includes the District of Columbia, the Commonwealth of Puerto Rico, and the possessions of the United States.

[80 AF C.B. 20]

27 CFR 178.11: MEANING OF TERMS
(Also 27 CFR 179.11)

A hand-held device with a hand grip bent at an angle to the bore and having a rifled bore which is designed to expel, by means of an explosive, two electrical contacts (barbs) connected by two wires to a high voltage source within the device is classified as a firearm. ATF Rul. 76-6 is amplified.

ATF Rul. 80-20

The Bureau has determined in AT Rul. 6 -6 thateth Taser Model TF1 was a firearm as that term is defined in Title 18, United States Code (U.S.C.), section 921(a)(3), and that the Model TF1 also met the "any other weapon" definition found in the National Firearms Act (NFA), Title 26, U.S.C., section 5845(e). This ruling was limited in its oapplication t Taser Models TF1 produced on or after May 1, 1976. Th Taser Models TF76 and TF76A were subsequently developed and differ from th Taser Model TF1 in that these models each have a hand grip bent at an angle to the bore and the bore of each is rifled.

The changes in the designeof th Taser Models TF76 and TF76A bring them within the exclusion found in the "any other weapon" definition of the NFA for pistols and revolvers having a rifled bore or rifled bores.

Held, th Taser Models TF76 and TF76A are not subject to the provisions of the NFA. However, they are firearms as defined in Title 18, U.S.C., section 921(a)(3) and are subject to the provisions of Title 18, U.S.C., Chapter 44 and Title 27, Code of Federal Regulations, Part 178.

ATF Rul. 6 -6, 1976, FAT C.B. 96 is hereby amplified.

[ATFB 198 -4 24]

27 CFR 178.11: MEANING OF TERMS

An out-of-State college student may establish residence in a State by residing and maintaining a home in a college dormitory or in a location off-campus during the school term.

ATF Rul. 80-21

"State of residence" is defined by regulation in 27 CFR 178.11 as the State in which an individual regularly resides or maintains a home. The regulation also provides an example of an individual who maintains a home in State X and a home in State Y. The individual regularly resides in State X except for the summer months and in State Y for the summer months of the year. The regulation states that during the time the individual actually resides in State X he is a resident of State X, and during the time he actually resides in State Y he is a resident of State Y.

Applying the above example to out-of-State college students it is held, that during the time the students actually reside in a college dormitory or at an of -campus location they are considered residents of the State where the dormitory or off-campus home is located. During the time out-of-State college students actually reside in their home State they are considered residents of their home State.

[ATFB 198 -4 25]

27 CFR 178.111: GENERAL

Nonresident U.S. citizens returning to the United States and nonresident aliens lawfully immigrating to the United States may obtain a permit to import firearms acquired outside of the United States, provided such firearms may be lawfully imported.

ATF Rul. 81-3

Section 922(a)(3) of Title 18, United States Code, makes it unlawful, with certain exceptions, for a person to bring into his State of residence a firearm which he acquired outside that State. An unlicensed resident of a State must, therefore, arrange for the importation of the firearm through a Federal firearms licensee.

The definition of "State of residence" in 27 CFR 178.11 provides that the State in which an individual regularly resides or maintains a home is the State of residence of that person. U.S. citizens who reside outside of the United States are not residents of a State while so residing. A person lawfully immigrating to the United States is not a resident of a State unless he is residing and has resided in a State for a period of at least 90 days. Therefore, such persons are not precluded by section 922(a)(3) from importing into the United States any firearms acquired outside of the United States that may be lawfully imported. The firearms must accompany such persons since once a person is in the United States and has acquired residence in a State he may import a firearm only by arranging for the importation through a Federal firearms licensee.

As applicable to this ruling, 18 U.S.C. 925(d) provides that firearms are importable if they are generally recognized as particularly suitable for, or readily adaptable to, sporting purposes, excluding National Firearms Act (NFA) firearms and surplus military firearms.

Held, a nonresident U.S. citizen returning to the United States after having resided outside of the United States, or a nonresident alien lawfully immigrating to the United States, may apply for a permit from ATF to import for personal use, not for resale, firearms acquired outside of the United States without having to utilize the services of a Federal firearms licensee. The application on ATF Form 6 Part I (7570.3A), Application and Permit for Importation of Firearms, Ammunition and Implements of War, should include a statement, on the application form or on an attached sheet, that:

(1) the applicant is a nonresident U.S. citizen who is returning to the United States from a residence outside of the United States or, in the case of an alien, is lawfully immigrating to the United States from a residence outside of the United States; and

(2) the firearms are being imported for personal use and not for resale.

[ATFB 198 -3 77]

27 CFR 179.11: MEANING OF TERMS

The AR15 auto sear is a machine gun as defined by 26 U.S.C. 5845(b).

ATF Rul. 81-4

The Bureau of Alcohol, Tobacco and Firearms has examined an auto sear known by various trade names including "AR15 Auto Sear," "Drop In Auto Sear," and "Auto Sear II," which consists of a sear mounting body, sear, return spring, and pivot pin. The Bureau finds that the single addition of this auto sear to certain AR15 type semiautomatic rifles, manufactured with M16 internal components already installed, will convert such rifles into machine guns.

The National Firearms Act, 26 U.S.C. 5845(b) defines "machine gun" to include any combination of parts designed and intended for use in converting a weapon to shoot automatically more than one shot, without manual reloading, by a single function of the trigger.

Held: The auto sear known by various trade names including "AR15 Auto Sear," "Drop In Auto Sear," and "Auto Sear II," is a combination of parts designed and intended for use in converting a weapon to shoot automatically more than one shot, without manual reloading, by a single function of the trigger. Consequently, the auto sear is a machine gun as defined by 26 U.S.C. 5845(b).

With respect to the machine gun classification of the auto sear under the National Firearms Act, pursuant to 26 U.S.C. 7805(b), this ruling will not be applied t auto sears manufactured before November 1, 1981. Accordingly, auto sears manufactured on or after November 1, 1981, will be subject to all the provisions of the National Firearms Act and 27 C.F.R. Part 179.

[ATFQB 1981-3 78]

Editor's Note

Regardless of the date of manufacture of a drop in auto sear, possession of such a sear and certain M-16 fire control parts is possession of a machine gun as defined by the NFA. Specifically, these parts are a combination of parts designed and intended for use in converting a weapon into a machine gun and are a machine gun as defined in the NFA. (See "Important Information Concerning AR15-Type Rifles" under "Items of Interest" in this publication.)

27 CFR 179.11: MEANING OF TERMS

The KG-9 pistol is a machine gun as defined in the National Firearms Act.

ATF Rul. 82-2

The Bureau of Alcohol, Tobacco and Firearms has examined a firearm identified as the KG-9 pistol. The KG-9 is a 9 millimeter caliber, semiautomatic firearm which i blowback operated and which fires from the open bolt position with the bolt incorporating a fixed firing pin. In addition, a component part of the weapon is disconnector which prevents more than one shot being fired with a single function of the trigger.

The disconnector is designed in the KG-9 pistol in such a way that a simple modification to it, such as cutting, filing, or grinding, allows the pistol to operate automatically Thus, this simple modification to the disconnector together with the configuration of the above design features (blowback operation, firing from the open bolt position, and fixed firing pin) in the GK -9 permits the firearm to shoot automatically, more than one shot, without manual reloading, by a single function of the trigger. The above combination of design features as employed in the KG-9 is normally not found in the typical sporting firearm.

The National Firearms Act, 26 U.S.C. 5845(b), defines a machine gun to include any weapon which shoots, is designed to shoot, or can be readily restored to shoot, automatically more than one shot, without manual reloading, by a single function of the trigger.

The "shoots automatically" definition covers weapons that will function automatically. The "readily restorable" definition define weapons which previously could shoot automatically but will not in their present condition. The "designed" definition includes those weapons which have not previously functioned as machine guns but possess design features which facilitate full automatic fire by simple modification or elimination of existing component parts.

Held: The GK -9 pistol is designed to shoot automatically more than one shot, without function of the trigger. Consequently, the KG-9 pistol is a machine gun as defined in section 5845(b) of the Act.

With respect to the machine gun classification of the KG-9 pistol under the National Firearms Act, pursuant to 26 U.S.C. § 7805(b), this ruling will not be applied to KG-9 pistols manufactured before January 19, 1982. Accordingly, KG-9 pistols manufactured on or after January 19, 1982, will be subject to all the provisions of the National Firearms Act and 27 C.F.R. Part 179.

[ATFB 1982 -1 18]

27 CFR 179.11: MEANING OF TERMS

The SM10 and SM11A1 pistols and SAC carbines are machine guns as defined in the National Firearms Act.

ATF Rul. 82-8

The Bureau of Alcohol, Tobacco and Firearms has reexamined firearms identified as SM10 pistols, SM11A1 pistols, and SAC carbines. The SM10 is a 9 millimeter or .45ACP caliber, semiautomatic firearm; the SM11A1 is a .380ACP caliber, semiautomatic firearm; and the SAC carbine is a 9 millimeter or .45ACP caliber, semiautomatic firearm. The weapons ar blowback operated, fire from the open bolt position with the bolt incorporating a fixed firing pin, and the barrels of the pistols are threaded to accept a silencer. In addition, component parts of the weapons are disconnector and trip which prevent more than one shot being fired with a single function of the trigger.

The disconnector and trip are designed in the SM10 and SM11A1 pistols and in the SAC carbine (firearms) in such a way that a simple modification to them, such as cutting, filing, or grinding, allows the firearms to operate automatically Thus, this simple modification to th disconnector or trip, together with the configuration of the above design features (blowback operating, firing from the open bolt position, and fixed firing pin) in the SM10 and SM11A1 pistols and in the SAC carbine, permits the firearms to shoot automatically, more than one shot, without manual reloading, by a single function of the trigger. The above combination of design features as employed in the SM10 and SM11A1 pistols and the SAC carbine are normally not found in typical sporting firearms.

The National Firearms Act, 26 U.S.C. § 5845(b), defines a machine gun to include any weapon which shoots, is designed to shoot, or can be readily restored to shoot, automatically more than one shot, without manual reloading, by a single function of the trigger.

The "shoots automatically" definition covers weapons that will function automatically. The "readily restorable" definition define weapons which previously could shoot automatically but will not in their present condition. The "designed" definition includes those weapons which have not previously functioned as machine guns but possess design features which facilitate full automatic fire by a simple modification or elimination of existing component parts.

Held: The SM10 and SM11A1 pistols and the SAC carbine are designed to shoot automatically more than one shot, without manual reloading, by a single function of the trigger. Consequently, the SM10 and SM11A1 pistols and SAC carbines are machine guns as defined in Section 5845(b) of the Act.

With respect to the machine gun classification of the SM10 and SM11A1 pistols and SAC carbines, under the National Firearms Act, pursuant to 26 U.S.C. 7805(b), this ruling will not be applied to SM10 and SM11A1 pistols and SAC carbines manufactured or assembled before June 21, 1982. Accordingly, SM10 and SM11A1 pistols and SAC carbines, manufactured or assembled on or after June 21, 1982, will be subject to all the provisions of the National Firearms Act and 27 C.F.R. Part 179.

[ATFB 1982-2 49]

27 CFR 179.11: MEANING OF TERMS

The YAC STEN MK II carbine is a machine gun as defined in the National Firearms Act.

ATF Rul. 83-5

The Bureau of Alcohol, Tobacco and Firearms has examined a firearm identified as the YAC STEN MK II carbine. The YAC STEN MK II carbine is 9 millimeter caliber firearm which has identical design characteristics to the original selective fire STEN submachine gun designed by Reginald Vernon Shepherd and Harold Joh Turpin. The weapon i blowback operated and fires from the open bolt position with the bolt incorporating a fixed firing pin. In addition, a component part of the weapon is a trip lever (disconnector) which has been modified to prevent more than one shot being fired with a single function of the trigger.

The trip lever (disconnector) is designed in such a way that a simple modification to it, such as bending, breaking or cutting, allows the weapon to operate automatically. Thus, this simple modification to the trip lever (disconnector), together with STEN submachine gun design features and components in the YAC STEN MK II carbine, permits the firearm to shoot automatically, more than one shot, without manual reloading by a single function of the trigger. The above combination of machine gun design features as employed in the YAC STEN MK II carbine are not normally found in the typical sporting firearm.

The National Firearms Act, 26 U.S.C. 5845(b), defines a machine gun to include any weapon which shoots, is designed to shoot, or can be readily restored to shoot, automatically more than one shot, without manual reloading, by a single function of the trigger.

The "shoots automatically" definition covers weapons that will function automatically. The "readily restorable" definition define weapons which previously could shoot automatically but will not in their present condition. The "designed" definition includes weapons which have not previously functioned as machine guns but possess specific machine gun design features which facilitate automatic fire by simple alteration or elimination of existing component parts.

Held: The YAC STEN MK II carbine is designed to shoot automatically more than one shot, without manual reloading, by a single function of the trigger. Consequently, the STEN MK II semiautomatic carbine is a machine gun as defined in Section 5845(b) of the Act.

[ATFB 1983-3 35]

27 CFR 179.111: IMPORTATION PROCEDURE

A National Firearms Act (NFA) firearm may not be imported for use as a sample for sales to law enforcement agencies if the firearm is a curio or relic unless it is established that the firearm is particularly suitable for use as a law enforcement weapon.

ATF Rul. 85-2

The Bureau of Alcohol, Tobacco and Firearms has approved a number of applications to import National Firearms Act (NFA) firearms for the use of registered importers to generate orders for such firearms from law enforcement agencies.

A review of the characteristics of the NFA firearms approved for importation as sales samples indicates that some of the firearms are not being imported for the purposes contemplated by the statute. Some of the NFA firearms imported are, in fact, curios or relics and are more suitable for use as collector's items than law enforcement weapons.

Importations of NFA firearms are permitted by 26 U.S.C. 5844, which provides in pertinent part:

"No firearms shall be imported or brought into the United States or any territory under its control or jurisdiction unless the importer establishes, under regulations as may be prescribed by the Secretary, that the firearm to be imported or brought in is:

(1) being imported or brought in for the use of the United States or any department, independent establishment, or agency thereof or any State or possession or any political subdivision thereof; or

(2) ***

(3) being imported or brought in solely for ... use as a sample by a registered importer or registered dealer;

except that, the Secretary may permit the conditional importation or bringing in of a firearm for examination and testing in connection with classifying the firearm."

The sole purpose of the statute permitting the importation of NFA firearms as sales samples is to permit registered importers to generate orders for firearms from government entities, primarily law enforcement agencies, on the basis of the sample.

The implementing regulation, 27 CFR Section 179.111, provides that the person importing or bringing a firearm into the United States or any territory under its control or jurisdiction has the burden of proof to affirmatively establish that the firearm is being imported or brought in for one of the authorized purposes. In addition, a detailed explanation of why the importation falls within one of the authorized purposes must be attached to the application to import. The mere statement that an NFA firearm is being imported as a sales sample for demonstration to law enforcement agencies does not meet the required burden of proof and is not a detailed explanation of why the importation falls within the import standards.

Held, an application to import a National Firearms Act firearm as a sample in connection with sales of such firearms to law enforcement agencies will not be approved if the firearm is determined to be a curio or relic unless it is established by specific information that the firearm is particularly suitable for use as a law enforcement weapon. For example, the importer must provide detailed information as to why a sales sample of a particular weapon is suitable for law enforcement purposes and the expected customers who would require a demonstration of the weapon. Information as to the availability of firearms to fill subsequent orders would help meet the burden of establishing use as a sales sample. Also, letters from law enforcement agencies expressing a need for a particular model or interest in seeing a demonstration of a particular firearm would be relevant.

[ATFB 85-2 62]

Editor's Note

The importation of machine guns for use as sales samples must also meet the requirements of 27 CFR 179.105(d).

27 CFR 178.118: IMPORTATION OF
CERTAIN FIREARMS CLASSIFIED
AS CURIOS OR RELICS
(Also 178.11 and 178.26)

Surplus military firearms frames or receivers alone not specifically classified as curios or relics by ATF will be denied importation.

ATF Rul. 85-10

Section 233 of the Trade and Tariff Act of 1984, 9 Stat. 2991, amended Title 18, United States Code, section 925 to allow licensed importers to import firearms listed by the Secretary as curios or relics, excluding handguns not generally recognized as particularly suitable for or readily adaptable to sporting purposes. The amendment had the effect of allowing the importation of surplus military curio or relic firearms that were previously prohibited from importation by 18 U.S.C. section 925(d)(3).

Congressional intent was expressed by Senator Robert Dole in 130 CONG. REC. S2234 (daily ed., Mar. 2, 1984), as follows:

First. This provision is aimed at allowing collectors to import fine works of art and other valuable weapons.

Second. This provision would allow the importation of certain military surplus firearms that are classified as curios and relics by regulations of the Secretary of the Treasury.

Third. In order for an individual or firm to import a curio or relic it must first be put on a list by petitioning the Secretary of the Treasury. The Secretary must find the firearm's primary value is that of being a collector's item.

Fourth. The only reason a person would purchase these firearms is because of their peculiar collector's status. And, in fact, they must be special firearms and classified as such in order to import.

This language clearly shows that Congress intended to permit the importation of surplus military firearms of special interest and value to collectors and recognized by ATF as meeting the curio or relic definition in 27 CFR 178.11. The regulation defines "curios or relics" as firearms of "special interest to collectors by reason of some quality other than is ordinarily associated with firearms intended for sporting use or as offensive or defensive weapons." The regulation further defines curios or relics to include "firearms which derive a substantial part of their monetary value from the fact that they are novel, rare, bizarre, or because of their association with some historical figure, period or event."

In classifying firearms as curios or relics under this regulation, ATF has recognized only assembled firearms as curios or relics.

Moreover, ATF's classification of surplus military firearms as curios or relics has extended only to those firearms in their original military configuration. Frames or receivers of curios or relics and surplus military firearms not in their original military configuration were not generally recognized as curios or relics by ATF since they were not of special interest or value as collector's items.

Specifically, they did not meet the definition of curio or reli in section 178.11 as firearms of special interest to collectors by reason of a quality other than is ordinarily associated with sporting firearms or offensive or defensive weapons.

Furthermore, they did not ordinarily have monetary value as novel, rare, or bizarre firearms; nor were they generally considered curios or relics because of their association with some historical figure, period or event.

It is clear from the legislative history that Congress did not intend for the frames or receivers alone of surplus military firearms, or any other surplus military firearms not in their original military configuration, to be importable under section 925(e). It is also clear that only those firearms classified by ATF as curios or relic were intended to be approved by ATF for importation.

Held, to be importable under 18 U.S.C. section 925(e), surplus military firearms must be classified as curios or relics by ATF. Applications by licensed importers to import frames or receivers alone of surplus military curio or relic firearms will not be approved under section 925(e). Surplus military firearms will not be classified as curios or relics unless they are assembled in their original military configuration, and applications for permits to import such firearms will not be approved.

[ATFB 85-3 46]

26 U.S.C. § 5845(f)(2): DESTRUCTIVE DEVICE (Nonsporting shotgun having a bore of more than one-half inch in diameter)

The USAS-12 shotgun has a bore of more than one-half inch in diameter and is not generally recognized as particularly suitable for sporting purposes. Therefore, it is classified as a destructive device for purposes of the National Firearms Act, 26 U.S.C. Chapter 53.

ATF Rul. 94-1

The Bureau of Alcohol, Tobacco and Firearms (ATF) has examined a firearm identified as the USAS-12 shotgun to determine whether it is a destructive device as that term is used in the National Firearms Act (NFA), 26 U.S.C. Chapter 53.

The USAS-12 is a 12-gauge, gas-operated autoloading semiautomatic shotgun which is chambered for 12-gauge 2 3/4-inch ammunition. It has an 18 1/4-inch barrel, is approximately 38 inches long, and weighs 12.4 pounds unloaded and approximately 15 pounds with a loaded magazine, depending on the capacity of the magazine. The USAS-12 is equipped with a 12-round detachable box magazine, but a 28-round detachable drum magazine is also available. The shotgun is approximately 11 inches deep with a box magazine. There is an integral carrying handle on top of the receiver, which houses a rifle-type aperture rear and adjustable post-type front sight. The USAS-12 has a separate combat-style pistol grip located on the bottom of the receiver, forward ofeth buttstock. An optional telescopic sight may be attached to the carrying handle. The barrel is located below the operating mechanism in such fashion that the barrel is in a straight line with the center of the buttstock.

Section 5845(f), Title 26, U.S.C., classifies certain weapons as "destructive devices" which are subject to the registration and tax provisions of the NFA. Section 5845(f)(2) provides as follows:

(f) **Destructive device.** - The term "destructive device" means * * *
(2) any type of weapon by whatever name known which will, or which may be readily converted to, expel a projectile by the action of an explosive or other propellant, the barrel or barrels of which have a bore of more than one-half inch in diameter, except a shotgun or shotgun shell which the Secretary or his delegate finds is generally recognized as particularly suitable for sporting purposes;

A "sporting purposes" test which is almost identical to that in section 5845(f)(2) appears in 18 U.S.C. § 925(d)(3). This provision of the Gun Control Act of 1968 (GCA) provides that the Secretary shall authorize a firearm to be imported into the United States if the firearm is "generally recognized as particularly suitable for or readily adaptable to sporting purposes." With the exception of the "readily adaptable" language, this provision is identical to the sporting shotgun exception to the destructive device definition. The definition of "destructive device" in the GCA (18 U.S.C. § 921(a)(4)) is identical to that in the NFA.

In determining whether shotguns with a bore of more than one-half inch in diameter are "generally recognized as particularly suitable for sporting purposes" and thus are not destructive devices under the NFA, we believe it is appropriate to use the same criteria used for evaluating shotguns under the "sporting purposes" test of section 925(d)(3). Congress used virtually identical language in describing the weapons subject to the two statutory schemes, and the language was added to the GCA and NFA at the same time.

In connection with the determination o importability, ATF determined that the USAS-12 shotgun was not eligible for importation under the sporting purposes test in section 925(d)(3). In reaching this determination, ATF evaluated the weight, size, bulk, designed magazine capacity, configuration, and other characteristics of the USAS-12. It was determined that the weight of the USAS-12, 12.4 pounds, made it much heavier than traditional 12-gauge sporting shotguns, which made it awkward to carry for extended periods, as in hunting, and cumbersome to fire at multiple small moving targets, as in skeet and trap shooting. The width of the USAS-12 with drum magazine, approximately 6 inches, and the depth with box magazine, in excess of 11 inches, far exceeded that of traditional sporting shotguns, which do not exceed 3 inches in width or 4 inches in depth. The large size and bulk of the USAS-12 made it extremely difficult to maneuver quickly enough to engage moving targets as is necessary in hunting, skeet, and trap shooting. The detachable box magazine with 12-cartridge capacity and the detachable drum magazine with 28-cartridge capacity were of a larger capacity than traditional repeating sporting shotguns, which generally contain tubular magazines with a capacity of 3-5 cartridges. Additionally, detachable magazines permit more rapid reloading than do tubular magazines. Finally, the combat-style pistol grip, the barrel-to-buttstock configuration, the bayonet lug, and the overall appearance and general shape of the weapon were radically different from traditional sporting shotguns and strikingly similar to shotguns designed specifically for or modified for combat and law enforcement use.

Section 7805(b), Title 26, U.S.C., provides that the Secretary may prescribe the extent, if any, to which any ruling relating to the internal revenue laws shall be applied without retroactive effect. Accordingly, all rulings issued under the Internal Revenue Code are applied retroactively unless they specifically provide otherwise. Pursuant to section 7805(b), the Director, as the delegate of the Secretary, may prescribe the extent to which any ruling will apply without retroactive effect.

Held: The USAS-12 is a shotgun with a bore of more than one-half inch in diameter which is not particularly suitable for sporting purposes. The weight, size, bulk, designed magazine capacity, configuration, and other factors indicate that the USAS-12 is a semiautomatic version of a military-type assault shotgun. Accordingly, the USAS-12 is a destructive device as that term is used in 26 U.S.C. § 5845(f)(2). Pursuant to section 7805(b), this ruling is applied prospectively effective March 1, 1994, with respect to the making, transfer, and special (occupational) taxes imposed by the NFA. All other provisions of the NFA apply retroactively effective March 1, 1994.

[ATFB 1993-94-1 21]

26 U.S.C. § 5845(f)(2): DESTRUCTIVE DEVICE (Nonsporting shotgun having a bore of more than one-half inch in diameter)

The Striker-12/Streetsweeper shotgun has a bore of more than one-half inch in diameter and is not generally recognized as particularly suitable for sporting purposes. Therefore, it is classified as a destructive for purposes of the National Firearms Act, 26 U.S.C. Chapter 53.

ATF Rul. 94-2

The Bureau of Alcohol, Tobacco and Firearms (ATF) has examined a firearm identified as the Striker-12/Streetsweeper shotgun to determine whether it is a destructive device as that term is used in the National Firearms Act (NFA), 26 U.S.C. Chapter 53.

The Striker-12d an Streetsweeper shotguns are virtually identical 12-gauge shotguns with a spring-driven revolving magazine. The magazine has a 12-round capacity. The shotgun has a fixed stock or folding shoulder stock and may be fired with the folding stock collapsed. The shotgun with an 18-inch barrel is 37 inches in length with the stock extended, and 26.5 inches in length with the stock folded. The shotgun is 5.7 inches in width and weighs 9.24 pounds unloaded. The Striker/Streetsweeper has two pistol grips, one in the center of the firearm below th buttstock, and one on the forearm. The Striker/Streetsweeper was designed and developed in South Africa as a military, security, and anti-terrorist weapon. Various types of 12-gauge cartridges can be fired from the shotgun, and a rapid indexing procedure allows various types of ammunition to be loaded into the cylinder and selected for firing. All 12 rounds can be fired from the shotgun in 3 seconds or less.

Section 5845(f), Title 26, U.S.C., classifies certain weapons as "destructive devices" which are subject to the registration and tax provisions of the NFA. Section 5845(f)(2) provides as follows:

> (f) **Destructive device.**--The term "destructive device" means * * *
> (2) any type of weapon by whatever name known which will, or which may
> be readily converted to, expel a projectile by the action of an
> explosive or other propellant, the barrel or barrels of which have a
> bore of more than one-half inch in diameter, except a shotgun or shotgun
> shell which the Secretary or his delegate finds is generally recognized
> as particularly suitable for sporting purposes; ..."

A "sporting purposes" test which is almost identical to that in section
5845(f)(2) appears in 18 U.S.C. § 925(d)(3). This provision of the Gun
Control Act of 1968 (GCA) provides that the Secretary shall authorize a
firearm to be imported into the United States if the firearm is "generally
recognized as particularly suitable for or readily adaptable to sporting
purposes." With the exception of the readily adaptable language, this
provision is identical to the sporting shotgun exception to the destructive
devices definition. The definition of "destructive device" in the GCA (18
U.S.C. § 921(a)(4)) is identical to that in the NFA.

In determining whether shotguns with a bore of more than one-half inch
in diameter are "generally recognized as particularly suitable for sporting
purposes" and thus are not destructive devices under the NFA, we believe it is
appropriate to use the same criteria used for evaluating shotguns under the
"sporting purposes" test of section 925(d)(3). Congress used virtually
identical language in describing the weapons subject to the two statutory
schemes, and the language was added to the GCA and NFA at the same time.

In 1984, ATF ruled that the Striker-12 was not eligible for importation
under section 925(d)(3) since it is not particularly suitable for sporting
purposes. In making this determination, the 198 letter-ruling notes that the
Striker was being used in a number of "combat" shooting events. In a letter
dated June 30, 1986, ATF again denied importation to the Striker-12, on the
basis that it did not meet the "sporting purposes" test of section 925(d)(3).
This letter states that, "We believe the weapon to have been specifically
designed for military and law enforcement uses."

In evaluating the physical characteristics of the Striker
12/Streetsweeper, ATF concludes that the weight, bulk, designed magazine
capacity, configuration, and other features indicate that it was designed
primarily for military and law enforcement use and is not particularly
suitable for sporting purposes.

The weight of the Striker-12/Streetsweeper, 9.24 pound unloaded, is on
the high end for traditional 12-gauge sporting shotguns, which generally weigh
between 7 and 10 pounds. Thus, the weight of the Striker-12/Streetsweeper
makes it awkward to carry for extended periods, as in hunting, and cumbersome
to fire at multiple small moving targets, as in skeet and trap shooting. The
width of the Striker-12/Streetsweeper, 5.7 inches, far exceeds that of
traditional sporting shotguns, which do not exceed three inches in width or
four inches in depth. The large size and bulk of the Striker-12/Streetsweeper
make it extremely difficult to maneuver quickly enough to engage moving
targets as is necessary in hunting, skeet, and trap shooting. The spring
driven revolving magazine with 12-cartridge capacity is a much larger capacity
than traditional repeating sporting shotguns, which generally contain tubular
magazines with a capacity of 3-5 cartridges. The folding shoulder stock and
the two pistol grips are not typical of sporting-type shotguns. Finally, the
overall appearance and general shape of the weapon are radically different
from traditional sporting shotguns and strikingly similar to shotguns designed
specifically for or modified for combat and law enforcement use.

Section 7805(b), Title 26, U.S.C., provides that the Secretary may prescribe the extent, if any, to which any ruling relating to the internal revenue laws shall be applied without retroactive effect. Accordingly, all rulings issued under the Internal Revenue Code are applied retroactively unless they specifically provide otherwise. Pursuant to section 7805(b), the Director, as the delegate of the Secretary, may prescribe the extent to which any ruling will apply without retroactive effect.

Held: The Striker-12/Streetsweeper is a shotgun with a bore of more than one-half imch i diameter which is not particularly suitable for sporting purposes. The weight, size, bulk, designed magazine capacity, configuration, and other factors indicate that the Striker-12/Streetsweeper is a military-type shotgun, as opposed to a shotgun particularly suitable for sporting purposes. Accordingly, the Striker-12/Streetsweeper is a destructive device as that term is used in 26 U.S.C. § 5845(f)(2). Pursuant to section 7805(b), this ruling is applied prospectively effective March 1, 1994, with respect to the making, transfer, and special (occupational) taxes imposed by the NFA. All other provisions of the NFA apply retroactively effective March 1, 1994.

[ATFB 1993-1994-1 23]

18 U.S.C. § 921(a)(4): DESTRUCTIVE DEVICE
26 U.S.C. § 5845(f)(2): DESTRUCTIVE DEVICE
(Firearm having a bore of more than one-half inch in diameter)

37/38 mm gas/flare guns possessed with cartridges containing wood pellets, rubber pellets or balls, or bean bags are classified as destructive devices for purposes of the Gun Control Act, 18 U.S.C. Chapter 44, and the National Firearms Act, 26 U.S.C. Chapter 53.

ATF Rul. 95-3

The Bureau of Alcohol, Tobacco and Firearms (ATF) has examined various 37/38 mm gas/flare guns in combination with certain types of ammunition to determine whether these are destructive devices as defined in the Gun Control Act (GCA), 18 U.S.C. Chapter 44, and the National Firearms Act (NFA), 26 U.S.C. Chapter 53.

Section 5845(f), Title 26, United States Code, classifies certain weapons as "destructive devices" which are subject to the registration and tax provisions of the National Firearms Act (NFA). Section 5845(f)(2) provides as follows:

> (f) **Destructive device.**—The term "destructive device" means * * * (2) any type of weapon by whatever name known which will, or which may be readily converted to, expel a projectile by the action of an explosive or other propellant, the barrel or barrels of which have a bore of more than one-half inch in diameter, except a shotgun or shotgun shell which the Secretary or his delegate finds is generally recognized as particularly suitable for sporting purposes; . . ."

Section 5845(f)(3) excludes from the term "destructive device" any device which is neither designed or redesigned for use as a weapon and any device, although originally designed for use as a weapon, which is redesigned for use as a signaling, pyrotechnic, line throwing, safety, or similar device.

The definition of "destructive device" in the GCA (18 U.S.C. § 921(a)(4)) is identical to that in the NFA.

ATF has previously held that devices designed for expelling tear gas or pyrotechnic signals are not weapons and are exempt from the destructive device definition. However, ammunition designed to be used against individuals is available for these 37/38 mm devices. This "anti-personnel" ammunition consists of cartridges containing wood pellets, rubber pellets or balls, and bean bags.

When a gas/flare gun is possessed with "anti-personnel" type ammunition, it clearly becomes an instrument of offensive or defensive combat and is capable of use as a weapon Since these gas/flare guns have a bore diameter of greater than one-half inch, fire a projectile by the means of an explosive, and, when possessed with "anti-personnel" ammunition, are capable of use as weapons, the combination of the gas/flare gun and "anti-personnel" ammunition is a destructive device as defined in the GCA and NFA. As a result, registration as a destructive device is required. Any person possessing a gas/flare gun with which "anti-personnel" ammunition will be used must register the making of a destructive device prior to the acquisition of any "anti-personnel" ammunition. In addition, the gas/flare guns ar classified as firearms as defined by the GCA when possessed with "anti-personnel" type ammunition.

Each gas/flare gun possessed with anti-personnel ammunition will be required to be identified as required by law and regulations (27 C.F.R. §§ 178.92 and 179.102), including a serial number. Any person manufacturing the gas/flare gun and the "anti-personnel" ammunition must, if selling them in combination, have the appropriate Federal firearms license as a manufacturer of destructive devices and must have paid the special (occupational) tax as a manufacturer of National Firearms Act firearms. Any person importing the gas/flare gun and the "anti-personnel" ammunition must, if importing them in combination, have the appropriate Federal firearms license as an importer of destructive devices and must have paid the special (occupational) tax as an importer of National Firearms Act firearms.

Further, the "anti-personnel" ammunition to be used in the gas/flare launchers is ammunition for destructive devices for purposes of the GCA. Any person manufacturing the "anti-personnel" ammunition must have the appropriate Federal firearms license as a manufacturer of ammunition for destructive devices. Any person importing the "anti-personnel" ammunition must have the appropriate Federal firearms license as an importer of ammunition for destructive devices.

HELD: 37/38 mm gas/flare guns possessed with "anti-personnel" ammunition, consisting of cartridges containing wood pellets, rubber pellets or balls, or bean bags, are destructive devices as that term is used in 18 U.S.C. § 921(a)(4) and 26 U.S.C. 5845(f)(2).

[ATFB 95-3 28]

18 U.S.C. § 923 (a): ENGAGING IN THE BUSINESS OF DEALING IN FIREARMS (Auctioneers)

Auctioneers who regularly conduct consignment-type auctions of firearms, for example, held every 1-2 months, on behalf of firearms owners where the auctioneer takes possession of the firearms pursuant to a consignment contract with the owner of the firearms giving the auctioneer authority to sell the firearms and providing for a commission to be paid by the owner upon sale of the firearms are required to obtain a license as a dealer in firearms.

ATF Rul. 96-2

An association of auctioneers has asked the Bureau of Alcohol, Tobacco and Firearms (ATF) for a ruling concerning the auctions conducted by their members and whether the sale of firearms at such auctions requires a Federal firearms license as a dealer in firearms.

The auctioneers' association stated that their members generally conduct two types of auctions: estate-type auctions and consignment auctions. In estate-typ auctions, articles to be auctioned, including firearms, are sold by the executor of the estate of an individual. In these cases the firearms belong to and are possessed by the executor. The auctioneer acts as an agent of the executor and assists the executor in finding buyers for the firearms. The firearms are possessed by the estate and their sale to third parties is controlled by the estate. The auctioneer is paid a commissio on the sale of each firearm by the estate at the conclusion of the auction.

The association states that, in consignment-type auctions, an auctioneer may take possession of firearms in advance of the auction. The firearms are inventoried, evaluated, and tagged for identification. The firearms belong to individuals or businesse who have entered into a consignment agreement with the auctioneer giving the auctioneer authority to sell the firearms. The agreement states that the auctioneer has the exclusive right to sell the items listed on the contract at a location, time, and date to be selected by the auctioneer. The consignment-type auctions generally involve accepting firearms for auction from more than one owner. Also, these auctions are held on a regular basis, for example, every 1-2 months.

Section 923(a), Title 18, U.S.C., provides that no person shall engage in the business of dealing in firearms until he has filed an application and received a license to do so. Section 922(a)(1), Title 18, U.S.C., provides that it is unlawful for any person, other than a licensee, to engage in the business of dealing in firearms. Licensees generally may not conduct business away from their licensed premises.

The term "dealer" is defined at 18 U.S.C. § 921(a)(11)(A) to include any person engaged in the business of selling firearms at wholesale or retail. The term "engaged in the business" as applied to a dealer in firearms means a person who devotes time, attention, and labor to dealing in firearms as a regular course of trade or business with the principal objective of livelihood and profit through the repetitive purchase and resale of firearms. A dealer can be "engaged in the business" without taking title to the firearms that are sold. However, the term does not include a person who makes occasional sales, exchanges, or purchases of firearms for the enhancement of a personal collection or for a hobby, or who sells all or part of his personal collection of firearms. 18 U.S.C. § 921(a)(21)(C).

In the case of estate-type auctions, the auctioneer acts as an agent of the executor and assists the executor in finding buyers for the estate's firearms. eTh firearms are possessed by the estate, and the sales of firearms are made by the estate. In these cases, the auctioneer does not meet the definition of "engaging in the business" as a dealer in firearms and would not require a license. An auctioneer engaged in estate-type auctions, whether licensed or not, may perform this function, including delivery of the firearms, away from the business premises.

In the case of consignment-type auctions held on a regular basis, for example, every 1-2 months, where persons consign their firearms to the auctioneer for sale pursuant to an agreement as described above, the auctioneer would be "engaging in the business" and would require a license. The auctioneer would be disposing of firearms as a regular course of trade or business within the definition of a "dealer" under § 921(a)(11)(A) and must comply with the licensing requirements of the law.

As previously stated, licensed auctioneers generally must engage in the business from their licensed premises. However, an auctioneer may conduct an auction at a location other than his licensed premises by displaying the firearms at the auction site, agreeing to the terms of sale of the firearms, then returning the firearms to the licensed premises for delivery to the purchaser.

Held: Persons who conduct estate-type auctions at which the auctioneer assists the estate in selling the estate's firearms, and the firearms are possessed and transferred by the estate, do not require a Federal firearms license.

Held further: Persons who regularly conduct consignment-type auctions, for example, held every 1-2 months, where the auctioneer takes possession of the firearms pursuant to a consignment contract giving the auctioneer the exclusive right and authority to sell the firearms at a location, time and date to be selected by the auctioneer and providing for a commission to be paid upon sale are required to obtain a license as a dealer in firearms pursuant to 18 U.S.C. § 923(a).

[ATFB 96-2 101]

PROCEDURES

PART 178: COMMERCE IN FIREARMS
(Also 27 CFR 178.94, 178.12)

Recordkeeping procedures for "drop shipments" of firearms are prescribed.

ATF Proc. 75-3

This ATF Procedure sets forth the recordkeeping procedures for "drop shipments" of firearms [other than National Firearms Act firearms as defined in section 5845(a) of Chapter 53, Title 26, U.S.C.] between federally licensed firearms dealers, importers, and manufacturers.

Where licensee "A" places an order for firearms with licensee "B", and "B" transmits the order to licensee "C" for direct shipment (drop shipment) to "A", a certified copy of the license/list of "A" must be forwarded to "C" prior to shipment of the order.

On shipment of the order to "A", "C" shall enter in his bound record the disposition of the firearms to "A". On receipt of the shipment by "A", he shall enter the acquisition of the firearms in his bound record. Both licensees shall make such entries in the manner prescribed by regulations.

Since the actual movement of the firearms is between "C" and "A", and since "B" does not take physical possession, "B" will make no entry in his bound record. However, "B" should make appropriate entries or notations in his commercial records to reflect the transaction.

For Example: Where a licensed dealer orders firearms from a wholesaler and the wholesaler requests drop shipment from a manufacturer to the dealer, a certified copy of the dealer's license (or a certified copy of the dealer's list, if he has a multi-licensed business organization) shall accompany the wholesaler's order to the manufacturer.

The manufacturer shall enter in his bound record the disposition of the firearms to the dealer, and the dealer shall enter the acquisition of the firearms in his bound record reflecting receipt from the manufacturer.

The wholesaler, although a part of the business transaction, neither acquires nor disposes of the firearms and would, therefore, enter nothing of the transaction in his bound record.

NFA Firearms: Transfer of National Firearms Act firearms may be accomplished only pursuant to the manner outlined in Subpart F, Part 179, Title 27, Code of Federal Regulations.

Inquiries: Inquiries concerning this procedure should refer to its number and be addressed to your nearest ATF office. (See ATF Publication 5300.5, State Laws and Published Ordinances-Firearms for the current address.)

[75 ATF C.B. 78] [Amended]

27 CFR 179.35: EMPLOYER IDENTIFICATION NUMBER
(Also see 179.34, 179.84 179.88, 179.90, 179.103 and 179.112)

Identification Number for Special (Occupational) Taxpayer

ATF Proc. 90-1

Purpose: The purpose of this ATF procedure is to inform Federal Firearms licensees who have paid the special (occupational) tax to import, manufacture, or deal in National Firearms Act (NFA) firearms of the discontinuance of the use of the ATF Identification Number and the replacement with the use of the Employer Identification Number (EIN) on all NFA transaction forms.

Background: Section 5801 of Title 26, U.S.C. provides that on first engaging in business, and thereafter on or before the first day of July of each year, every importer, manufacturer, and dealer in NFA firearms shall pay the appropriate special (occupational) tax. In addition, section 5802 requires each importer, manufacturer, and dealer to register with the Secretary his name and the address of each location where he will conduct business. The filing of ATF Form 5630.5, with payment of the appropriate tax required by section 5801, also accomplishes registration requirements under section 5802.

The regulations at 27 CFR 179.34 require that the special tax be paid by return (ATF Form 5630.5, Special Tax Registration and Return) and require that all information called for on the return be provided, including the Employee Identification Number. 27 CFR 179.35 provides the instruction for applying for an EIN.

The regulations in 27 CFR 179.84, 179.88, and 179.90 provide require that the application to transfer an NFA firearm identify the special tax stamp, if any, of the transferor and transferee. The regulations in 27 CFR 179.103 and 179.112 require that the notice submitted to register NFA firearms identify the special tax stamp of the manufacturer or importer respectively. Identification of the tax stamp is necessary to ensure the tax liability has been satisfied, that the parties are qualified to import, manufacture, or deal in NFA firearms, and, in certain instances, is necessary to ensure that both parties in a transfer application are entitled to an exemption form the transfer tax.

In 1980, because of delays in the issuance of special tax stamps resulting in the inability of special taxpayers to conduct business operations, ATF Procedure 80-6 was implemented. Thi procedure notified taxpayers that they could obtain an ATF identification number which should be used in lieu of the IRS special tax stamp number on all NFA transaction forms. This procedure was established to facilitate the processing of NFA forms and to eliminate the delay caused by the time period required for IRS processing of the special tax stamp.

ATF has recently taken over the collection of special tax from the Internal Revenue Service, and is now issuing the special tax stamps. The number used to identify the special tax stamp is the EIN.

Because the number used to identify the special tax stamp is the EIN, this number must appear on all forms (applications, notices, and returns) involving NFA firearms. The problems that caused the implementation of the procedure in 1980 have been resolved. In fact, the assignment of an ATF identification number is now duplicative and requires more paperwork of the taxpayer. Accordingly, the use of the ATF identification number is no longer necessary and is discontinued.

ATF Procedure 80-6 is cancelled.

Inquiries: Inquiries regarding this ATF procedure should refer to its number and be addressed to the Bureau of Alcohol, Tobacco and Firearms, Chief, National Firearms Act Branch, Washington, DC 20226.

[ATFB 1990-1 55]

INDUSTRY CIRCULARS

Industry Circular 72-23

SHIPMENT OR DELIVERY OF FIREARMS
BY LICENSEES TO EMPLOYEES,
AGENTS, REPRESENTATIVES, WRITERS,
AND EVALUATORS

Purpose. The purpose of this circular is to clarify the provisions of 18 U.S.C. Chapter 44, and Subpart F of the regulations thereunder (27 C.F.R. 178) pertaining to the shipment of firearms in interstate commerce by a firearms licensee to its own nonlicensed employees, agents, and representatives, for the use and benefit of the licensee's business. The position of the Bureau is set out in Revenue Ruling 69-248.

Background. Revenue Ruling 69-248 provides as follows: [See RR 69-248].

Scope. Included within the category of agents and representatives discussed in the Revenue Ruling are professional writers, consultants and evaluators who in the course of their professions acquire firearms from a licensee for research or evaluation. The Revenue Ruling applies only to firearms acquired from a licensee for limited lengths of time and where the title to and ultimate control of the firearm remains in the licensee. Should the writer or evaluator desire to permanently keep the examined firearm, prior arrangements must be made to acquire the firearm through a licensee in such writer's or evaluator's State of residence and the Revenue Ruling would have no application.

Restriction. This Revenue Ruling also does not apply to firearms and ammunition within the purview of the National Firearms Act (26 U.S.C. Chapter 53).

Records. The licensee should enter in his firearms records the shipment or delivery of firearms to the employee, agent, representative, writer, consultant, or evaluator in accordance with Subpart H of the regulations. Upon the completion of the business purpose for which the firearms were received the firearms must be returned to the licensee who should enter their receipt in his records.

Editor's Note

Rev. Rul. 69-248 and ATF Rul. 73-19 discuss this matter in greater detail and can be found in this publication.

Industry Circular 72-30

IDENTIFICATION OF PERSONAL
FIREARMS ON LICENSED PREMISES
NOT OFFERED FOR SALE

Purpose. The purpose of this circular is to urge licensed firearms dealers to identify their personal collection of firearms kept at the business premises.

Scope. The provisions of Section 923(g), 18 U.S.C. Chapter 44, and Subpart H of the regulations (27 CFR 178) require all licensed firearms dealers to maintain records of their receipt and disposition of all firearms at the licensed premises. Section 178.121(b) of the regulations and the law further provide for the examination and inspection during regular business hours or other reasonable times of firearms kept or stored on business premises by licensees and any firearms record or document required to be maintained.

Guidelines for Identifying Personal Firearms on the Business Premises of Licensed Dealers. A presumption exists that alls firearm on a business premises are for sale and accordingly must be entered in the records required to be maintained under the law and regulations. However, it is recognized that some dealers may have personal firearms on their business premises for purposes of display or decoration and not for sale Firearms dealers who have such personal firearms on licensed premises should not intermingle such firearms with firearms held for sale. Such firearms should be segregated from firearms held for sale and appropriately identified (for example, by attaching a tag) as being "not for sale". Personal firearms on license premises which are segregated from firearms held for sale and which are appropriately identified as not being for sale need not be entered in the dealer's records.

There may be occasions where firearms dealer utilizes his license to acquire firearms for his personal collection. Such firearms must be entered in his permanent acquisition records and subsequently be recorded as a disposition to himself in his private capacity. If such personal firearms remain on the licensed premises, the procedures described above with respect to segregation and identification must be followed.

The above procedures will facilitate the examination and inspection of the records fo firearms dealers and result in less inconvenience to licensees.

Industry Circular 74-13

GUIDELINES FOR VERIFYING IDENTITY
AND LICENSED STATUS OF
TRANSFEREE

General: The Bureau urges all firearms licensees to require whatever information they deem necessary and within reason in order to verify the identity and licensed status of transferee licensees with whom they do business.

Personal Appearance: A licensee who appears in person at another licensee's business premises for the purpose of acquiring firearms should be required to furnish, to the transferor, positive identification in addition to a certified copy of his license [or in addition to a copy of his certified list, if a multi-licensed entity]. Such identification should prove to the satisfaction of the transferor that the person receiving firearms is, in fact, the same person to whom the license was issued.

Mail Order Sales: When the shipment is to be made to an address other than the transferee's premises as listed on his license or on his certified list, it is suggested that the transferor verify the address as being that of the transferee.

Industry Circular 77-20

DUPLICATION OF SERIAL NUMBERS
BY LICENSED IMPORTERS

ATF has noted cases where some licensed importers have adopted the same serial number for more than one firearm. These instances of duplication have generally occurred when firearms are received from more than one source.

Title 27, CFR section 178.92 requires that the serial number affixed to a firearm must not duplicate the number affixed to any other firearm that you import into the United States. Those of you who import destructive devices are under the same requirement due to the inclusion of destructive devices in the definition of firearm as used in 27 CFR 178.11. ATF Ruling 75-28 stated that a serial number affixed by the foreign manufacturer may be adopted to fulfill this unique serial number requirement. However, the manufacturer's serial number must be affixed in the manner set forth in 27 CFR 178.92 and must not duplicate a number previously adopted by you for another firearm.

If you receive two or more firearms with the same serial number, it is your responsibility to affix additional markings to make each serial number unique.

ATF Ruling 75-28 also reminds you of the other identifying marks required by 27 CFR 178.92 In addition to a unique serial number, each firearm must be marked to show the model (if any); the caliber or gauge; the name of the manufacturer and importer, or recognizable abbreviations; the country of manufacture; and the city and State (or recognized abbreviations) in which your licensed premises are located.

GENERAL INFORMATION

TABLE OF CONTENTS

PAGE

GENERAL INFORMATION

Items of Interest

1. INFORMATION ABOUT ATF

For assistance with applications, records, transactions, and other regulatory matters, contact your local ATF office. Information about criminal violations of Federal firearms laws should also be referred to that office.

Check ATF local listings in your telephone directory under **"U.S. Government."** If unavailable, see the listings contained in ATF Publication 5300.5, State Laws and Published Ordinances - Firearms, which is furnished to all Federal firearms licensees.

Requests for ATF forms and publications should be directed to the ATF Distribution Center, P.O. Box 5950, Springfield, Virginia 22150-5950, (703) 455-7801.

ATF publishes the ATF Quarterly Bulletin, by which the agency informs interested persons about current alcohol, tobacco, firearms, and explosives matters, including regulatory, procedural, and administrative information; items of general interest; and, excerpts from public laws and congressional committee reports. The ATF Quarterly Bulletin may be purchased from the Superintendent of Documents, U.S. Government Printing Office, Washington, DC 20402, (202) 512-1800. Checks should be made payable to the Superintendent of Documents.

2. WEB SITE ADDRESSES

Further information pertaining to ATF and other Federal agencies referenced in this publication can now be obtained at the following web site addresses:

ATF: www.atf.treas.gov
FBI (NICS): www.fbi.gov/programs.htm
Department of State, Office of Defense Trade Controls: www.pmdtc.org
Department of Commerce, Bureau of Export Administration: www.bxa.doc.gov

3. IMPORTANT INFORMATION CONCERNING AR-15 TYPE RIFLES

ATF has encountered various AR-15 type assault rifles such as those manufactured by Colt, E.A. Company, SGW, Sendra and others, which have been assembled with fire control components designed for use in M16 machine guns. The vast majority of these rifles which have been assembled with an M16 bolt carrier, hammer, trigger, disconnector and selector will fire automatically merely by manipulation of the selector or removal of the disconnector. Many of these rifles using less than the 5 M16 parts listed above also will shoot automatically by manipulation of the selector or removal of the disconnector.

Any weapon which shoots automatically, more than 1 shot, without manual reloading, by a single function of the trigger, is a machine gun as defined in 26 U.S.C. 5845(b), the National Firearms Act (NFA). In addition, the definition of a machine gun also includes any combination of parts from which a machine gun may be assembled, if such parts are in possession or under the control of a person. An AR-15 type assault rifle which fires more than 1 shot by a single function of the trigger is a machine gun under the NFA. Any machine gun is subject to the NFA and the possession of an unregistered machine gun could subject the possessor to criminal prosecution.

Additionally, these rifles could pose a safety hazard in that they may fire automatically without the user being aware that the weapon will fire more than 1 shot with a single pull of the trigger.

In order to avoid violations of the NFA, M16 hammers, triggers, disconnectors, selectors and bolt carriers must not be used in assembly of AR-15 type semiautomatic rifles, unless the M16 parts have been modified to AR-15 Model SP1 configuration. Any AR-15 type rifles which have been assembled with M16 internal components should have those parts removed and replaced with AR-15 Model SP1 type parts which are available commercially. The M16 components also may be modified to AR-15 Model SP1 configuration.

It is important to note that any modification of the M16 parts should be attempted by fully qualified personnel only.

Should you have any questions concerning AR-15 type rifles with M16 parts, please contact your nearest ATF office. Our telephone numbers are listed in the **"United States Government"** section of your telephone directory under the **"United States Treasury Department."**

4. FEDERAL AGE RESTRICTIONS

Federal law prohibits Federal firearms licensees from selling or delivering any firearm or ammunition to any individual who the licensee knows or has reasonable cause to believe is less than 18 years of age, and, if the firearm, or ammunition is other than a shotgun or rifle, or ammunition for a shotgun or rifle, to any individual who the licensee knows or has reasonable cause to believe is less than 21 years of age. (18 U.S.C. 922(b)(1), 27 CFR 178.99(b)(1).)

[**Note:** Ammunition interchangeable between rifles and handguns (such as .22 caliber rimfire) may be sold to an individual 18 years of age, but less than 21, if the licensee is satisfied that the ammunition is being acquired for use in a rifle.]

Additionally, it is unlawful for a person to transfer a handgun to a juvenile (a person less than 18 years of age) and it is unlawful for a juvenile to possess a handgun. Exceptions are provided for the transfer of a handgun to and possession by a juvenile for the purposes of employment, ranching, farming, target practice or hunting as provided for in 18 U.S.C. 922(x).

Pursuant to regulations at 27 CFR 178.103, ATF distributes posters to licensed importers, manufacturers, and dealers that caution against the transfer of handguns to juveniles, as well as the possession of handguns by juveniles, in violation of 18 U.S.C. 922(x). The regulations require licensees to display the posters on their premises. Some licensees erroneously interpreted the poster to mean that that they may now lawfully transfer handguns to any person over 17 years of age. Enactment of section 922(x) making it unlawful to sell handguns to juveniles (persons under 18 years of age) and for juveniles to possess handguns did not alter section 922(b)(1) of the GCA that continues to prohibit licensees from transferring handguns to persons under 21 years of age.

5. SALES OF FIREARMS TO LAW ENFORCEMENT OFFICERS

Section 925(a)(1) of the GCA exempts law enforcement agencies from the transportation, shipment, receipt, or importation controls of the GCA when firearms are to be used for the official business of the agency. An individual law enforcement officer's receipt and possession of firearms for use in carrying out official duties on behalf of an agency is also exempt, unless the officer has been convicted of a misdemeanor crime of domestic violence.

If a law enforcement officer is issued a certification letter on the agency's letterhead, signed by a person in authority within the agency stating the officer will use the firearm in performance of official duties and that a records check reveals that the purchasing officer has not been convicted of a misdemeanor crime of domestic violence, the officer specified in the certification may purchase a firearm from you regardless of the State in which the officer resides, or in which the agency is located.

You (the seller) are not required to prepare a Form 4473 covering this particular sale, as the certification letter is evidence of the transaction. Moreover, you are not required to comply with the Brady law where the purchase is made pursuant to such letter. However, disposition to the officer is to be entered in your permanent records and the certification letter kept in your files. The permanent records should show the residence address of the purchasing officer, not the address of the officer's employing agency.

If a law enforcement officer desiring to purchase a firearm does not have a certification letter, you (the seller) may still make the sale if the requirements of the Brady law (18 U.S.C. 922(t)), including a background check of the purchaser, are met. A Form 4473 covering such a sale must be prepared, the transaction must be entered in your permanent records, and all other applicable requirements of the law and regulations must be met.

ATF considers the following as persons having authority to certify that law enforcement officers purchasing firearms will use the firearms in performance of official duties:

a. In a city or county police department, the director of public safety or the chief or commissioner of police.

b. In a sheriff's office, the sheriff.

c. In a State police or highway patrol department, the superintendent or the supervisor in charge of the office to which the State officer or employee is assigned.

d. In Federal law enforcement offices, the supervisor in charge of the office to which the Federal officer or employee is assigned.

The Bureau would also recognize someone signing on behalf of a person in authority, provided there is a proper delegation of authority and overall responsibility has not changed in any way. (See **Question O11** for sales of semiautomatic assault weapons and large capacity ammunition feeding devices to law enforcement officers.) If the purchasing officer is a supervisory officer, the certification must be made by that officer's supervisor. In other words, the purchasing officer and the certifying officer may not be the same person.

6. SALE OF FIREARMS TO ALIENS IN THE UNITED STATES

ATF frequently receives questions about the legality of selling firearms to aliens. The purpose of this Item of Interest is to clarify the status of aliens who wish to purchase firearms and provide some of the exportation guidelines.

a. Domestic Sales

In order to purchase firearms in the United States from a Federal firearms licensee, an alien must:

(1) Be 18 years of age (21 for handguns);
(2) Provide the licensee with a government-issued photo identification document;
(3) Complete ATF Form 4473, Firearms Transaction Record;
(4) Comply with the Brady law, 18 U.S.C. 922(t);
(5) Be a resident of the State in which the firearm purchase is made for a period of 90 days and substantiate residency by documentation (for example, utility bills or a lease agreement);
(6) Not be admitted to the United States under a nonimmigrant visa (for exceptions, see 18 U.S.C. 922(y)(2)); and
(7) Not be a felon or within any other category of prohibited person.

An alien who is legally in the United States will be considered to be a resident of a State for the purpose of complying with the GCA if he or she is residing in that State and has resided in that State continuously for at least 90 days before purchasing a firearm. **Note**, however, that even a legal resident alien who has lived in the United States for many years will have to wait 90 days before purchasing a firearm if the alien changes his or her State of residence.

Editor's Note

A 1997 amendment to the definition of State of residence in 27 CFR 178.11 removed the written statement of an alien's embassy or consulate as a means of acquiring a residence for the purpose of obtaining a firearm.

b. Export Sales

Removal of a firearm or ammunition from the U.S. by an alien is an **exportation**. With few exceptions, the firearms licensee must obtain an export license (Form DSP-5) from the State Department's **Office of Defense Trade Controls (ODTC)** prior to exportation. Also, a licensee may export a firearm or ammunition to an alien if an export license is obtained from ODTC.

ODTC takes the position that when a dealer knows or believes that a foreign customer intends to take a rifle or handgun out of the U.S., the dealer is legally obligated to notify ODTC that the firearm was sold for the purpose of exportation.

Where the licensee exports the firearms to the alien, the licensee need only record the name and address of the foreign customer in his or her bound book. Therefore, Form 4473 need not be completed.

Exportation Guidelines

Exportation of firearms other than sporting shotguns is regulated by the Department of State, Office of Defense Trade Controls.

For further information about obtaining an export license as well as detailed rifle and handgun exportation information, contact:

**OFFICE OF DEFENSE TRADE CONTROLS
PM/DTC, SA-6, ROOM 228
DEPARTMENT OF STATE
WASHINGTON, DC 20522-0602
TELEPHONE (703) 875-6644**

Web site: www.pmdtc.org

Exportation Guidelines (Shotguns)

The Department of Commerce oversees the exportation of shotguns with barrels between 18 and 28 inches in length. The Department of Commerce requires a general license to export these items. There is no fee for a general license. For further information, contact:

BUREAU OF EXPORT ADMINISTRATION
EXPORT COUNSELING DIVISION
DEPARTMENT OF COMMERCE
14th ST. & PENNSYLVANIA AVE. N.W.
WASHINGTON, DC 20230
TELEPHONE (202) 482-4811
Web site: www.bxa.doc.gov

Certain Shipments to Canada

Certain shipments to Canada are exempt from the Form DSP-5 licensing requirement. For all exportations to Canada, the person exporting the firearm must present a Shipper Export Declaration (Department of Commerce Form 7525-V) to the U.S. Customs Service prior to, or at the time of, exportation (22 CFR 123.25).

Licensees located near the Canadian border have asked ATF and ODTC if a Canadian resident could visit the shop, negotiate a purchase, and then arrange to receive the firearm in Canada. The dealer could bring or ship the firearm to the border, present a Form 7525-V to Customs, and then arrange to have the Canadian receive the firearm in Canada. The dealer records the transaction in his or her bound book, with the disposition showing sale to the Canadian with the Canadian address listed. No ATF Form 4473 is required since delivery was not made in the U.S. See also **Item 6** below containing information on Canadian firearms laws and regulations.

If the dealer performs this type of transaction more than occasionally, the dealer is required to register with ODTC as an exporter even though Forms DSP-5 are not required.

Under the provisions of the Arms Export Control Act of 1976, an export license must be obtained from ODTC before exporting a rifle or a handgun anywhere other than to Canada. Application for an export license is made to ODTC on Form DSP-5. There is no fee for the license. For all exportations, the person exporting the firearm must present a Shipper Export Declaration (Department of Commerce Form 7525-V) to the U.S. Customs Service prior to, or at the time of, exportation (22 CFR 123.25). Anyone who **regularly** exports firearms is required to register with ODTC as an exporter on Form DSP-9 and pay a registration fee, as follows:

> 1 year..............$250
> 2 years.............$500
> 3 years.............$700
> 4 years.............$850
> 5 years.............$1000
> [22 CFR 122.2]

c. Sales to Diplomats

Diplomats, as individuals, are not exempt from Federal, State or local firearms laws. Sales to individuals, including diplomats and embassy personnel, must comply with all requirements of the GCA and the firearms regulations (27 CFR Part 178).

d. Sales to Foreign Embassies or Consulates

Special provisions have been made to allow for the sale of small quantities of firearms to foreign missions for the purpose of the physical security of embassy grounds. The arms become the property of the government whose embassy made the purchase, not the private property of an individual.

The dealer should obtain documentation which will show that the sale was a bona fide sale to a foreign mission and not a sale to an individual diplomat. Documentation should contain the following:

(1) A purchase order or invoice from the foreign mission;

(2) Payment out of government funds rather than from private funds; or

(3) A written statement by the principal officer of the embassy or consulate that the weapons are being purchased by, and will be the property of, the mission.

Once the dealer has documented that a sale is to a foreign mission, he or she may complete the transaction by shipping or delivering the firearms directly to the foreign mission. Form 4473 need not be completed since the sale is considered to be an exportation.

ATF views the transaction as an exportation, as embassy grounds are regarded as foreign territory.

ODTC does not view the sale of **"reasonable quantities"** of firearms to a foreign embassy to be an exportation. Consequently, the dealer need not obtain an export license from ODTC to deliver firearms to the embassy. **ODTC should be contacted for further information.**

7. CANADIAN FIREARMS INFORMATION

a. General

Implementation of the **Firearms Act** on December 1, 1998, brought about extensive changes to Canadian firearm regulations. **Most changes affecting visitors bringing firearms into Canada will not come into effect until January 1, 2001.**

b. Individuals Bringing Firearms Into Canada

- An individual must be at least 18 years of age to bring a firearm into Canada.
- Prohibited firearms may not be brought into Canada or transported through Canada.
- Restricted firearms, including handguns, may only be imported with prior authorization from the Chief Firearms Officer of the province or territory to which you are traveling.
- Anyone entering Canada must declare all firearms to Canada Customs.

For more information on bringing firearms into Canada prior to January 1, 2001, contact Revenue Canada Customs at (613) 946-0240.

1. Prohibited firearms

The following firearms are classified as prohibited firearms and cannot be brought into Canada:

- short-barreled handguns (handguns with a barrel length equal to or less than 105 mm)
- .25 caliber handguns
- .32 caliber handguns

(No handgun listed above is prohibited if it is prescribed by regulation for use in competitions governed by the rules of the International Shooting Union.)

- "sawed-off" rifles or "sawed-off" shotguns less than 660 mm in overall length
- "sawed-off" rifles or "sawed-off" shotguns which have a barrel length of less than 457 mm and are equal to or more than 660 mm in overall length
- all automatic firearms
- automatic firearms that have been converted to semiautomatic or single shot
- all firearms that the Governor in Council classifies as prohibited in the regulations (military and paramilitary firearms and firearms deemed to have no legitimate sporting or recreational use).

Cartridge magazines are also regulated. Generally, magazines used in semiautomatic, centerfire rifles and shotguns, or handguns not commonly available in Canada, with a capacity to contain more than 5 cartridges are prohibited. Magazines for other handguns that can contain more than 10 cartridges are also prohibited.

Where a prohibited firearm is declared at Canada Customs, a customs officer may allow the firearm to be exported back to its country of origin. Firearms that are not immediately exported are forfeited.

For information on firearms prohibited by regulations made by the Governor in Council or on firearms prescribed as International Shooting Union handguns, contact the Canadian Firearms Centre.

2. Restricted firearms

The following firearms are classified as restricted firearms requiring special authority to bring into Canada:

- all handguns which are not prohibited firearms
- semiautomatic centerfire rifles and shotguns that have a barrel length less than 470 mm and that have not been made prohibited
- rifles and shotguns that can fire after being reduced to an overall length of less than 660 mm, by any temporary means such as folding or telescoping
- all firearms that the Governor in Council classifies as restricted in the regulations (for information on these regulations, contact the Canadian Firearms Centre).

Anyone bringing a restricted firearm into Canada must have an Authorization to Transport for the restricted firearm. This authorization will permit transport of the restricted firearm between specified places within Canada. **This authorization must be obtained in advance from the Chief Firearms Officer of the Canadian province or territory to be visited.**

An applicant for an Authorization to Transport must show good and sufficient reason to bring restricted firearms to Canada, such as for use in target practice, or a target shooting competition at an approved shooting club or range. Restricted firearms cannot be used for hunting.

For more information on Authorizations to Transport, call the Chief Firearms Officer of the Canadian province or territory that you will be visiting. You can obtain the address, telephone and fax numbers from the Canadian Firearms Centre.

3. January 1, 2001: Confirmed Firearms Declarations

All firearms users in Canada must have a license authorizing possession of firearms as of January 1, 2001. Visitors to Canada will require authorization to possess firearms in the form of a confirmed Firearms Declaration. When firearms are declared on entry into Canada, the declaration will be confirmed by a customs officer.

Firearms Declarations must normally be made in writing. In certain cases, a customs officer can accept an oral declaration. In most cases, it will be possible to complete and mail in a pre-declaration form, which can be confirmed when you arrive at the border. As 2001 approaches, Firearms Declaration forms will be available through outfitters, hunting and shooting clubs, Canadian tourism offices, and the Canadian Firearms Centre.

The declaration will require basic information about the visiting firearm user, the destination in Canada, and the reason for bringing a firearm into Canada. Background checks, including a criminal history search, will be conducted. Descriptive information about each firearm being brought into the country will also be required (i.e., make, model, serial number, caliber/gauge).

Certain items are not considered firearms for purposes of these regulations, including firearms classified as antiques under the Criminal Code and pellet guns designed to fire a projectile at no more than 152.4 meters per second (500 feet per second). While the firearms import/export regulations do not apply, these items should still be declared at Canada Customs.

4. Confirmation

On arrival at the border, a customs officer may review the Firearms Declaration and examine the firearms. If everything checks out, the customs officer will confirm the Firearms Declaration and, where restricted firearms are being imported, the Authorization to Transport. The visitor will receive a confirmation number either in writing or orally.

A confirmed Firearms Declaration will serve two purposes:

- it will act as a temporary license authorizing possession of the firearm(s) listed thereon; and
- it will serve as a temporary registration certificate for the firearms imported.

A Firearms Declaration will expire after 60 days. A Firearms Declaration for restricted firearms will expire on the earlier of 60 days or the expiration date of the Authorization to Transport. A new Firearms Declaration will be required for every re-entry into Canada. A Firearms Declaration may be renewed from within Canada by contacting the Chief Firearms Officer of the province or territory visited.

Firearms Declarations must be turned in when leaving Canada. In many cases, visitors will be able to leave Firearms Declarations in a drop-off box. In some cases, however, a Canadian customs officer at the border crossing may inspect it and the firearms to ensure that all the firearms previously brought into Canada are being exported.

5. Refusal to confirm

A customs officer can refuse to confirm a Firearms Declaration if:

- the declaration form is not completed truthfully, or required information is not provided;
- the requirements set out in the **Firearms Act** and regulations are not met; or
- he or she concludes that it is desirable, in the interests of the visitor's safety and/or the safety of others, that the firearm not be allowed to enter Canada.

If the customs officer refuses to confirm the Firearms Declaration, then, depending on the situation, the officer may:

- require that the firearm be exported from Canada; or
- detain the firearm and give the visitor a reasonable amount of time to comply with requirements.

6. Fee

The fee for a 60-day Firearms Declaration will be $50 in Canadian funds. This amount will be payable only **once every 12 months.** If, within the 12-month period, the Firearms Declaration is renewed, or a new declaration is issued on re-entry to Canada, the fee will be waived. The $50 fee covers all firearms imported by the same individual. There is no fee for the Authorization to Transport restricted firearms.

c. Borrowing Firearms while in Canada

On April 1, 1999, a temporary license authorizing non-residents to borrow firearms will be introduced. The Non-Resident's Sixty-Day Possession Licence will allow a holder to borrow **non-restricted** firearms (ordinary rifles and shotguns) in Canada. Persons who are **18 years of age or over** may get a license to borrow firearms for the following purposes:

- hunting under the supervision of an outfitter or other person authorized to provide organized hunting services in Canada;
- hunting with a Canadian resident who has the proper firearms license and hunting license;
- competing in a shooting competition;
- target practice at an approved shooting club or range;
- participating in a historical re-enactment or display;
- engaging in a business or scientific activity being carried on in a remote area where firearms are necessary for the control of predators;
- participating in a parade, pageant or other similar event; or
- using firearms for movie, television, video or theatrical productions or publishing activities.

Visitors are advised to apply for this temporary license well in advance of arrival at the Canadian border. License application forms are available from the Canadian Firearms Centre.

Fee: The fee for a Non-Resident's Sixty-Day Possession Licence for borrowed firearms will be $30 in Canadian funds. This license may be renewed once in a 12-month period at no extra charge. Subsequent renewals cost $30.

d. Storing, Transporting and Handling Firearms in Canada

Canadian federal law requires that firearms be transported unloaded. Restricted firearms must also be rendered inoperable with a secure locking device and locked inside an opaque container that cannot be readily broken open or into or accidentally opened during transportation. When left in an unattended vehicle, the firearms must be in a locked trunk or a similar compartment. Where the vehicle does not have a trunk or similar compartment, the firearms must be locked inside the vehicle and out of sight. Provincial and municipal regulations may require a non-restricted firearm to be transported in a case in certain areas and at certain times.

When stored, firearms must be unloaded and rendered incapable of being fired, by using a secure locking device, by removing the bolt or bolt carrier or by storing the firearm in a sturdy, secure, locked container or room. Restricted firearms must be stored in a sturdy, secure, locked container or room **and** rendered inoperable with a secure locking device. Ammunition must be stored separately, unless it is in a securely locked container (ammunition and firearms may be stored in the same secure, locked container).

More information on requirements for the safe storage, transportation and handling of firearms in Canada is available from the Canadian Firearms Centre.

e. Canadian Firearms Centre

The Canadian Firearms Centre is an agency of the Government of Canada responsible for the implementation of the **Firearms Act**. For more information or to order a copy of the **Firearms Act** and regulations, or other Canadian Firearms Centre publications, contact:

1-800-731-4000 (Toll Free in the United States and Canada)
Web site: www.cfc-ccaf.gc.ca
e-mail: canadian.firearms@justice.x400.gc.ca
Mail: Canadian Firearms Centre
284 Wellington Street
Ottawa, Ontario, Canada
K1A 0H8

This article is intended to provide general information only. For legal references, refer to the Firearms Act and its regulations. Provincial, territorial and municipal laws, regulations, and policies may also apply.

8. OPERATIONS BY FEDERALLY LICENSED FIREARMS COLLECTORS

a. Licensing

A collector of curios or relics may obtain a collector's license under the GCA. The privileges conferred by this license extend only to curio or relic transactions, as discussed in detail below. In transactions involving firearms not classified as curios or relics, the licensed collector has the same status as a nonlicensee. A person need not be federally licensed to collect curios or relics. However, the individual must be licensed in order to lawfully receive curios or relics from outside his or her State of residence. Federal law, regulations, and general information pertaining to licensed collectors and curios or relics can be found in this publication.

Recordkeeping requirements for licensed collectors are discussed in detail in 27 CFR Part 178.

b. What are Curios or Relics?

As set out in the regulations (27 CFR 178.11), curios or relics include firearms which have special value to collectors because they possess some qualities not ordinarily associated with firearms intended for sporting use or as offensive or defensive weapons.

Please note that ammunition is no longer classified as curios or relics since the Congress in 1986 removed the interstate controls over ammunition under the GCA.

To be recognized as curios or relics, firearms must:

1. Have been manufactured at least 50 years prior to the current date, but not including replicas thereof; or

2. Be certified by the curator of a municipal, State or Federal museum which exhibits firearms to be curios or relics of museum interest; or

3. Derive a substantial part of their monetary value from the fact that they are novel, rare, or bizarre, or from the fact of their association with some historical figure, period, or event.

Collectors wishing to obtain a determination whether a particular firearm qualifies for classification as a curio or relic in accordance with 27 CFR 178.26 should submit a written request for a ruling. The letter should include:

(1) A complete physical description of the item;

(2) Reasons the collector believes the item merits the classification;

(3) Data concerning the history of the item, including production figures, if available, and market value.

In some cases, actual submission of the firearm may be required prior to a determination being made. Requests should be sent to the Bureau of ATF, Firearms Technology Branch, Washington, DC 20226.

ATF's classifications of curios or relics are published in ATF P 5300.12, Firearms Curios or Relics List. Curios or relics are listed in the publication under the following headings:

Section I. Ammunition Classified as Curios or Relics: As noted above, Congress ended the recognition of ammunition curios or relics. Thus, no ammunition has received curio or relic classification since August 1986.

Section II. Firearms Classified as Curios or Relics Under the GCA: Licensed collectors may acquire, hold or dispose of these firearms as curios or relics. However, they are still firearms as defined in 18 U.S.C. 921(a)(3) and are, therefore, subject to all GCA controls. Generally, this category includes commemorative handguns, semiautomatic pistols, revolvers and rifles.

Section III. NFA Firearms Removed From the NFA as Collectors' Items and Classified as Curios or Relics Under the GCA: Weapons in this section are excluded entirely from the provisions of the NFA. Thus, approval from ATF to transfer these weapons is not required. They need not be registered in the National Firearms Registration and Transfer Record and they are not subject to the transfer tax. These weapons are still firearms under the GCA and remain subject to regulation under 27 CFR Part 178.

Section IV. NFA Firearms Classified as Curios or Relics Under the GCA: These weapons (e.g., machine guns) are firearms within the scope of the NFA and are subject to all the Act's provisions. Accordingly, these weapons cannot be lawfully transferred or received unless they are registered with ATF in the National Firearms Registration and Transfer Record.

c. Licensed Collector's Activities

Subject to other applicable provisions of the law and regulations, a collector's license entitles its holder to transport, ship, receive and acquire curios or relics in interstate or foreign commerce and to dispose of curios or relics in interstate or foreign commerce to any other Federal firearms licensee. Dispositions of curios or relics by licensed collectors are not subject to the requirements of the Brady law; however, dispositions should not be made to any person whom the collector knows or has reasonable cause to believe is a felon or is within any other category of persons to whom sales are prohibited by 18 U.S.C. 922(d).

However, ATF has recognized only assembled firearms as curios or relics. Moreover, ATF's classification of surplus military firearms as curios or relics has extended only to those firearms in their original military configuration.

Frames or receivers of curios or relics are not generally recognized as curios or relics by ATF since they are not of special interest or value as collectors' items. Specifically, they do not meet the definition of curio or relic in 27 CFR 178.11 as firearms of special interest to collectors by reason of a quality other than is ordinarily associated with sporting firearms or offensive or defensive weapons.

Those collectors having questions concerning the importability of specific curio or relic firearms should contact the Bureau of ATF, Firearms and Explosives Imports Branch, Washington, DC 20226.

The principal advantage of a collector's license is that the collector can lawfully acquire curios or relics from both licensees and nonlicensees without regard to his/her State of residence. A licensed collector may acquire and dispose of curios or relics at any location, the only limitation being that a disposition made to a nonlicensee is to be made to a resident of the same State in which the collector is licensed.

d. Restrictions on Licensed Collector's Activities

As stated earlier, the collector's license covers only transactions in curios or relics. A licensed collector has the same status as a nonlicensee with respect to transactions in firearms that are not curios or relics.

While a licensed collector may acquire curios or relics and dispose of same from a personal collection, the collector is not authorized to engage in a firearms dealing business in curios or relics pursuant to a collector's license. As stated in 27 CFR 178.41(d), "...if the acquisition and disposition of curios or relics by a collector bring the collector within the definition of a manufacturer, importer or dealer under this part, he shall qualify as such." For example, if a collector acquires curios or relics for the purpose of sale rather than to enhance a collection, the collector would have to be licensed as a dealer in firearms under the GCA. Additionally, if the collector is dealing in NFA firearms, the collector would be liable for the special (occupational) tax prescribed by the NFA. **The sole intent and purpose of the collector's license is to enable a firearms collector to obtain a curio or relic from outside his or her State of residence.**

9. ANTIQUE FIREARMS UNDER THE GCA, NFA AND THE ARMS EXPORT CONTROL ACT

Under section 921(a)(16) of the GCA, the term **antique firearm** means:

"**(A)** any firearm (including any firearm with a matchlock, flintlock, percussion cap, or similar type of ignition system) manufactured in or before 1898; and

(B) any replica of any firearm described in subparagraph (A) if such replica—

(i) is not designed or redesigned for using rimfire or conventional centerfire fixed ammunition, or

(ii) uses rimfire or conventional centerfire fixed ammunition which is no longer manufactured in the United States and which is not readily available in the ordinary channels of commercial trade; or

(C) any muzzle loading rifle, muzzle loading shotgun, or muzzle loading pistol, which is designed to use black powder, or a black powder substitute, and which cannot use fixed ammunition."

For purposes of subparagraph (C), a muzzle loading rifle, shotgun, or pistol is not an "antique firearm" for purposes of the GCA if it incorporates a firearm frame or receiver, if it is a firearm which is converted into a muzzle loading weapon, or if it can be readily converted to fire fixed ammunition by replacing the barrel, bolt, breechblock, or any combination of such parts.

Under section 5845(g) of the NFA, **antique firearm** means:

"...Any firearm not designed or redesigned for using rim fire or conventional center fire ignition with fixed ammunition and manufactured in or before 1898 (including any matchlock, flintlock, percussion cap, or similar type of ignition system or replica thereof, whether actually manufactured before or after 1898) and also any firearm using fixed ammunition manufactured in or before 1898, for which ammunition is no longer manufactured in the United States and is not readily available in the ordinary channels of commercial trade."

To illustrate the distinction between the two definitions of antique firearm under the GCA and NFA, a rifle manufactured in or before 1898 would not come under the provisions of the GCA, even though it uses conventional ammunition. However, if such rifle has a barrel of less than 16 inches in length AND uses conventional fixed ammunition which is available in the ordinary channels of commercial trade, it would still be a **firearm** subject to the provisions of the NFA.

An antique firearm as defined in **both** the GCA and NFA is exempt from all of the provisions and restrictions contained in both laws. Consequently, such an antique firearm may be bought, sold, transported, shipped, etc., without regard to the requirements of these laws.

Under the Arms Export Control Act certain "antique firearms" are not subject to the import controls under that Act. These "antique firearms" are defined as "muzzle loading (black powder) firearms (including any firearm with a matchlock, flintlock, percussion cap, or similar type of ignition system) or firearms covered by Category I(a) established to have been manufactured in or before 1898." No all-inclusive list of antique firearms is published by ATF.

10. IMPORTATION BY NONLICENSEES

a. Nonlicensed U.S. Residents

A permit must be obtained to import or bring into the United States any firearm or ammunition. The firearm or ammunition must be generally recognized as particularly suitable for, or readily adaptable to, sporting purposes.

Surplus military firearms are generally excluded from importation into the United States except for certain curio or relic surplus military firearms imported by licensed importers only.

A federally licensed firearms dealer located in the nonlicensee's State of residence may act as an agent to import the nonlicensee's personal firearm, provided that the firearm is lawfully importable. The form to be used by the licensee is ATF Form 6, Part I, Application and Permit For Importation of Firearms, Ammunition and Implements of War, and may be obtained from the Bureau of ATF, Firearms and Explosives Imports Branch, Washington, DC 20226.

A nonlicensee may obtain a permit to import sporting ammunition for personal use (excluding armor piercing handgun ammunition, tracer, incendiary, or large capacity ammunition feeding devices), and firearm parts (other than frames, receivers, or actions) without engaging the services of a Federal firearms licensee. Silencer parts and certain machine gun parts are subject to the NFA and may not be imported. If the nonlicensee chooses to have a licensee handle the importation, the licensee should complete and send to ATF an ATF Form 6, Part I, in accordance with the instructions on the form. The nonlicensee's name, address, and telephone number should appear in Item 9, "Specific purpose of importation."

No permit or authorization from ATF is required to bring into the United States a firearm or ammunition that was **previously taken out** of the U.S. by the person bringing it in. The U.S. Customs Service is authorized to release a firearm or ammunition without a permit from ATF upon a proper showing of proof that the firearm or ammunition was taken out of the country by the person bringing it in. This proof is best established by having registered the item or items with U.S. Customs on Customs Form 4457, Certificate of Registration, at the point and time of departure.

For further information, see ATF Rul. 81-3 and ATF Rul. 85-10, set out within the Rulings, Procedures, and Industry Circulars portion of this publication.

b. Nonresident U.S. Citizens Returning to the United States and Nonresident Aliens Immigrating to the United States

A nonresident U.S. citizen returning to the United States after having resided outside of the United States, or a nonresident alien lawfully immigrating to the United States, may apply for a permit from ATF to import for personal use, not for resale, firearms and ammunition acquired outside of the United States without having to utilize the services of a federally licensed firearms dealer. The application on ATF Form 6, Part I should include a statement, on the application form or on an attached sheet, that:

(1) the applicant is a nonresident U.S. citizen who is returning to the United States from a residence outside of the United States or, in the case of an alien, is lawfully immigrating to the United States from a residence outside of the United States, and

(2) the firearms and ammunition are being imported for personal use and not for resale.

No permit will be issued to import firearms or ammunition which are not generally recognized as particularly suitable for, or readily adaptable to, sporting purposes, surplus military firearms, or National Firearms Act (NFA) firearms (e.g., machine guns, silencers, destructive devices, short-barreled rifles, short-barreled shotguns, etc.).

The firearms must accompany such persons, since once a person is in the United States and has acquired residence in a State, he or she may import a firearm only by arranging for the importation through a federally licensed firearms dealer.

Section 922(a)(3) of Title 18, U.S.C. makes it unlawful, with certain exceptions, for a person to bring into his State of residence a firearm which he or she acquired outside that State. An unlicensed resident of a State must, therefore, arrange for the importation of the firearm through a federally licensed firearms dealer.

The definition of "State of residence" in 27 CFR 178.11 provides that the State in which an individual is present with the intention of making a home in that State is the State of residence of that person. U.S. citizens who reside outside of the United States are not residents of a State while so residing. A person lawfully immigrating to the United States is not a resident of a State unless he or she is residing and has resided in a State for a period of at least 90 days.

c. Members of the Armed Forces

(1) Import Permit Requirements

Section 925(a)(4) of the GCA provides that:

"When established to the satisfaction of the Secretary to be consistent with the provisions of this chapter [the GCA] and other applicable Federal and State laws and published ordinances, the Secretary may authorize the transportation, shipment, receipt or importation into the United States to the place of residence of any member of the United States Armed Forces who is on active duty outside the United States (or who has been on active duty outside the United States within the 60 day period immediately preceding the transportation, shipment, receipt, or importation), of any firearm or ammunition which is:

(A) determined by the Secretary to be generally recognized as particularly suitable for sporting purposes, or determined by the Department of Defense to be a type of firearm normally classified as a war souvenir, and

(B) intended for the personal use of such member."

Applications to import firearms are filed on ATF Form 6, Part II and should include a detailed description of each firearm to be imported. Incomplete information will cause return of your application. Applications should be completed in triplicate and mailed to the Bureau of ATF, Firearms and Explosives Imports Branch, Washington, DC 20226.

A member of the Armed Forces who does not meet the above criteria must obtain the services of a Federal firearms licensee located in his or her State of residence to import a firearm on behalf of the member. The licensee would submit an application on ATF Form 6, Part I.

A permit must be obtained for all firearms to be imported, regardless of the date purchased. However, this does not apply to a firearm previously taken out of the United States by the person bringing it in, nor to a firearm shipped by a licensee in the United States to a serviceman on active duty outside the United States or to an authorized rod and gun club abroad specifically for the serviceman importing the firearm.

The U.S. Customs Service is authorized to release a firearm without a permit from ATF upon presentation of proof to Customs that the firearm was taken out of the United States by the person bringing it in, or shipped from the United States to the serviceman under one of the aforementioned conditions.

If your application is approved, the original will be returned to you. This will be your authorization to import the firearm(s) described on the form. The permit is valid for 1 year from the date of approval. If disapproved, your application will be stamped **disapproved** and returned to you with the reason for disapproval stated.

Authorization will not be given to import a machine gun, or any other firearm as defined in the NFA, regardless of the degree of serviceability.

Authorization will not be given to import any surplus military firearm (unless it has been listed as a curio or relic in accordance with 18 U.S.C. 921(a)(13) and 925(e)) or any firearm not recognized as particularly suitable for sporting purposes. However, such firearm may be brought into the United States without a permit, provided the person bringing it in can establish to the satisfaction of U.S. Customs officials that he or she previously owned the firearm in the United States and is now returning it to the United States.

To determine whether or not a handgun may be authorized for importation **as particularly suitable for sporting purposes**, the "factoring" criteria for pistols and revolvers (ATF Form 4590) is used.

(2) Importation of War Souvenirs or War Trophy Firearms

The regulations (27 CFR 178.114(c)) provide that firearms determined by the Department of Defense to be war souvenirs may be imported into the United States by members of the U.S. Armed Forces under such provisions and procedures as the Department of Defense may issue.

For information regarding the classification of war souvenirs or trophies by the Department of Defense, see DOD regulations AR 608-4, OPNAVINST 3460.7A, AFR 125-13, and MCO 5800.6A, describing articles and material that are not considered war trophies and may not be kept or imported into the United States by members of the U.S. Armed Forces.

The aforementioned Department of Defense regulations list machine guns and other firearms coming within the purview of the NFA, regardless of the degree of serviceability, among the items which are prohibited from being retained and introduced into the United States by Armed Forces personnel.

The Customs Service is authorized to release a firearm without an import permit from ATF where a properly executed DD Form 603, **Registration of War Trophy Firearms**, is presented certifying that the firearm to be brought in has been classified as a war souvenir under DOD regulations. To be valid, the DOD Form 603 must have been issued during a period authorized by DOD.

(3) Revenue Ruling 69-309

The requirement that an import permit be obtained for each firearm to be imported was relaxed somewhat by Rev. Rul. 69-309. This ruling allows members of the U.S. Armed Forces, under specified conditions, to import up to 3 rifles or shotguns, excluding surplus military, and up to 1,000 rounds of ammunition without obtaining an import permit. The waiver provided by this ruling does not include handguns. A permit for each handgun to be imported must be obtained.

[Rev. Rul. 69-309 is set out within the Rulings, Procedures, and Industry Circulars portion of this publication. A sample copy of the certification required by the ruling is set forth below.]

Sample of Certification under
Revenue Ruling 69-309
(Attaches to Form 6A)

Under the penalties provided by law, I hereby declare that I now am, or have been, on active duty outside the United States within 60 days immediately preceding this importation; that I am returning to the United States from a permanent overseas duty station; that the transportation to, and the receipt and possession by, me at my place of residence or new permanent duty station located at [City]_____ [State]_____of the firearm(s) and/or ammunition described on the attached ATF Form 6A would not constitute any violation of the Gun Control Act (18 U.S.C. Chapter 44) or of Section 38 of the Arms Export Control Act of 1976 (22 U.S.C. 2778), or of any applicable State law, or of any published, local ordinance.

_____	_____	_____
Signature	Date	Rank
_____	_____	_____
Date of Birth	Branch/Service	Serial Number

(4) ATF Ruling 74-13

ATF was informed by State and local authorities that handguns were transported, shipped, received, or imported into the United States by members of the U.S. Armed Forces to their State of residence without such members having obtained the required permits or other authorizations required by the State for lawful possession or ownership of handguns in that State.

The ruling holds that a member of the U.S. Armed Forces who is a resident of any State or territory which requires that a permit or other authorization be issued prior to possessing or owning a handgun shall submit evidence of compliance with State law before an application to import a handgun may be approved.

[ATF Rul. 74-13 is set out within the Rulings, Procedures, and Industry Circulars portion of this publication.]

11. SPECIAL TAXPAYERS AND NFA FIREARMS

a. General

Anyone wishing to manufacture, import, or deal in firearms as defined in the NFA must:

1. Be properly licensed as a Federal firearms licensee;

2. Have an employer identification number (even if you have no employees); and

3. Pay the Special (Occupational) Tax required of those manufacturing, importing, or dealing in NFA firearms.

Those weapons defined as NFA firearms can be found in sections 5845(a)-(f) of the NFA.

After payment of the tax, you will receive a Special (Occupational) Tax Stamp as evidence you have paid the required occupational tax as a NFA manufacturer, importer, or dealer.

b. What You Need To Proceed

If you do not already have an employer identification number (EIN), you must obtain and complete a Form SS-4 application to obtain such a number. This number must appear on all registration documents when you apply to receive or transfer any NFA firearm. You may obtain the Form SS4 from any Social Security Administration Office, any IRS Service Center, or IRS District Office.

Federal firearms licensees who wish to engage in business of importing, manufacturing, or dealing in NFA firearms are required to pay Special (Occupational) Tax for each business location. The tax year begins July 1st and ends June 30th of the following year. If you begin business any time during the tax year, you are responsible for the full amount of tax for the entire year, _i.e._, the taxes are not prorated.

CLASS OF SPECIAL TAX	ACTIVITY COVERED	ANNUAL TAX AMOUNT	TYPE OF FIREARMS LICENSE
Class 1	Importer	$1,000*	Type 08 or 11
Class 2	Manufacturer	$1,000*	Type 07 or 10
Class 3	Dealer	$ 500	Type 01

* =If your gross receipts for the prior Fiscal Year were less than $500,000, the tax is $500.

If you want to be a Class 3 dealer, you could have a Type 01, Type 02, Type 07, or Type 08 Federal firearms license. The tax you would pay (Class 3) allows you to deal, only, in NFA firearms. Being a Class 3 dealer will not, however, have any effect on your business activity involving non-NFA firearms.

Submit ATF Form 5630.7, Annual Special Tax Registration and Return, along with your check or money order [not cash] to:

BUREAU OF ALCOHOL, TOBACCO AND FIREARMS
P.O. BOX 371970
PITTSBURGH, PENNSYLVANIA 15250-7970

Upon receipt of your properly completed ATF Form 5630.7, together with your remittance, a Special Tax Stamp will be mailed directly to you.

c. Permanent Change of Address, Location or Name

If you change your address, location, or trade name, you must file a new

ATF Form 5630.7 advising us of that change. You may accomplish this easily by attaching ATF F 5630.7 to your Special Tax Stamp and mailing them to the Pittsburgh address shown.

To change your trade name, you must also obtain an amended Federal firearms license. This is done by sending a copy of your license (with the changes noted thereon) to the Firearms and Explosives Licensing Center, P.O. Box 2994, Atlanta, Georgia 30301.

To change your location, you must file an Application for an Amended Federal Firearms License, ATF Form 5300.38, with the Chief, Firearms and Explosives Licensing Center, not less than 30 days prior to the move. You must obtain the amended license before commencing business at the new location.

We suggest that you contact the NFA Branch at (202) 927-8330 if there is a change in who controls the business or in business structure.

d. Applications to Transfer or Make NFA Firearms

All applications to transfer or make NFA firearms must be submitted in duplicate, with both copies bearing original signatures. Extra care in ensuring that the applications are completed accurately will expedite the flow of your paperwork. Particular attention should be given to the serial number of the weapon to ensure that it does not have suffixes or prefixes.

With regard to transfers going to individuals, please ensure that the law enforcement certification is signed by someone acceptable to sign, and that the certifying officer does, in fact, have jurisdiction where the transferee resides. See also Question **M19** on acceptable certifying officials.

All ATF Form 4 applications must be accompanied by 2 properly completed sets of fingerprint cards (FBI Form FD-258). Fingerprint classification can take as long as 4-8 weeks. In some cases (particularly when fingerprints have not been properly taken), fingerprint classification can, and does, take several months. For your Form 4 applications to be expeditiously acted upon, it is imperative that the fingerprint cards you submit be complete in all respects.

All applications for taxpaid making or transfer (ATF Forms 1 and ATF Forms 4) should be forwarded, together with proper remittance, to the following address:

BUREAU OF ATF
P.O. BOX 73201
CHICAGO, ILLINOIS 60673

All other applications and correspondence should be forwarded to the National Firearms Act Branch at the address appearing at the end of this item.

e. Machine guns

Machine guns produced, imported, or registered after May 19, 1986, the effective date of 18 U.S.C. 922(o), are restricted for use by a government agency or for exportation. We will allow Class 3 dealers to receive necessary "sales samples" of these firearms if they obtain a letter from a local law enforcement agency, on the agency's letterhead, indicating a **bona fide** need to see the weapon.

If we, through an error in processing, fail to note on your transfer document(s) that certain weapons are restricted, such error will not exempt you from complying with the restrictions of 18 U.S.C. 922(o).

f. Forms

Forms you may need in the conduct of your business (but not bound books which are privately sold) are available from:

ATF DISTRIBUTION CENTER
PO BOX 5950
SPRINGFIELD, VIRGINIA 22150-5950
(703) 455-7801

g. Going Out Of Business

(1) NFA Activities Only

Machine guns. If you, as a Special (Occupational) Taxpayer, decide not to renew your payment of the special tax, all machine guns which you possess that are restricted under 18 U.S.C. 922(o) must be transferred to a qualified Special (Occupational) Taxpayer having a legitimate need for the weapon(s), to a government agency, or be exported. Such transfer must occur before you allow your Federal firearms license and special tax status to expire. Otherwise, these firearms must be abandoned to ATF or be subject to seizure.

When you, as a Special (Occupational) Taxpayer, go out of business as a manufacturer, importer, or dealer in NFA firearms, you may, **if you are a sole proprietor**, retain those machine guns imported or manufactured prior to May 19, 1986, the effective date of 18 U.S.C. 922(o). You should record your possession of the weapon(s) as an individual in your acquisition/disposition book. Machine guns imported under 26 U.S.C. 5844 for use as samples may only be transferred to government agencies or to a Federal firearms licensee who has paid the special (occupational) tax to manufacture, import, or deal in NFA firearms. **This provision allowing retention of machine guns imported or manufactured prior to May 19, 1986, applies to corporations or partnerships only if the weapons remain the property of the corporation or partnership after the special (occupational) tax stamp expires.** If the machine guns are distributed to others by the corporation or partnership when business as a special taxpayer ceases (for example, to an individual officer, stockholder, or partner), the transaction will be treated as a transfer subject to the transfer, registration, and tax provisions of the NFA.

Other NFA weapons. When you, as a Special (Occupational) Taxpayer, go out of business as a manufacturer, importer, or dealer in NFA firearms and have on hand firearms other than machine guns, you may, **if you are a sole proprietor**, retain those firearms. You should record your possession of the weapon(s) as an individual in your acquisition/disposition book. **This provision allowing retention of firearms applies to corporations or partnerships only if the weapons remain the property of the corporation or partnership after the special (occupational) tax stamp expires.** If the firearms are distributed to others by the corporation or partnership when business as a special taxpayer ceases (for example, to an individual officer, stockholder, or partner), the transaction will be treated as a transfer subject to the transfer, registration, and tax provisions of the NFA. Any NFA firearms retained that were imported under 26 U.S.C. 5844 for use as samples may only be transferred to government agencies or to a Federal firearms licensee who has paid the special (occupational) tax to manufacture, import, or deal in NFA firearms.

CAUTION

The mere possession of a license and a special (occupational) tax stamp as a dealer in NFA firearms does not qualify a person to receive firearms free of transfer tax. Where a person who possesses a license and tax stamp is not actually engaged in the business of selling NFA firearms and receives such firearms by a tax-free transfer on ATF Form 3, an unlawful transfer in violation of the NFA has occurred because the transfer tax was not paid. In such cases, the firearms involved are subject to seizure and forfeiture. (Also, see ATF Rul. 76-22, set out within the Rulings, Procedures, and Industry Circulars portion of this publication.)

Should your special tax status lapse, your continued possession of certain firearms may place you in violation of various State laws and local ordinances. We urge you to carefully consider the consequences of possessing NFA firearms in your particular city, county, and State without being a Special (Occupational) Taxpayer.

(2) Disposition of Records

If someone is taking over the business, the licensee will underline the final entry in each bound book, note the date of transfer, and deliver all records and forms kept by the licensee to the successor (who must apply for and receive his own license before lawfully engaging in business) or deliver the records and forms to the ATF Out-of-Business Records Center. If there is no business successor, within 30 days of business discontinuance the licensee must ship the required records and forms to the ATF Out-of-Business Records Center, 2029 Stonewall Jackson Drive, Falling Waters, WV 25419.

h. Questions or Problems

In the event you have any inquiries relating to your NFA business activity, please contact:

BUREAU OF ALCOHOL, TOBACCO AND FIREARMS
NATIONAL FIREARMS ACT BRANCH
WASHINGTON, DC 20226
(202) 927-8330

12. MOVING REGISTERED NFA FIREARMS INTERSTATE

A person who desires to transport a machine gun, short-barreled rifle, short-barreled shotgun, or destructive device interstate must first apply to ATF for permission to do so. ATF Form 5320.20, Application to Transport Interstate or to Temporarily Export Certain National Firearms Act (NFA) Firearms, can be used for this purpose. Only after the person receives ATF approval can the firearm be taken into another State, even for a short period of time. This requirement does not apply to a licensee qualified under the National Firearms Act to engage in business with respect to the weapon or device to be transported or to a licensed collector if the device or weapon to be transported is a curio or relic.

Alternatively, the lawful owner of the firearm may write a letter, in duplicate, giving:

   a. A complete description and identification of the device or weapon to be transported;

   b. A statement whether such transportation involves a transfer of title;

   c. The need for such transportation;

   d. The approximate date such transportation is to take place;

e. The present location of such device or weapon, the place to which it is to be transported, and the transportation to be used (including, if by common carrier, the name and address of the carrier); and

f. Evidence that the transportation or possession of such device or weapon is not inconsistent with the laws at the place of destination.

An application will not be approved if possession of the firearm at the place of destination would place the possessor in violation of State or local law.

If you have any questions regarding this subject, please contact:

BUREAU OF ALCOHOL, TOBACCO AND FIREARMS
NATIONAL FIREARMS ACT BRANCH
WASHINGTON, DC 20226
(202) 927-8330

13. LISTS OF LICENSEES/PERMITTEES

Current lists of Federal firearms licensees and Federal explosives licensees and permittees are available. Prices are quoted on request.

For a copy of the order form, contact:

BUREAU OF ALCOHOL, TOBACCO AND FIREARMS
DISCLOSURE BRANCH
WASHINGTON, DC 20226
(202) 927-8480

We must advise, however, that Federal law generally prohibits the disclosure (either affirmatively or negatively) of information concerning the registration of NFA firearms. Therefore, please do not ask us for the names of persons engaged in business with respect to machine guns or other NFA firearms.

14. IDENTIFYING FIREARMS

To insure that firearms are properly identified, ATF wishes to remind licensees that it is their responsibility to ensure that firearms are properly identified in accordance with the law. The markings on the firearms are vital to our tracing program. Violations of the law and regulations may result in criminal and administrative action, including license revocation, denial of license renewal applications, and recall. For required markings see 27 CFR 178.92 and 179.104.

15. NFA FIREARMS IN DECEDENT'S ESTATES

Possession of an NFA firearm not registered to the possessor is a violation of Federal Law and the firearm is subject to seizure and forfeiture. However, a reasonable time is allowed for transfer of lawfully registered firearms in a decedent's estate.

It is the responsibility of the executor or the administrator of an estate to transfer firearms registered to a decedent. ATF Form 5, Application for Tax Exempt Transfer and Registration of a Firearm, is used to apply for a tax-exempt transfer to a lawful heir. A lawful heir is anyone named in the decedent's will or, in the absence of a will, anyone entitled to inherit under the laws of the State in which the decedent last resided. NFA firearms may be transferred directly interstate to a beneficiary of the estate. However, if any Federal, State or local law prohibits the heir from receiving or possessing the firearm, ATF will not approve the application. When a firearm is being transferred to an individual heir, his or her fingerprints on FBI Forms FD-258 must accompany the transfer application.

ATF Form 4 is used to apply for the taxpaid transfer of a serviceable NFA firearm to a person outside the estate (not a beneficiary). ATF Form 5 is also used to apply for the tax-exempt transfer of an unserviceable NFA firearm to a person outside the estate. As noted above, all requirements, such as fingerprint cards for transfers to individuals and compliance with State or local law, must be met before an application could be approved.

If the NFA firearm in the estate was imported for use as a "sales sample," this restriction on the firearm's possession remains. The NFA firearm may only be transferred to a Federal firearms licensee who has paid the special (occupational) tax to deal in NFA firearms or to a government agency.

For further information, contact:

BUREAU OF ALCOHOL, TOBACCO AND FIREARMS
NATIONAL FIREARMS ACT BRANCH
WASHINGTON, DC 20226
(202) 927-8330

16. "STRAW PURCHASES"

Questions have arisen concerning the lawfulness of firearms purchases from licensees by persons who use a "straw purchaser" (another person) to acquire the firearms. Specifically, the actual buyer uses the straw purchaser to execute the Form 4473 purporting to show that the straw purchaser is the actual purchaser of the firearm. In some instances, a straw purchaser is used because the actual purchaser is prohibited from acquiring the firearm. That is to say, the actual purchaser is a felon or is within one of the other prohibited categories of persons who may not lawfully acquire firearms or is a resident of a State other than that in which the licensee's business premises is located. Because of his or her disability, the person uses a straw purchaser who is not prohibited from purchasing a firearm from the licensee. In other instances, neither the straw purchaser nor the actual purchaser is prohibited from acquiring the firearm.

In both instances, the straw purchaser violates Federal law by making false statements on Form 4473 to the licensee with respect to the identity of the actual purchaser of the firearm, as well as the actual purchaser's residence address and date of birth. The actual purchaser who utilized the straw purchaser to acquire a firearm has unlawfully aided and abetted or caused the making of the false statements. The licensee selling the firearm under these circumstances also violates Federal law if the licensee is aware of the false statements on the form. It is immaterial that the actual purchaser and the straw purchaser are residents of the State in which the licensee's business premises is located, are not prohibited from receiving or possessing firearms, and could have lawfully purchased firearms from the licensee.

An example of an illegal straw purchase is as follows: Mr. Smith asks Mr. Jones to purchase a firearm for Mr. Smith. Mr. Smith gives Mr. Jones the money for the firearm. If Mr. Jones fills out Form 4473, he violates the law by falsely stating that he is the actual buyer of the firearm. Mr. Smith also violates the law because he has unlawfully aided and abetted or caused the making of false statements on the form.

Where a person purchases a firearm with the intent of making a gift of the firearm to another person, the person making the purchase is indeed the true purchaser. There is no straw purchaser in these instances. In the above example, if Mr. Jones had bought a firearm with his own money to give to Mr. Smith as a birthday present, Mr. Jones could lawfully have completed Form 4473. The use of gift certificates would also not fall within the category of straw purchases. The person redeeming the gift certificate would be the actual purchaser of the firearm and would be properly reflected as such in the dealer's records.

17. FEDERAL EXCISE TAX ON FIREARMS AND AMMUNITION

A Federal excise tax is imposed by 26 U.S.C. 4181 on the sale by the manufacturer, importer or producer of firearms, shells, and cartridges. The tax is 10 percent of the sale price for pistols and revolvers, 11 percent of the sale price for firearms and 11 percent of the sale price for shells and cartridges.

The excise tax attaches only to the sale of complete firearms and ammunition or firearms that, although in a knockdown condition, are complete as to all component parts.

The term **firearm** for excise tax purposes includes all portable weapons, such as rifles, carbines, machine guns, shotguns, and fowling pieces from which a shot, bullet, or projectile may be discharged by an explosive. The term firearm also includes pistols and revolvers. Antique firearms are also subject to the excise tax.

Shells and cartridges include any article consisting of a projectile, explosive, and container that is designed, assembled, and ready for use without further manufacture in firearms, pistols or revolvers.

Reloading of used shells or cartridges is considered manufacturing for purposes of excise tax. Sale of such shells by the reloader is subject to the excise tax. However, if the reloader merely reloads shells belonging to a customer and is paid for labor and materials, the reloading service is not a taxable sale, as long as the reloader returns the identical shells provided by the customer to that same customer. In such instances the customer is the manufacturer and would not be liable for tax if the shells are manufactured for personal use. If the customer sells reloaded shells or uses them in a business, e.g., shooting range, the customer would be liable for the tax.

Returns and Deposits

Regulations in 27 CFR Part 53 require that taxpayers incurring a tax liability on the sale or use of firearms and ammunition file excise tax returns quarterly on ATF Form 5300.26, Federal Firearms and Ammunition Excise Tax Return. In addition, taxpayers are required to make semimonthly deposits of tax on ATF Form 5300.27.

Further Information

For more detailed information regarding firearms and ammunition excise taxes, refer to ATF Publication 5300.16, Firearms or Ammunition Manufacturers and Importers Information Booklet.

18. ARMOR PIERCING AMMUNITION

For purposes of the prohibitions imposed upon manufacture, importation, and transfer of armor piercing ammunition in 18 U.S.C. 922(a)(7)-(8) and

923(e), armor piercing ammunition includes the following:

KTW AMMUNITION, all calibers. Identified by a green coating on the projectile.

ARCANE AMMUNITION, all calibers. Identified by a pointed bronze or brass projectile.

THV AMMUNITION, all calibers. Identified by a brass or bronze projectile and a head stamp containing the letters SFM and THV.

CZECHOSLOVAKIAN manufactured 9mm Parabellum (Luger) ammunition having an iron or steel bullet core. Identified by a cupronickel jacket and a head stamp containing a triangle, star, and dates of 49, 50, 51, or 52. This bullet is attracted to a magnet.

GERMAN manufactured 9mm Parabellum (Luger) having an iron or steel bullet core. Original packaging is marked Pistolenpatronen 08 m.E. May have black colored bullet. This bullet is attracted to a magnet.

MSC AMMUNITION, caliber .25. Identified by a hollow point brass bullet. **NOTE:** MSC ammunition, caliber .25 identified by a hollowpoint copper bullet is **not** armor piercing.

BLACK STEEL ARMOR PIERCING AMMUNITION, all calibers, as produced by National Cartridge, Atlanta, Georgia.

BLACK STEEL METAL PIERCING AMMUNITION, all calibers, as produced by National Cartridge, Atlanta, Georgia.

7.62mm **NATO AP,** identified by black coloring in the bullet tip. This ammunition is used by various NATO countries. The U.S. military designation is M61 AP.

7.62mm **NATO SLAP.** Identified by projectile having a plastic sabot around a hard penetrator. The penetrator protrudes above the sabot and is similar in appearance to a Remington accelerator cartridge.

PMC ULTRAMAG, .38 Special caliber, constructed entirely of a brass type material, and a plastic pusher disc located at the base of the projectile. **NOTE:** **PMC ULTRAMAG 38J** late production made of copper with lead alloy projectile is **not** armor piercing.

OMNISHOCK. A .38 Special cartridge with a lead bullet containing a mild steel core with a flattened head resembling a wad cutter. **NOTE:** **OMNISHOCK** cartridges having a bullet with an aluminum core are **not** armor piercing.

7.62x39mm with steel core. These projectiles have a steel core. **NOTE:** Projectiles having a lead core with steel jacket or steel case are **not** armor piercing.

In addition, the Violent Crime Control and Law Enforcement Act of 1994 added to the definition of armor piercing ammunition the following:

"... a full jacketed projectile larger than .22 caliber designed and intended for use in a handgun and whose jacket has a weight of more than 25 percent of the total weight the projectile."

Exemptions: The following articles are exempted from the definition of armor piercing ammunition.

5.56 mm (.223) SS 109 and M855 Ammunition, identified by a green coating on the projectile tip.

U.S. .30-06 M2AP, identified by a black coating on the projectile tip.

19. ASSEMBLY OF NONSPORTING SEMIAUTOMATIC RIFLES AND SHOTGUNS FROM IMPORTED PARTS

Section 922(r), Title 18, U.S.C., makes it unlawful for any person to assemble from imported parts any semiautomatic rifle or any shotgun which is identical to any rifle or shotgun prohibited from importation under section 925(d)(3) of the GCA. Regulations implementing the law in 27 C.F.R. 178.39 provide that a violation of section 922(r) will result if a semiautomatic rifle or shotgun is assembled with more than 10 of the following imported parts:

(1) Frames, receivers, receiver castings, forgings, or stampings
(2) Barrels
(3) Barrel extensions
(4) Mounting blocks (trunnions)
(5) Muzzle attachments
(6) Bolts
(7) Bolt carriers
(8) Operating rods
(9) Gas pistons
(10) Trigger housings
(11) Triggers
(12) Hammers
(13) Sears
(14) Disconnectors
(15) Buttstocks
(16) Pistol grips
(17) Forearms, handguards
(18) Magazine bodies
(19) Followers
(20) Floorplates

Section 922(r) does not prohibit the importation, sale, or possession of parts which may be used to assemble a semiautomatic rifle or shotgun in violation of the statute. However, 18 U.S.C. § 2 provides that a person who aids or abets another person in the commission of an offense is also responsible for the offense. Therefore, a person who sells parts knowing that the purchaser intends to use the parts in assembling a firearm in violation of section 922(r) would also be responsible for the offense.

It should be noted that a violation of section 922(r) may result even if the assembled rifle or shotgun does **not** meet the definition of "semiautomatic assault weapon" in the GCA. For example, a person who assembled an SKS rifle with a folding stock and fixed magazine would violate section 922(r), since the SKS would have more than 10 of the imported parts specified in the regulation. However, the SKS rifle with a fixed magazine would still not be regulated as a "semiautomatic assault weapon."

20. FIREARMS THEFT/LOSS REPORTING

Federal firearms licensees are required to report the theft or loss of firearms from their inventory or collection to local authorities as well as to ATF within 48 hours after the theft or loss is discovered.

The notification to ATF must be made through our toll free theft hotline number which is **1-800-800-3855**. The hotline is operational 24 hours a day, 7 days a week. The caller should indicate that he or she is a Federal firearms licensee and be ready to furnish his or her Federal firearms license number. The hotline representative will provide the licensee with an incident number which should be recorded in the licensee's bound book.

The verbal notification must be followed up by a written notification to ATF within the same 48-hour period. The written notification should be made on ATF F 3310.11, Federal Firearms Licensee Theft/Loss Report, in accordance with the instructions on the form. This form is available from the ATF Distribution Center.

Theft or loss of NFA firearms should be reported to the NFA Branch at (202) 927-8330 immediately upon discovery.

21. IMPORTATION OF SEMIAUTOMATIC ASSAULT RIFLES
AND MODIFIED VERSIONS OF SUCH RIFLES

Section 925(d)(3), Title 18, U.S.C., states that the Secretary shall authorize a firearm or ammunition to be imported or brought into the United States if it is of a type that does not fall within the definition of a firearm in section 5845(a) of the Internal Revenue Code of 1954 and is generally recognized as particularly suitable for or readily adaptable to sporting purposes, excluding surplus military firearms.

In 1989, ATF identified a type of rifle known as a semiautomatic assault rifle and found that weapons of this type were not generally recognized as particularly suitable for or readily adaptable to sporting purposes under the sporting purposes test for importation in 18 U.S.C. 925(d). Accordingly, on July 6, 1989, ATF determined that rifles of this type were not importable into the United States. ATF's finding was based, in part, on the determination that these rifles have certain characteristics that are common to modern military assault rifles and that distinguish them from traditional sporting rifles. These characteristics include the ability to accept a detachable magazine, folding/telescoping stocks, separate pistol grips, ability to accept a bayonet, flash suppressors, bipods, grenade launchers, and night sights. It was decided that any of these military features, other than the ability to accept a detachable magazine, would make a semiautomatic assault rifle not importable. The particular rifles barred from importation were as follows:

AKS Variants

AK 47 type
AK47S type
AK/4 type
AKS type
AKM type
AKMS type
ARM type
84S type
84S1 type
84S3 type
86S type
867S type
Galil type
Type 56 type
Type 56S type
Valmet M76 type
Valmet M78 type
M76 counter sniper type

Other

AUG type
FNC type
Uzi carbine
Algimec AGM1 type
AR180 type
Australian Automatic Arms SAR type
Beretta AR70 type
Beretta BM59 type
CIS SR88 type
HK91 type
HK93 type
HK94 type
G3SA type
K1 type
K2 type
AR100 type
M14S type
MAS223 type
SIG 550SP type
SIG 551SP type
SKS with detachable magazine

FAL Variants

FAL type
L1A1A type
SAR 48

Subsequent to the 1989 decision, certain semiautomatic assault rifles that failed the 1989 sporting purposes test for importation were modified to remove all of their military features other than the ability to accept a detachable magazine. They could still accept the large capacity magazines originally designed and produced for the military assault rifles from which they were derived. The modified weapons were permitted to be imported because they met the 1989 sporting purposes test for importation.

On November 14, 1997, the President and the Secretary of the Treasury ordered a review of the importation of modified versions of semiautomatic assault rifles into the United States. Consequently, it was decided on April 6, 1998, that the following modified versions of semiautomatic assault rifles could no longer be imported under the sporting purposes test.

AK47 Variants

MAK90
314
56V
89
EXP56A
SLG74
NHM90
NHM90-2
NHM91
SA85M
SA93
A93
AKS762
SA2000
ARM
MISR
MISTR
SA85M
Mini PSL
ROMAK 1
ROMAK 2
ROMAK 4
Hunter Rifle
386S
PS/K
Galil Sporter
Haddar
Haddar II
WUM 1
WUM 2
SLR95
SLR96
SLR97
SLG94
SLG95
SLG96

FN-FAL Variants

L1A1 Sporter
FAL Sporter
FZSA
SAR4800
XFAL
C3
C3A
LAR Sporter

HK Variants

BT96
Centurian 2000
SR9
PSG1
MSG90
G3SA
SAR8

Uzi Variants

Officers 9
320 carbine
Uzi Sporter

SIG SG550 Variants

SG550-1
SG550-2

It should also be noted that on September 13, 1994, Congress passed the Violent Crime Control and Law Enforcement Act which made it unlawful, with certain exceptions, to manufacture, transfer, or possess semiautomatic assault weapons as defined by the statute (18 U.S.C. 922(v)). Therefore, any weapon banned under section 922(v) would also be denied importation into the United States because its possession would be illegal. Weapons covered by section 922(v) are defined in 18 U.S.C 921(a)(30).

22. IMPORTATION OF HANDGUNS

Section 925(d)(3), Title 18, U.S.C., states that the Secretary shall authorize a firearm or ammunition to be imported or brought into the United States if it is of a type that does not fall within the definition of a firearm in section 5845(a) of the Internal Revenue Code of 1954 and is generally recognized as particularly suitable for or readily adaptable to sporting purposes, excluding surplus military firearms. The statute also prohibits the importation of any frame, receiver, or barrel of a firearm that would be prohibited if assembled.

In 1968, the Secretary established the Treasury Department Firearms Evaluation Panel. The purpose of this panel was to assist in the establishment of guidelines for use in determining if a particular firearm would be importable under the sporting purposes test prescribed by section 925(d)(3). This panel was composed of representatives from the firearms industry, law enforcement, and the Federal Government.

The panel developed objective numerical criteria with minimum qualifying scores to determine if a handgun is particularly suitable for or readily adaptable to sporting purposes. The criteria, ATF Form 4590, Factoring Criteria for Weapons, assigns point values to handguns based on dimensions, material used in construction, weight, caliber, safety features and miscellaneous equipment. The criteria also have prerequisite requirements concerning safeties and minimum dimensions.

Form 4590 is divided into two parts. The right side of the form is used to evaluate revolvers and the left side is used to evaluate pistols. The minimum qualifying score for a revolver is 45 points and the minimum qualifying score for a pistol is 75 points. The form also provides that revolvers must pass a safety test.

Any handgun being imported into the United States must pass these criteria. The fact that a particular weapon may be of domestic manufacture or classified as a curio or relic does not exempt it from the factoring criteria.

If you have any questions concerning Form 4590, or the score a particular handgun achieves, please contact the ATF Firearms Technology Branch at (202) 927-7910.

QUESTIONS AND ANSWERS

TABLE OF CONTENTS

2. Multiple bound books
3. Bound book availability
4. Approved variations
5. Time for entries into bound book
6. Ammunition
7. Rental firearms
8. Sale of personal firearms
9. Computer records

F. CONDUCT OF BUSINESS--LICENSEES

1. Compliance with State and local laws
2. Sales to out-of-State nonlicensees
3. Out-of-State sales to law enforcement officers
4. Employees under age 21
5. Handguns--reporting multiple sales
6. Firearm purchasers--age requirements
7. Ammunition purchasers--age requirements
8. Verification--licensees' status
9. Verification--multi-licensed entity
10. Deliveries after license expiration
11. License to sell ammunition
12. Sales at gun shows
13. Sales at out-of-State gun shows
14. Mailing firearms
15. Consignment sales
16. Reporting stolen firearms
17. Stolen firearms--records
18. Distribution of multiple sale forms
19. Responses to trace requests
20. Written notification on juvenile handgun possession
21. Posting signs on juvenile handgun possession

G. COLLECTORS

1. Collector's license
2. Collector's limitation
3. Collector cannot engage in business
4. Collector's need for dealer's license
5. Form 4473
6. Brady law
7. Notification and posting signs concerning juveniles
8. Disposition of records

H. MANUFACTURERS

1. License for each type of business
2. Firearms not covered by ammunition license
3. Ammunition covered by firearms license
4. Reloaders of ammunition
5. Excise taxes

I. GUNSMITHS

1. Activity requires license--when
2. Records
3. Form 4473
4. Immediate repairs
5. NFA weapon repair
6. Notification and posting signs concerning juveniles
7. Brady law

J. PAWNBROKERS

1. Disposition records
2. Transactions
3. Prohibited categories of persons
4. Notification and posting signs concerning juveniles
5. Brady law

K. AUCTIONEERS

1. License requirements
2. Location of sales

L. IMPORTING AND EXPORTING

1. Importation by licensees other than importers
2. Exportation

M. FIREARMS--NATIONAL FIREARMS ACT (NFA)

1. Firearms requiring registration
2. Lawful acquisitions
3. Tax on making
4. Paying tax on making
5. Unserviceable NFA firearms
6. NFA firearms acquired by police
7. Unregistered NFA firearms
8. Penalties for unlawful possession
9. Disposal of unregistered firearm
10. Exemptions from making/transfer tax
11. Qualified to import/manufacture/deal
12. Payment--special (occupational) tax
13. Special tax required for each activity--exception
14. Licensed collector--NFA firearms
15. Required transfer procedures
16. Authorization to make
17. Conversion part(s) subject to registration
18. Out-of-State transfer
19. Acceptable "law enforcement certifications"
20. Absence of duty to sign "law enforcement certification"
21. Refusal to sign "law enforcement certification"
22. Approval to transfer interstate--when required
23. Disapproved application to transfer interstate--options
24. Application to transfer prior to receipt not allowed
25. Pistol and attachable shoulder stock
26. Evidence of registration required
27. Status--empty grenades, shells, etc.
28. Status--muzzle-loading cannons
29. Status--grenade and rocket launcher attachments
30. Definition of "conversion kit"

N. MACHINE GUNS--NFA

1. Manufacture by unlicensed person
2. Transfers by registered owners

O. SEMIAUTOMATIC ASSAULT WEAPONS AND LARGE CAPACITY AMMUNITION FEEDING DEVICES

1. Restrictions on semiautomatic assault weapons
2. Definition of semiautomatic assault weapon
3. Restrictions on large capacity ammunition feeding devices

4. Definition of large capacity ammunition feeding devices
5. Exceptions to prohibitions
6. Semiautomatic firearms also regulated under the NFA
7. Replacement parts
8. Sales to law enforcement officers
9. Retention by officers upon retirement
10. Assembly of semiautomatic assault weapon from previously owned parts
11. Documentation for sales to law enforcement officers
12. Stocking of semiautomatic assault weapons
13. Stocking of large capacity ammunition feeding devices
14. Markings on semiautomatic assault weapons
15. Markings on large capacity ammunition feeding devices
16. Magazines for Appendix A weapons not subject to restrictions
17. Evidence that weapon not subject to restrictions
18. Semiautomatic assault weapons classified as curios or relics
19. Modifying weapons for sale to the public

P. BRADY LAW

1. Waiting periods
2. Persons who must comply
3. Effective date of permanent Brady law
4. Operation of NICS
5. Agencies that perform NICS checks
6. States as points of contact for NICS checks
7. Licensees advised of points of contact
8. Charge for NICS checks
9. Enrolling with FBI for access to NICS
10. Firearms subject to NICS checks
11. Antique firearms
12. Transfers between licensees
13. Transfers of curios or relics by licensed collectors
14. Transfers by licensed dealers to licensed collectors
15. Licensees' sales of personal firearms
16. Loan or rental of firearms
17. Sales at gun shows
18. Redemption of pawned firearms
19. Pawnbroker's disposition of pawned firearm when transfer "denied"
20. Repeated pawn of same firearm
21. Pawnbroker's NICS checks prior to accepting firearms in pawn
22. Pawnbroker's completion of Form 4473 on optional NICS check
23. Action of pawnbroker when NICS check results in "denial"
24. Action of pawnbroker when NICS check results in "approval"
25. Return of repaired firearm; transfer of replacement firearm
26. Return of consignment firearm
27. Sales to law enforcement officers for official use
28. Sales to law enforcement officers for personal use
29. Exceptions to requirement to initiate NICS check
30. Required steps on transfer
31. Identification of purchaser
32. Non-over-the-counter sales subject to NICS
33. Non-over-the-counter sales not subject to NICS
34. Brady form
35. Form 4473
36. Social security numbers on Form 4473
37. NICS responses--"proceed", "denied", "delayed"
38. "Business day" for purposes of Brady law
39. Transfer of firearm after "delayed" NICS response
40. Action of licensee after "denied" NICS response
41. Advice by licensee to proposed transferee denied by NICS

A. GENERAL QUESTIONS

(A1) Does the law regulate who can be in the business?

Yes. The Gun Control Act (GCA), administered by the Bureau of Alcohol, Tobacco and Firearms (ATF) of the Department of the Treasury, contains Federal licensing standards for various firearms businesses (manufacturers, importers, and dealers). An example of these standards is that the applicant must have a business premises.

[18 U.S.C. 923(d), 27 CFR 178.47]

(A2) Who can get a license?

ATF will approve the application if the applicant:

·Is 21 years or more of age;

·Is not prohibited from shipping, transporting, receiving or possessing firearms or ammunition;

·Has not willfully violated the GCA or its regulations;

·Has not willfully failed to disclose material information or willfully made false statements concerning material facts in connection with his application;

·Has premises for conducting business or collecting; and,

·The applicant certifies that--

(1) the business to be conducted under the license is not prohibited by State or local law in the place where the licensed premises is located;

(2) within 30 days after the application is approved the business will comply with the requirements of State and local law applicable to the conduct of the business;

(3) the business will not be conducted under the license until the requirements of State and local law applicable to the business have been met;

(4) the applicant has sent or delivered a form to the chief law enforcement officer where the premises is located notifying the officer that the applicant intends to apply for a license; and

(5) secure gun storage or safety devices will be available at any place in which firearms are sold under the license to persons who are not licensees ("secure gun storage or safety device" is defined in 18 U.S.C. 921(a)(34)).

Editor's Note

The requirement to certify to the availability of gun storage or safety devices was enacted on October 22, 1998, by Public Law 105-277. The requirement is codified in section 923(d)(1)(G) of the GCA and becomes effective 180 days after the date of enactment.

[18 U.S.C. 923(d)(1), 27 CFR 178.47(b)]

(A3) Does the Federal Government issue a license or permit to carry a concealed weapon?

No. Neither ATF nor any other Federal agency issues such a permit or license. Carrying permits may be issued by a State or local government.

(A4) Do antique firearms come within the purview of the GCA?

No.

[18 U.S.C. 921(a)(3) and (16), 27 CFR 178.11 and 178.141(d)]

(A5) What kinds of ammunition are covered by the GCA?

Ammunition includes cartridge cases, primers, bullets or propellant powder designed for use in any firearm other than an antique firearm. Items **NOT** covered include blank ammunition, tear gas ammunition, pellets and nonmetallic shotgun hulls without primers.

Generally, no records are required for ammunition transactions. However, information about the disposition of armor piercing ammunition is required to be entered into a record by importers, manufacturers, and collectors. A license is not required for dealers in ammunition only.

[18 U.S.C. 921(a)(17) and 922(b)(5), 27 CFR 178.11]

(A6) Does the GCA control the sale of firearms parts?

No, except that frames or receivers of firearms are "firearms" as defined in the law and subject to the same controls as complete firearms. Silencer parts are also firearms under the GCA, as well as under the National Firearms Act (NFA). Certain machinegun parts, such as conversion parts or kits, are also subject to the NFA.

The GCA generally prohibits the transfer and possession of large capacity ammunition feeding devices manufactured after September 13, 1994. "Large capacity ammunition feeding devices" are those that can accept more than 10 rounds of ammunition.

[18 U.S.C. 921(a)(3), (24), and (31), 922(w), 27 CFR 178.11 and 178.40a]

(A7) Does the GCA prohibit anyone from making a handgun, shotgun or rifle?

With certain exceptions a firearm may be made by a nonlicensee provided it is not for sale and the maker is not prohibited from possessing firearms. However, a person is prohibited from making a semiautomatic assault weapon or assembling a nonsporting semiautomatic rifle or nonsporting shotgun from imported parts. In addition, the making of an NFA firearm requires a tax payment and approval by ATF. An application to make a machinegun will not be approved unless documentation is submitted showing that the firearm is being made for a Federal or State agency.

[18 U.S.C. 922(o), (r), (v), and 923, 27 CFR 178.39, 178.40, 178.41 and 179.105]

(A8) Are black powder dealers required to be licensed as an ammunition dealer under the GCA?

No. However, black powder dealers are subject to the provisions of 27 CFR Part 55, Commerce in Explosives, which requires that a dealer in any quantity of black powder must have a license as a dealer in low explosives.

B. UNLICENSED PERSONS

(B1) To whom may an unlicensed person transfer firearms under the GCA?

A person may sell a firearm to an unlicensed resident of his or her State, if the buyer is not prohibited by law from receiving or possessing a firearm, or to a licensee in any State. A firearm other than a curio or relic may not be transferred interstate to a licensed collector.

[18 U.S.C 922(a)(3) and (5), 922(b)(3), 27 CFR 178.29]

(B2) From whom may an unlicensed person acquire a firearm under the GCA?

A person may only buy a firearm within the person's own State, except that he or she may buy a rifle or shotgun, in person, at a licensee's premises in any State, provided the sale complies with State laws applicable in the State of sale and the State where the purchaser resides.

[18 U.S.C 922(a)(3) and (5), 922(b)(3), 27 CFR 178.29]

(B3) May an unlicensed person obtain a firearm from an out-of-State source if the person arranges to obtain the firearm through a licensed dealer in the purchaser's own State?

A person not licensed under the GCA and not prohibited from acquiring firearms may purchase a firearm from an out-of-State source and obtain the firearm if an arrangement is made with a licensed dealer in the purchaser's State of residence for the purchaser to obtain the firearm from the dealer.

[18 U.S.C 922(a)(3) and (5), 922(b)(3), 27 CFR 178.29]

(B4) May an unlicensed person obtain ammunition from an out-of-State source?

Yes, provided he or she is not a person prohibited from receiving firearms and ammunition.

[18 U.S.C. 922(g) and (n)]

(B5) Are there certain persons who cannot legally receive or possess firearms?

Yes, a person who −

(1) Has been convicted in any court of, a crime punishable by imprisonment for a term exceeding 1 year;

(2) Is a fugitive from justice;

(3) Is an unlawful user of or addicted to any controlled substance;

(4) Has been adjudicated as a mental defective or has been committed to a mental institution;

(5) Is an alien illegally or unlawfully in the United States or an alien admitted to the United States under a nonimmigrant visa;

(6) Has been discharged from the Armed Forces under dishonorable conditions;

(7) Having been a citizen of the United States, has renounced his or her citizenship;

(8) Is subject to a court order that restrains the person from harassing, stalking, or threatening an intimate partner or child of such intimate partner; or

(9) Has been convicted of a misdemeanor crime of domestic violence

cannot lawfully receive, possess, ship, or transport a firearm.

A person who is under indictment or information for a crime punishable by imprisonment for a term exceeding 1 year cannot lawfully receive a firearm. Such person may continue to lawfully possess firearms obtained prior to the indictment or information.

[18 U.S.C. 922(g) and (n), 27 CFR 178.32(a) and (b)]

(B6) Do law enforcement officers who are subject to restraining orders and who receive and possess firearms for purposes of carrying out their official duties violate the law?

Not if the firearms are received and possessed for official use only. The law prohibits persons subject to certain restraining orders from receiving, shipping, transporting or possessing firearms or ammunition. To be disabling, the restraining order must:

1. specifically restrain the person from harassing, stalking, or threatening an "intimate partner" of the person (e.g., spouse);

2. be issued after a hearing of which notice was given to the person and at which the person had an opportunity to participate; and

3. include a finding that the person subject to the order represents a credible threat to the "intimate partner" or child of the "intimate partner" OR explicitly prohibits the use, attempted use, or threatened use of force against the partner.

However, the GCA has an exception for the receipt and possession of firearms and ammunition on behalf of a Federal or State agency. Therefore, the GCA does not prohibit a law enforcement officer under a restraining order from receiving or possessing firearms or ammunition for use in performing official duties. Possession of the firearm off duty would be lawful if such possession is authorized by the officer's department. An officer subject to a disabling restraining order would violate the law if the officer received or possessed a firearm or ammunition for other than official use. (See Question **Q15** on officers' receipt and possession of firearms and ammunition after a conviction of a misdemeanor crime of domestic violence.)

[18 U.S.C. 922(g)(8), 925(a)(1)]

(B7) May a nonlicensee transport firearms for sporting or other lawful purposes?

Yes. Federal law provides a person, who is not prohibited by the GCA from receiving or transporting firearms, the right to transport a firearm under certain conditions, notwithstanding State or local law to the contrary. The firearms must be unloaded and in a locked trunk or, in a vehicle lacking a trunk, in a locked container other than the glove compartment or console. Also, the carrying and possession must be lawful at the place of origin and destination.

[18 U.S.C. 926A, 27 CFR 178.38]

(B8) May a nonlicensee ship a firearm through the U.S. Postal Service?

A nonlicensee may mail a shotgun or rifle to a resident of his or her own State or to a licensee in any State. Handguns are not mailable. A common or contract carrier must be used to ship a handgun. A nonlicensee may not transfer any firearm to a nonlicensed resident of another State.

The Postal Service recommends that longguns be sent by registered mail and that no marking of any kind which would indicate the nature of the contents be placed on the outside of any parcel containing firearms.

(B9) May a nonlicensee ship a firearm by carrier?

A nonlicensee may ship a firearm by carrier to a resident of his or her own State or to a licensee in any State. A common or contract carrier must be used to ship a handgun. In addition, Federal law requires that the carrier be notified that the shipment contains a firearm and prohibits common or contract carriers from requiring or causing any label to be placed on any package indicating that it contains a firearm.

[18 U.S.C. 922(a)(2)(A) and 922(e), 27 CFR 178.31]

(B10) May a nonlicensee ship firearms interstate for his or her use in hunting or other lawful activity?

Yes. A person may ship a firearm to himself or herself in care of another person in the State where he or she intends to hunt or engage in any other lawful activity. The package should be addressed to the owner. Persons other than the owner should not open the package and take possession of the firearm.

(B11) May a person who is relocating out-of-State move firearms with other household goods?

Yes. A person who lawfully possesses a firearm may transport or ship the firearm interstate when changing his or her State of residence.

Certain NFA firearms must have prior approval from the Bureau of ATF, NFA Branch, Washington, DC 20226, before they may be moved interstate. The person must notify the mover that firearms are being transported. He or she should also check State and local laws where relocating to ensure that movement of firearms into the new State does not violate any State law or local ordinance.

[18 U.S.C. 922(a)(4), 27 CFR 178.28 and 178.31]

(B12) What constitutes residency in a State?

The State of residence is the State in which an individual is present with the intention of making a home in that State. A member of the Armed Forces on active duty is a resident of the State in which his or her permanent duty station is located. If a member of the Armed Forces maintains a home in one State and the member's permanent duty station is in a nearby State to which he or she commutes each day, then the member may purchase a firearm in either the State where the duty station is located or the State where the home is maintained. An alien who is legally in the United States is considered to be a resident of a State only if the alien is residing in that State and has resided in that State continuously for a period of at least 90 days prior to the date of sale of the firearm (See also **Item 5**, "Sale of Firearms to Aliens in the United States", under Items of Interest).

[18 U.S.C. 921(b) and 922(b)(3), 27 CFR 178.11]

(B13) May a person who resides in one State and owns property in another State purchase a handgun in either State?

If a person maintains a home in 2 States and resides in both States for certain periods of the year, he or she may, during the period of time the person actually resides in a particular State, purchase a handgun in that State. But simply owning property in another State does not qualify the person to purchase a handgun in that State.

(B14) May foreign visitors and other aliens legally in the United States buy firearms?

An alien legally in the United States who has been admitted into the country under a nonimmigrant visa is generally prohibited from receiving or possessing firearms and a licensee may not lawfully transfer firearms to such alien. In addition, a foreign visitor is not a resident of a State and, therefore, may not purchase **and** take delivery of a firearm in the United States. A foreign visitor may purchase a firearm and have it exported by a licensee. The licensee must obtain an export license from the State Department for this type of transaction.

In addition, an alien legally in the United States would have a State of residence and may acquire firearms in that State only if he or she is residing in that State and has resided in that State continuously for at least 90 days prior to the purchase. Aliens acquiring firearms from licensees are required to prove their identity and residency by presenting government-issued photo identification and substantiating documentation showing that he or she has resided in the State continuously for the 90-day period, for example, utility bills, lease agreements, credit card statements, pay stubs or other documents from the purchaser's place of employment.

See also **Item 6**, "Sale of Firearms to Aliens in the United States", under Items of Interest.

[18 U.S.C. 921, 922(b)(3), (d), (g), 27 CFR 178.11, 178.99(a)]

(B15) May a parent or guardian purchase firearms or ammunition as a gift for a juvenile (less than 18 years of age)?

Yes. However, possession of handguns by juveniles (less than 18 years of age) is generally unlawful. Juveniles may only receive and possess handguns with the written permission of a parent or guardian for limited purposes, e.g., employment, ranching, farming, target practice or hunting.

[18 U.S.C. 922(x)]

(B16) Are curio or relic firearms exempt from the provisions of the GCA?

No. Curios or relics are still firearms subject to the provisions of the GCA; however, curio or relic firearms may be transferred in interstate commerce to licensed collectors or other licensees.

C. LICENSING

(C1) How does one get a license?

Submit ATF Form 7, Application for License, or ATF Form 7CR, Application for License (Collector of Curios or Relics), with the appropriate fee in accordance with the instructions on the form. These forms may be obtained from the Firearms and Explosives Licensing Center in Atlanta, Georgia or your local ATF office.

[18 U.S.C. 923, 27 CFR 178.44]

(C2) May one license cover several locations?

No. A separate license must be obtained for each location. Storage facilities are not required to be covered by a separate license. However, the records maintained on licensed premises must reflect all firearms held in the separate storage facility. Firearms may be shipped directly to separate storage facilities as long as they are properly recorded as an acquisition in the licensee's records.

[27 CFR 178.50]

(C3) Does an importer or manufacturer of firearms also need a dealer's license?

No, as long as the importer or manufacturer is engaged in business at the licensed importing or manufacturing premises in the same type of firearms authorized by the license.

[27 CFR 178.41(b)]

(C4) If a person timely files an application for renewal of a license and the present license expires prior to receipt of the new license, may the person continue to conduct the business covered by the expired license?

Yes. A person who timely files an application for renewal of a license may continue operations authorized by the expired license until the application is finally acted upon. An application is timely filed when it is received at the appropriate P.O. Box in Dallas, Texas with the appropriate renewal fee prior to the expiration date of the license.

If a person does not timely file a license renewal application and the license expires, the person must file ATF Form 7, Application for License, or an ATF Form 7CR, Application for License (Collector of Curios or Relics), as required by 27 CFR 178.44, submit the application fee applicable to a new business, and obtain the required license before continuing business activity.

[27 CFR 178.45]

(C5) Must a licensed importer's, manufacturer's, or dealer's records be surrendered to ATF if the licensee discontinues business?

If the business is being discontinued completely, the licensed dealer, manufacturer or importer is required, within 30 days, to forward the business records to the following address:

Bureau of ATF
Out-of-Business Records Center
2029 Stonewall Jackson Drive
Falling Waters, West Virginia 25419

Failure to surrender your required records is a felony and could result in the licensee being fined up to $250,000, imprisoned up to 5 years, or both. A licensee discontinuing business must immediately notify the Licensing Center in Atlanta, Georgia.

If someone is taking over the business, the original licensee will underline the final entry in each bound book, note the date of transfer, and forward all records and forms to the successor (who must apply for and receive his or her own license before lawfully engaging in business) or forward the records and forms to the ATF Out-of-Business Records Center. If the successor licensee receives records and forms from the original licensee, the successor licensee may choose to forward these records and forms to the ATF Out-of-Business Record Center. The successor licensee will begin business with a new set of records reflecting any beginning inventory on hand.

[18 U. S. C. 923(g)(4), 27 CFR 178.127]

(C6) What records am I required to forward to ATF upon discontinuance of my business?

The records consist of the licensee's bound acquisition/disposition (A/D) records, ATF Forms 4473, ATF Forms 5300.35 (the Brady forms), ATF Forms 3310.4 (Report of Multiple Sale or Other Disposition of Pistols and Revolvers), ATF Forms 3310.11 (Federal Firearms Licensee Theft/Loss Report), records of transactions in semiautomatic assault weapons, and law enforcement certification letters. If the licensee was granted a variance to use a computerized recordkeeping system, the licensee is required to provide a complete print-out of the entire A/D records, and an ASCII text file (conforming to common industry standards) along with a file description.

[27 CFR 178.127]

(C7) May a successor owner of a business entity, other than one who is a successor under the provisions of 27 CFR 178.56 (for example, the surviving spouse or child, or a receiver or trustee in bankruptcy), commence a firearms business prior to receiving a Federal firearms license in the successor's name?

No. Each person intending to engage in business as a firearms dealer, importer or manufacturer or an ammunition importer or manufacturer must obtain the required Federal firearms license prior to commencing business.

[27 CFR 178.41]

(C8) Does a Federal firearms license allow the licensee to carry a firearm in the course of business?

No. A Federal firearms license confers no right or privilege to carry a firearm, concealed or otherwise. Permits to carry are issued by State or local authorities.

[27 CFR 178.58]

(C9) May a person obtain a dealer's license to engage in business only at gun shows?

No. A license may only be issued for a permanent premises at which the license applicant intends to do business. A person having such license may conduct business at gun shows located in the State in which the licensed premises is located and sell and deliver curio or relic firearms to other licensees at any location.

[18 U.S.C. 923(a) and (j)]

(C10) May a licensee change the location of the licensed business or activity?

To change your location, you must file an application for an amended license, ATF Form 5300.38, not less than 30 days prior to the move. You must obtain the amended license before commencing business at the new location.

The application for an amended license would include the certification of compliance with State and local laws and notification of local law enforcement officials as outlined in Question **A2**. Tear-out Forms 5300.38 for changing the location of a licensed business are in the back of this publication.

[27 CFR 178.52]

D. ATF FORM 4473 - FIREARMS TRANSACTION RECORD

(D1) Where can a dealer get ATF Forms 4473?

They are available free of charge from the ATF Distribution Center. The current address is P.O. Box 5950, Springfield, VA 22150-5950. Please order a quantity of forms estimated for 6 months use.

(D2) Is an ATF Form 4473 needed in the transfer of a firearm by a nonlicensee?

No. ATF Form 4473 is required only for transfers by a licensee.

[27 CFR 178.124]

(D3) Does a dealer have to execute ATF Form 4473 to take a weapon out of the dealer's inventory for his or her own use?

No. However, the "bound book" must be properly posted to reflect the disposition of the firearm from business inventory to personal use.

[27 CFR 178.124, 178.125a]

(D4) Who signs ATF Form 4473 for the seller?

ATF Form 4473 must be signed by the person who verified the identity of the buyer.

[27 CFR 178.124(c)]

(D5) Is a Social Security card a proper means of identification?

No. A Social Security card, alien registration card, or military identification alone does not contain sufficient information to identify a firearms purchaser. However, a purchaser may be identified by any combination of documents which together establish all of the required information: Name, residence address, date of birth or age, signature, and photograph of the holder. In addition, the documents used to establish the purchaser's identity must have been issued by a government agency.

[27 CFR 178.124(c)]

(D6) When must the ATF Form 4473 be signed?

Part I (yellow) used for over-the-counter sales must be completed, signed and dated by the buyer prior to delivery of the firearm.

Part II (green) used for intrastate non-over-the-counter sales must be completed, signed and dated in duplicate by the buyer at the time of sale.

[27 CFR 178.124(c), 178.124(f)]

E. RECORDS REQUIRED - LICENSEES

(E1) What is a "bound book?"

A "bound book" is a permanently bound book or an orderly arrangement of loose-leaf pages which must be maintained on the business premises. In either event, the format must follow that prescribed in the regulations and the pages must be numbered consecutively.

[27 CFR 178.125]

(E2) May a dealer keep more than one "bound book" at the same time?

Yes. A dealer in firearms is not limited to using only one "bound book." A few dealers account for different brands or types of firearms in separate "bound books." Many maintain a separate repair "bound book."

[27 CFR 178.125]

(E3) Does the Government sell a record book for licensees to use in recording their receipts and dispositions of firearms?

No. Certain trade associations have them available at nominal cost. Your supplier should be able to tell you about this.

(E4) What is the dealer's responsibility where a variation from normal regulatory practice has been authorized?

The ATF letter authorizing the variation must be kept at the licensed premises and available for inspection. For businesses with more than a single licensed outlet, each outlet covered by the variation must have a copy of the letter authorizing the change.

[27 CFR 178.22, 178.125(h)]

(E5) How much time does a dealer have to record acquisitions and dispositions of firearms in his or her "bound book"?

If commercial records are kept containing the required information, are available for inspection, and are separate from other commercial documents, dealers have 7 days from the time of receipt or disposition to record the receipt or disposition in the "bound book."

Receipts not covered by these records must be entered in the "bound book" by the close of the next business day after the acquisition or purchase. If a disposition is made before the acquisition has been entered in the "bound book," the acquisition entry must be made at the same time as the disposition entry.

[27 CFR 178.125(d)-(i)]

(E6) Are the ammunition recordkeeping requirements the same as for firearms?

No. No records are required for ammunition other than armor piercing ammunition. Disposition records must be kept by licensed manufacturers, importers, and collectors for transactions in armor piercing ammunition.

[27 CFR 178.125]

(E7) Are rental firearms subject to recordkeeping control?

Yes, but the recordkeeping is not imposed on the loan or rental of firearms for use only on the premises.

[27 CFR 178.97]

(E8) May a licensee who has firearms in his or her private collection sell any of these firearms without making firearms record entries?

No. A licensee may sell a firearm from his or her personal collection, subject only to the restrictions on firearm sales by unlicensed persons, provided the firearm has been entered in the licensee's bound book and transferred to the licensee's private collection at least 1 year prior to the sale. On selling the personal firearm after 1 year, the sale must be recorded in a "bound book" for disposition of personal firearms, but no ATF Form 4473 is required.

[27 CFR 178.125a]

(E9) May a licensee maintain computer records in lieu of the "bound book"?

Yes. The Regional Director (Compliance) or other designated ATF official must approve a request for a recordkeeping variance before the licensee may use a computer system in lieu of the "bound book" record required by the regulations.

[27 CFR 178.22 and 178.125(h)]

F. CONDUCT OF BUSINESS - LICENSEES

(F1) Does the Federal firearms law require licensees to comply with State laws and local published ordinances which are relevant to the enforcement of the GCA?

Yes. It is unlawful for any licensed importer, licensed manufacturer, licensed dealer, or licensed collector to sell or deliver any firearm or ammunition to any person if the person's purchase or possession would be in violation of any State law or local published ordinance applicable at the place of sale or delivery. (See Question **A2** for requirement to comply with State and local law to qualify for a license.)

[18 U.S.C. 922(b)(2), 27 CFR 178.99(b)(2)]

(F2) May a licensed dealer sell a firearm to a nonlicensee who is a resident of another State?

Generally, a firearm may not be lawfully sold by a licensed dealer to a nonlicensee who resides in a State other than the State in which the seller's licensed premises is located. However, the sale may be made if the firearm is shipped to a licensed dealer whose business is in the purchaser's State of residence and the purchaser takes delivery of the firearm from the dealer in his or her State of residence. In addition, a licensee may sell a rifle or shotgun to a person who is not a resident of the State where the licensee's business premises is located in an over-the-counter transaction, provided the transaction complies with State law in the State where the licensee is located and in the State where the purchaser resides and provided the sale complies with all applicable Federal laws.

[18 U.S.C. 922(b)(3)]

(F3) May a dealer sell firearms to law enforcement agencies and individual officers in another State?

Yes. Sales and deliveries of firearms to out-of-State police and sheriff departments are not prohibited by the GCA. A dealer may also sell or ship firearms, other than NFA firearms, to an individual law enforcement officer, regardless of age, if the dealer has a signed statement of the officer's agency, stating that the items are to be used in the buyer's official duties and that the officer has not been convicted of a misdemeanor crime of domestic violence. ATF Form 4473 need not be executed, and the Brady law is not applicable; however, the bound book must be properly posted, and the signed statement included in the dealer's records. (For information on sales of semiautomatic assault weapons to individual law enforcement officers, see Question O11.) For further information on sales of firearms to law enforcement officers, see **Item 4**, "Sales of Firearms to Law Enforcement Officers", under Items of Interest.

[27 CFR 178.141]

(F4) May an employee of a licensed dealer, such as a manager or clerk, who is under 21 years of age, sell handguns and ammunition suitable for use in handguns for the licensee?

Yes, if the employee is not a prohibited person (e.g., a felon). However, to sell handguns, a person less than 18 years of age must have the prior written consent of a parent or guardian and the written consent must be in the person's possession at all times. Also, the parent or guardian giving the written consent may not be prohibited by law from possessing a firearm.

[18 U.S.C. 922(x)]

(F5) As a licensed dealer, must I advise ATF if I sell more than one handgun to an individual?

If you sell more than one handgun to any nonlicensee during a period of 5 consecutive business days, the sale must be reported on ATF Form 3310.4, Report of Multiple Sale or Other Disposition of Pistols and Revolvers, and forwarded to the ATF office specified on the form no later than the close of business on the day the second handgun was sold. A copy of the form must also be sent to the State police or the local law enforcement agency where the sale occurred. A copy must also be kept in the records of the dealer.

[18 U.S.C. 923(g)(3), 27 CFR 178.126a]

(F6) Does a customer have to be a certain age to buy firearms or ammunition from a licensee?

Yes. Longguns and longgun ammunition may be sold only to persons 18 years of age or older. Sales of handguns and ammunition for handguns are limited to persons 21 years of age and older. Although some State and local ordinances have lower age requirements, dealers are bound by the minimum age requirements established by the GCA. If State law or local ordinances establish a higher minimum age, the dealer must observe the higher age requirement.

[18 U.S.C. 922(b)(1), 27 CFR 178.99(b)]

(F7) May a licensee sell interchangeable ammunition such as .22 cal. rimfire to a person less than 21 years old?

Yes, provided the buyer is 18 years of age or older, and the dealer is satisfied that it is for use in a rifle. If the ammunition is intended for use in a handgun, the 21 year old minimum age requirement is applicable.

[18 U.S.C. 922(b)(1), 27 CFR 178.99(b)]

(F8) In transactions between licensees, how is the seller assured that a

purchaser of a firearm is a licensed dealer?

Verification must be established by the transferee furnishing to the transferor a certified copy of the transferee's license and by any other means the transferor deems necessary.

[27 CFR 178.94]

(F9) Must a multi-licensed business submit a certified copy of each of its licenses when acquiring firearms?

No. It need only provide the seller a list, certified to be true, correct and complete, containing the name, address, and license number and expiration date for each location.

[27 CFR 178.94]

(F10) May a licensee continue to deliver to a business whose license has expired?

Yes, for a period of 45 days following the expiration date of the license. After the 45-day period, the transferor is required to verify the licensed status of the transferee with the Chief, Firearms and Explosives Licensing Center. If the transferee's license renewal application is still pending, the transferor must obtain evidence from the Regional Director (Compliance) that a license renewal application has been timely filed by the transferee and is still pending.

[27 CFR 178.94]

(F11) Is a license required to engage in the business of selling small arms ammunition?

No. A license is not required for a dealer in ammunition only, but a manufacturer or an importer must be licensed.

(F12) May licensed dealers sell firearms at gun shows?

Generally, a licensee may sell firearms at a gun show located only in the same State as that specified on the seller's license. However, a licensee may sell curio or relic firearms to another licensee at any location. The transfer of NFA firearms may be lawfully made only upon an ATF approved transfer application.

[18 U.S.C. 923(j), 27 CFR 178.100]

(F13) What may a licensed dealer do at an out-of-State gun show?

A licensed dealer may sell and deliver curio or relic firearms to another licensee at an out-of-State gun show. With respect to other firearms transactions, a licensed dealer may only display and take orders for firearms at an out-of-State gun show. In filling any orders for firearms, the dealer must return the firearms to his or her licensed premises and deliver them from that location. Any firearm ordered by a nonlicensee must be delivered or shipped from the licensee's premises to a licensee in the purchaser's State of residence, and the purchaser must obtain the firearm from the licensee in the purchaser's State. Except for sales of curio or relic firearms to other licensees, sales of firearms and simultaneous deliveries at the gun show, whether to other licensees or to nonlicensees, violate the law because the dealer would be unlawfully engaging in business at an unlicensed location.

[18 U.S.C. 922(a)(1), (b)(3), 923(a), (j)]

(F14) Who may ship firearms through the U.S. Postal Service?

Federal firearm licensees may deposit an unloaded firearm in the mails for conveyance to any officer, employee, agent, or watchman who is eligible under 18 U.S.C. 1715 to receive pistols, revolvers, and other firearms capable of being concealed on the person for use in connection with his or her official duties.

However, any person proposing to mail a handgun must file with the postmaster, at the time of mailing, an affidavit signed by the addressee stating that the addressee is qualified to receive the firearm, and the affidavit must bear a certificate stating that the firearm is for the official use of the addressee. See the current Postal Manual for details.

The Postal Service recommends that all firearms be sent by registered mail and that no marking of any kind which would indicate the nature of the contents be placed on the outside of any parcel containing firearms. (See also Question **B8**.)

(F15) Must a dealer record firearms received on consignment?

Yes. Firearms received for sale on consignment must be entered in the dealer's "bound book."

Sales of the firearms are handled in the same manner as other firearm sales. Return of the remaining firearms by the licensee to the consignor is entered in the dealer's disposition record, and the consignor must complete an ATF Form 4473 if the consignor is a nonlicensee. It should also be noted that the sale of such firearms or their return to the consignor must also comply with the Brady law.

(F16) To whom does a dealer report stolen firearms?

A theft or lost of firearms must be reported to your local police as well as to ATF within 48 hours after the discovery. Licensees should notify ATF on the 24-hour, 7 days a week toll free line at 1-800-800-3855 and by preparing and submitting ATF Form 3310.11, Federal Firearms Licensee Theft/Loss Report.

Theft or loss of NFA firearms should also be reported to the NFA Branch, (202) 927-8330, immediately upon discovery.

[18 U.S.C. 923(g)(6), 27 CFR 178.39a and 179.141]

(F17) If my firearms are stolen, what do I do about my records?

Take an inventory of stock on hand and enter "stolen" and the date in the disposition section of the "bound book" for those stolen firearms. In addition, at the time a licensee reports the theft on the ATF toll free line, the licensee will be provided a control number that should be placed in the records as well as on ATF Form 3310.11, Federal Firearms Licensee Theft/Loss Report.

[18 U.S.C. 923(g)(6), 27 CFR 178.39a]

(F18) How many copies of the ATF Form 3310.4, Report of Multiple Sale or Other Disposition of Pistols and Revolvers, must be completed and what becomes of each copy?

ATF Form 3310.4 must be completed in triplicate (3 copies). The original is sent to ATF's National Tracing Center by FAX at 1-877-283-0288 or by mail to Box 1061, Falling Waters, West Virginia 25419-1061. A copy is to be sent to the designated State police or the local law enforcement agency in the jurisdiction where the sale took place. The remaining copy is to be retained in the records of the dealer and held for not less than 5 years.

[27 CFR 178.126a, 178.129]

(F19) What is my responsibility to respond to a request to trace a firearm?

A licensee must provide the requested information immediately and in no event later than 24 hours after receipt of a request by ATF for information required to be kept. Failure to respond to the request for trace information can result in monetary fines, imprisonment, and/or revocation of the licensee's Federal firearms license.

[18 U.S.C. 923(g)(7), 27 CFR 178.25a]

(F20) Does the requirement to give written notification to handgun transferees about juvenile handgun possession apply to a licensed dealer who returns firearms to their owners, for example, handguns that the dealer repaired?

Yes. The requirement to give written notification to nonlicensees to whom handguns are transferred applies to the return of handguns, as well as to their sale. However, the requirement does not apply where a handgun is delivered to a customer through another licensee. The licensee who delivers the handgun must provide the notification.

[18 U.S.C. 922(x), 27 CFR 178.103]

(F21) Does the requirement to post a sign on the licensed premises about juvenile handgun possession apply to a licensed dealer who only disposes of handguns to nonlicensees who do not appear at the dealer's premises?

No. The sign posting requirement does not apply where the licensee only disposes of handguns to nonlicensees who do not appear at the licensed premises, for example, the licensee ships repaired or replacement handguns to nonlicensees.

[18 U.S.C. 922(x), 27 CFR 178.103]

G. COLLECTORS

(G1) Is there a specific license which permits a collector to acquire firearms in interstate commerce?

Yes. The person may obtain a collector's license; however, this license applies only to transactions in curio or relic firearms.

[27 CFR 178.41(c), (d), 178.50(b) and 178.93]

(G2) Does a collector's license afford any privileges to the licensee with respect to acquiring or disposing of firearms other than curios or relics in interstate or foreign commerce?

No. A licensed collector has the same status under the GCA as a nonlicensee except for transactions in curio or relic firearms.

[27 CFR 178.93]

(G3) Does a license as a collector of curio or relic firearms authorize the collector to engage in the business of dealing in curios or relics?

No. A dealer's license must be obtained to engage in the business of dealing in any firearms, including curios or relics. A collector's license only enables the collector to obtain curio and relic firearms interstate.

[18 U.S.C. 922(a) and 923(a)(1), 27 CFR 178.41]

(G4) Since a licensed firearms dealer may legally receive firearms interstate, including curios or relics, is there any reason why a dealer would need both a dealer's license and collector's license?

No.

[27 CFR 178.50(b)]

(G5) Are licensed collectors required to execute ATF Form 4473 for transactions in curio or relic firearms?

No. Licensed collectors are only required to keep a "bound book" record.

[27 CFR 178.125(f)]

(G6) Are licensed collectors' transfers of curio or relic firearms subject to the Brady law, including the provision for making background checks on transferees?

No, but it is unlawful to transfer a firearm to any person knowing or having reasonable cause to believe that such person is a felon or is within any other category of person prohibited from receiving or possessing firearms. (See also Questions **P13** and **P14**.)

[18 U.S.C. 922(d), 27 CFR 178.32(d)]

(G7) Are licensed collectors required to comply with the requirements that written notification be given to handgun transferees and signs be posted on juvenile handgun possession?

The requirement that written notification concerning juvenile handgun possession be given by licensees to a nonlicensee to whom a handgun is delivered applies to curio or relic handguns transferred by licensed collectors. However, the sign posting requirement does not apply to licensed collectors. In the case of collectors, a requirement to post signs at the licensed premises would serve no purpose because the premises is not a business premises open to the public and licensed collectors may lawfully dispose of curio or relic handguns away from their licensed premises.

[18 U.S.C. 922(x), 27 CFR 178.103]

(G8) Are licensed collectors required to turn in their acquisition/disposition records to ATF if their collector's license is not renewed or they discontinue their collecting activity?

No. The GCA requires the delivery of required records to the Government within 30 days after a firearms "business" is discontinued. A license as a collector of curios or relics does not authorize any business with respect to firearms. This is in contrast to firearms importers, manufacturers, and dealers who are licensed to engage in a firearms business. Therefore, the records required to be kept by licensed collectors under the law and regulations are not business records and are not required to be turned in to ATF when collector's licenses are not renewed or collecting activity under such licenses is discontinued.

H. MANUFACTURERS

(H1) Must a person who engages in the business of manufacturing and importing firearms have a separate license to cover each type of business?

Yes. A separate license is required to cover each of these types of businesses.

[27 CFR 178.41]

(H2) May a person licensed as a manufacturer of ammunition also manufacture firearms?

No. A person licensed as a manufacturer of ammunition may not manufacture firearms unless he or she obtains a license as a firearms manufacturer.

(H3) May a person licensed as a manufacturer of firearms also manufacture ammunition?

Yes. The person may also manufacture ammunition (not including destructive device ammunition or armor piercing ammunition) without obtaining a separate license as a manufacturer of ammunition.

(H4) Is one who reloads ammunition required to be licensed as a manufacturer?

Yes, if the person engages in the business of selling or distributing reloads for the purposes of livelihood or profit. No, if the person reloads only for personal use.

[27 CFR 178.41]

(H5) Must a licensed manufacturer pay excise taxes?

Yes. Licensed manufacturers incur excise tax on the sale of firearms and ammunition manufactured. (See Item **16** under "Items of Interest".)

I. GUNSMITHS

(I1) Is a license needed to engage in the business of engraving, customizing, refinishing or repairing firearms?

Yes. A person conducting such activities as a business is considered to be a gunsmith within the definition of a dealer.

[27 CFR 178.11]

(I2) Does a gunsmith need to enter in a permanent "bound book" record every firearm received for adjustment or repair?

If a firearm is brought in for repairs and the owner waits while it is being repaired or if the gunsmith is able to return the firearm to the owner during the same business day, it is not necessary to list the firearm in the "bound book" as an "acquisition." If the gunsmith has possession of the firearm from one business day to another or longer, the firearm must be recorded as an "acquisition" and a "disposition" in the permanent "bound book" record.

[27 CFR 178.125(e)]

(I3) Is ATF Form 4473 required when a gunsmith returns a repaired firearm?

No, provided the firearm is returned to the person from whom received.

[27 CFR 178.124(a)]

(I4) May a gunsmith make immediate repairs at locations other than his or her place of business?

Yes.

(I5) May a licensed gunsmith receive an NFA firearm for purposes of repair?

Yes, for the sole purpose of repair and subsequent return to its owner. It is suggested that the owner obtain permission from ATF for the transfer by completing and mailing ATF Form 5 to the NFA Branch and receive approval prior to the delivery. The gunsmith should do the same prior to returning the firearm.

Only the face of the form need be completed in each instance. ATF Forms 5 may be obtained from the Bureau of ATF, NFA Branch, Washington, DC 20226, (202) 927-8330.

(I6) Is a licensed gunsmith required to comply with the requirements to give written notification to handgun transferees and post signs on juvenile handgun possession?

The requirement that written notification on juvenile handgun possession be given to a nonlicensee to whom a handgun is delivered applies to all Federal firearms licensees. It also applies to the return of handguns to their owners, as well as to their sale. Thus, a gunsmith who repairs or customizes a nonlicensee's handgun must provide the notification to the nonlicensee when the handgun is returned. The sign posting requirement also applies to gunsmiths, unless the gunsmith only disposes of handguns to nonlicensees who do not appear at the gunsmith's licensed premises, for example, when repaired handguns are shipped to nonlicensees.

[18 U.S.C. 922(x), 27 CFR 178.103]

(I7) Is a licensed gunsmith's return of repaired or customized firearms to their owners subject to the Brady law, including the provision for making background checks on transferees?

No, but it is unlawful to transfer a firearm to any person knowing or having reasonable cause to believe that such person is a felon or is within any other category of person prohibited from receiving or possessing firearms. (See also Question **P25**.)

[18 U.S.C. 922(d), 27 CFR 178.32(d)]

J. PAWNBROKERS

(J1) What disposition records must be kept by a pawnbroker upon the redemption of a pawned firearm?

The redemption of a pawned firearm is a "disposition" of a firearm under Federal firearms law and is subject to all the recordkeeping requirements under the GCA. Disposition must be properly entered in the pawnbroker's "bound book," and ATF Form 4473 must be executed in connection with the redemption. (See also Question **J4**.)

[27 CFR 178.124 and 178.125]

(J2) What is the procedure for a licensed pawnbroker to return a firearm?

The procedure varies, depending upon the firearm and the situation.

Some Examples -

(1) Pawnbroker and nonlicensee are residents of the same State: The pawnbroker may return a handgun or longgun to either the person who pawned it or a holder of the pawn ticket who resides in the pawnbroker's State. Use ATF Form 4473, Part I (yellow) at the time of redemption.

(2) Pawnbroker and nonlicensee are not residents of the same State:

a. The pawnbroker may return a handgun only to the person who pawned it, using ATF Form 4473, Part I (yellow) at the time of redemption.

b. The pawnbroker may return a rifle or shotgun to the person who pawned it.

c. The pawnbroker may transfer a rifle or shotgun to the holder of a pawn ticket who did not pawn it at the licensed premises, provided that the transaction complies with the law of the State where the pawnbroker's business is located and the law of the State where the pawn ticket holder resides. An ATF Form 4473, Part I (yellow) is used for this transaction.

[18 U.S.C. 922(a)(2) and 922(b)(3)]

(J3) Are there prohibited categories of persons from whom a pawnbroker should not accept firearms?

Yes. The pawnbroker cannot lawfully return a firearm to a person who is underage or within a prohibited category of persons to whom the sale or other disposition of the firearm would be unlawful. For example, a pawnbroker cannot lawfully return a pawned handgun to a person who is less than 21 years of age, nor can he or she return a firearm to a convicted felon or to anyone else who is prohibited from receiving the firearm. (See also Question **J5**.)

[18 U.S.C. 922(d), 27 CFR 178.99]

(J4) Are licensed pawnbrokers required to comply with the requirements that written notification be given to handgun transferees and signs be posted on juvenile handgun possession?

The requirement that written notification on juvenile handgun possession be given to a nonlicensee to whom a handgun is delivered applies to all types of Federal firearms licensees. It also applies to the return of handguns to their owners, as well as to their sale. Thus, a pawnbroker who returns a handgun to its owner upon its redemption from pawn must provide the notification to the owner. The sign posting requirement also applies to licensed pawnbrokers.

[18 U.S.C. 922(x), 27 CFR 178.103]

(J5) Are licensed pawnbrokers' firearms sales or return of firearms redeemed from pawn subject to the Brady law, including the provision for making

background checks of transferees?

Yes. As provided by Public Law 105-277, enacted on October 21, 1998, a licensed pawnbroker may also contact the National Instant Criminal Background Check System (NICS) for a background check on a person at the time the person offers to pawn a firearm. If NICS advises the pawnbroker that receipt or possession of the firearm by the person attempting to pawn the firearm would violate the law, the pawnbroker must advise local law enforcement within 48 hours after receipt of the information. A pawnbroker who contacts NICS about a person prior to accepting the person's firearm in pawn must still comply with the requirements of the Brady law at the time of the firearm's redemption, i.e., NICS must again be contacted for a background check on the person at the time of redemption. (See also questions **P18** through **P24.**)

K. AUCTIONEERS

(K1) Does an auctioneer who is involved in firearms sales need a dealers license?

Generally speaking, there are two types of auctions: estate-type auctions and consignment auctions. In estate-type auctions, the articles to be auctioned (including firearms) are being sold by the executor of the estate of an individual. In these cases the firearms belong to and are possessed by the executor. The auctioneer is acting as an agent of the executor and assisting the executor in finding buyers for the firearms. The firearms are controlled by the estate, and the sales of firearms are being made by the estate. In these cases, the auctioneer does not meet the definition of engaging in business as a dealer in firearms and would not need a license. An auctioneer who has a license may perform this function away from his or her licensed premises.

In consignment-type auctions, an auctioneer often takes possession of firearms in advance of the auction. These firearms are generally inventoried, evaluated, and tagged for identification. The firearms belong to individuals who have entered into a consignment agreement with the auctioneer giving that auctioneer authority to sell the firearms. The auctioneer has possession and control of the firearms. Under these circumstances, an auctioneer would generally need a license.

An auctioneer who buys firearms for purposes of resale will also need a license.

(K2) If a licensed auctioneer is making sales of firearms, where may those sales be made?

Firearms may be displayed at an auction site away from the auctioneer's licensed premises and sales of the firearms can be agreed upon at that location, but delivery may only be made to purchasers after the firearms have been returned to the auctioneer's licensed premises. The simultaneous sale and delivery of the auctioned firearms away from the licensed premises would violate the law, i.e., engaging in business at an unlicensed location. However, if the auctioneer is assisting an estate dispose of firearms, the estate is the seller of the firearms, and the estate is in control and possession of the firearms, the firearms would not have to be returned to the licensed premises prior to their delivery. (See also Question **K1.**)

L. IMPORTING AND EXPORTING

(L1) May a licensed dealer who does not have an importer's license make an occasional importation?

Yes. A licensee may make an occasional importation of a firearm for a nonlicensee or for the licensee's personal use (not for resale). The licensee must first submit an ATF Form 6, Part I to the Imports Branch for approval. The licensee may then present the approved Form 6 and completed ATF Form 6A to the U.S. Customs Service. Contact the Bureau of ATF, Imports Branch, Washington, DC 20226, (202) 927-8320 for forms.

(L2) Does a licensee need an export license to export a firearm?

The GCA does not provide for an export license. However, firearms and ammunition shall be exported in accordance with provisions of the Arms Export Control Act of 1976 and a license must be obtained from the Office of Defense Trade Controls, PM/DTC, SA-6, Room 228, U.S. Department of State, Washington, DC 20522-0602; (703) 875-6644. In the case of exporting NFA firearms, a permit, ATF Form 9, must be obtained from ATF.

The export of sporting shotguns is regulated by the U.S. Department of Commerce. For further information, contact them at their nearest district office or the Bureau of Export Administration, Export Counseling Division, U.S. Department of Commerce, 14th St. & Pennsylvania Ave. N.W., Washington, DC 20230, (202) 482-4811.

[22 U.S.C. 2778, 27 CFR 179.114-179.116]

M. FIREARMS - NATIONAL FIREARMS ACT (NFA)

(M1) The types of firearms that must be registered in the National Firearm Registration and Transfer Record are defined in the NFA and in 27 CFR Part 179. What are some examples?

Some examples of the types of firearms that must be registered are:

Machineguns;

The frames or receivers of machineguns;

Any combination of parts designed and intended for use in converting weapons into machineguns;

Any part designed and intended solely and exclusively for converting a weapon into a machinegun;

Any combination of parts from which a machinegun can be assembled if the parts are in the possession or under the control of a person;

Silencers and any part designed and intended for fabricating a silencer;

Short-barreled rifles;

Short-barreled shotguns;

Destructive devices; and,

"Any other weapons."

A few examples of destructive devices are:

Molotov cocktails;

Anti-tank guns (over caliber .50);

Bazookas; and,

Mortars.

A few examples of "any other weapon" are:

H&R Handyguns;

Ithaca Auto-Burglar guns;

Cane guns; and,

Gadget-type firearms and "pen" guns which fire fixed ammunition.

(M2) How can an individual legally acquire NFA firearms?

Basically, there are 2 ways that an individual (who is not prohibited by Federal, State, or local law from receiving or possessing firearms) may legally acquire NFA firearms:

(1) By lawful transfer of a registered weapon from its lawful owner residing in the same State as the transferee. Obtain any forms needed from the Bureau of ATF, NFA Branch, Washington, DC 20226.

(2) By obtaining prior approvals to make NFA firearms.

[27 CFR 179.84-179.87 and 179.62-179.67]

(M3) What is the tax on making an NFA firearm?

The tax is $200 for making any NFA firearm, including "any other weapon."

(M4) How is this tax paid?

A money order or check made payable to the Bureau of ATF together with the application forms are to be mailed to the Bureau of ATF, NFA Branch, Washington, DC 20226.

(M5) What is an unserviceable firearm?

An unserviceable firearm is defined as one which is incapable of discharging a shot by means of an explosive and which is incapable of being readily restored to a firing condition.

An acceptable method of rendering most firearms unserviceable is to fusion weld the chamber closed and fusion weld the barrel solidly to the frame. Certain unusual firearms require other methods to render the firearms unserviceable.

An unserviceable NFA firearm is still subject to the controls of the NFA, but may be transferred tax free as a curio or ornament. Contact the Bureau of ATF, Firearms Technology Branch, Washington, DC 20226, (202) 927-7910 for instructions.

[27 CFR 179.11 and 179.91]

(M6) What is the status of an unregistered NFA firearm acquired through seizure or abandonment by a State?

When NFA firearms are desired for official use, they must be registered by filing ATF Form 10 with the Bureau of ATF, NFA Branch, Washington, DC 20226.

Since approval is conditioned on an "official use only" basis, subsequent transfers on ATF Form 5 cannot be approved except to a government agency for official use.

[27 CFR 179.104]

(M7) May a private citizen who owns an NFA firearm which is not registered have the firearm registered?

No. An unregistered NFA firearm is a contraband firearm and it is unlawful to possess the weapon. The possessor should contact the nearest ATF office to arrange for its disposition.

[26 U.S.C. 5861(d)]

(M8) What can happen to someone who has an NFA firearm which is not registered to him?

Violators may be fined not more than $250,000, and imprisoned not more than 10 years, or both. In addition, any vessel, vehicle or aircraft used to transport, conceal or possess an unregistered NFA firearm is subject to seizure and forfeiture, as is the weapon itself.

[49 U.S.C. 781-788, 26 U.S.C. 5861, 26 U.S.C. 5872]

(M9) What should a person do if he or she comes into possession of an unregistered NFA firearm?

Contact the nearest ATF office immediately.

(M10) Are there any exemptions from the making or transfer tax provisions of the NFA?

Yes. These are noted below, along with the required form number.

You will have to contact the Bureau of ATF, NFA Branch, Washington, DC 20226, (202) 927-8330. Completed forms must be approved by the NFA Branch prior to the making or transfer:

(1) Tax exempt transfer and registration of a firearm between special (occupational) taxpayers: ATF Form 3.

(2)(a) Tax-exempt making of a firearm on behalf of a Federal or State agency: ATF Form 1. Tax-exempt transfer and registration of the firearm: ATF Form 5.

(b) A licensed manufacturer under contract to make NFA firearms for the U.S. Government may be granted exemption from payment of the special (occupational) tax as a manufacturer of NFA firearms and exemption from all other NFA provisions (except importation) with respect to the weapons made to fulfill the contract. Exemptions are obtained by writing the NFA Branch, stating the contract number(s) and the anticipated date of termination. This exemption must be renewed each year prior to July 1.

(3) Tax-exempt transfer and registration of an unserviceable firearm which is being transferred as a curio or ornament; tax exempt transfer of a firearm to a lawful heir: ATF Form 5.

[26 U.S.C. 5851, 27 CFR 179.69, 179.70, 179.88, 179.89, 179.90 and 179.91]

(M11) How does a person qualify to import, manufacture, or deal in NFA firearms?

The person must be licensed under the GCA and pay the required special (occupational) tax imposed by the NFA. In addition, an importer (except importers of sporting shotguns and shotgun ammunition) must also be registered with ATF under the Arms Export Control Act of 1976.

After becoming licensed under the GCA, he or she must file ATF Form 5630.7 with the appropriate tax payment in the entire amount with ATF.

[26 U.S.C. 5801, 18 U.S.C. 923, 27 CFR 47.31, 178.41, 179.34 and 179.193]

(M12) When must firearms special (occupational) taxes be paid, how much are the taxes, and how are they paid?

On first engaging in business and thereafter on or before the first day of July, these taxes must be paid in full. The current taxes are set out in the following table. Taxes are paid in the manner discussed in Question **M11**, above.

SPECIAL (OCCUPATIONAL) TAX RATES
UNDER THE NFA

CLASS OF TAXPAYER		ANNUAL FEE
1	Importer of Firearms (Including "Any Other Weapons")	$1000.00
2	Manufacturer of Firearms (Including "Any Other Weapons")	$1000.00
3	Dealer of Firearms (Including "Any Other Weapons")	$ 500.00
1	Importer of Firearms (Including "Any Other Weapons") REDUCED*	$ 500.00
2	Manufacturer of Firearms (Including "Any Other Weapons") REDUCED*	$ 500.00

*   REDUCED = Rates which apply to certain taxpayers whose total gross receipts in the last taxable year are less than $500,000.

(M13) Does a single special (occupational) tax payment entitle a person or firm to import and manufacture firearms?

No. A separate special (occupational) tax payment must be made for each of these activities. However, Class 1 and Class 2 special (occupational) taxpayers are qualified to deal in NFA firearms without also having to pay special (occupational) tax as a Class 3 dealer.

[27 CFR 179.39]

(M14) May a licensed collector obtain NFA firearms in interstate commerce?

Only if the firearms are classified as curios or relics, are registered, and are transferred in accordance with the provisions of the NFA. In addition, the collector must meet the requirements set forth under Question **M15**.

(M15) What are the required transfer procedures for an individual who is not qualified as a manufacturer, importer, or dealer of NFA firearms?

ATF Form 4 (5320.4) must be completed, in duplicate. The transferor must first complete the face of the form. The transferee must complete the transferee's certification on the reverse of the form and have the "Law Enforcement Certification" completed by the chief law enforcement officer.

The transferee is to affix, on each copy of the form, a 2-inch by 2-inch photograph of the transferee taken within the past year (proofs, group photographs or photocopies are unacceptable). The transferee's address must be a street address, not a post office box. If there is no street address, specific directions to the residence must be included.

If State or local law requires a prior permit or license to purchase, possess, or receive NFA firearms, a copy of the transferee's permit or license must accompany the application. A check or money order for $200 ($5 for transfer of "any other weapon") shall be made payable to ATF by the transferor. All signatures on both copies must be in ink.

Submit fingerprints on FBI Form FD-258, in duplicate. Fingerprints must be taken by a person qualified to do so, and must be clear and classifiable. If wear or damage to the fingertips do not allow clear prints, and if the prints are taken by a law enforcement official, a statement on his or her official letterhead giving the reason why good prints are unobtainable should accompany the fingerprints.

Forward completed information and appropriate tax payment to the Bureau of ATF, P.O. Box 73201, Chicago, IL 60673. Transfer of the NFA firearm may be made only upon approval of the ATF Form 4 by the NFA Branch. If the application is approved, the original of the form with the cancelled stamp affixed showing approval will be returned to the applicant. Otherwise, the tax will be refunded.

Upon approval of the ATF Form 4, the transferor should transfer the firearm as soon as possible, since the firearm is now registered to the transferee.

[26 U.S.C. 5812, 27 CFR 179.83-179.86]

(M16) How does an individual obtain authorization to make an NFA firearm?

Prior to making the firearm, the individual must submit ATF Form 1, Application to Make and Register a Firearm, to the Bureau of ATF, NFA Branch, Washington, DC 20226, and receive approval. The applicant must follow the procedures described in Question M**15** concerning completion of the form, including photographs, fingerprints and certifications. The applicant must forward the original and a duplicate of the form along with a check or money order for $200 made payable to the Bureau of ATF. If the application is approved, the original of the form with the cancelled stamp affixed showing approval will be returned to the applicant. Otherwise, the tax will be refunded.

[26 U.S.C 5822, 27 CFR 179.62-179.65]

(M17) Are parts which would convert a firearm into an NFA firearm subject to registration?

Yes. Examples:

An M-2 conversion kit;

Any part designed and intended solely and exclusively to convert a weapon into a machinegun. (See Question **M1**.)

(M18) May a licensed firearms dealer, qualified to deal in NFA firearms, transfer a firearm to an unlicensed person in another State?

No. The GCA generally prohibits the interstate transfer of a firearm from a licensed dealer to a nonlicensee, except for an over-the-counter sale of a longgun to an unlicensed person where the sale complies with the legal requirements in the States of both buyer and seller.

[18 U.S.C. 922(a)(2) and (b)(3), 27 CFR 178.29-178.30]

(M19) What law enforcement officials' certifications on an application to transfer or make an NFA weapon are acceptable to ATF?

As provided by regulations, certifications by the local chief of police, sheriff of the county, head of the State police, or State or local district attorney or prosecutor are acceptable. The regulations also provide that certifications of other officials are appropriate if found in a particular case to be acceptable to the Director. Examples of such other officials include State attorneys general and judges of State courts having authority to conduct jury trials in felony cases.

[27 CFR 179.63, 179.85]

(M20) Is the chief law enforcement officer required to sign the law enforcement certification?

No. Although ATF cannot approve an application to make or transfer an NFA weapon without a law enforcement certification, no official is required to sign the certification.

(M21) If the chief law enforcement official whose jurisdiction includes the proposed transferee's residence refuses to sign the "law enforcement certification," will the signature of an official in another jurisdiction be acceptable?

No.

(M22) Does the registered owner of a destructive device, machinegun, short barreled shotgun, or short barreled rifle need authorization to lawfully transport such items interstate?

Yes, unless the owner is a qualified dealer, manufacturer or importer, or a licensed collector transporting only curios or relics. Prior approval must be obtained, even if the move is temporary, and is requested by either submitting a letter application containing all necessary information, or by submitting ATF Form 5320.20 to the Bureau of ATF, NFA Branch, Washington, DC 20226. Possession of the firearms must still comply with State and local laws.

[27 CFR 178.28]

(M23) If an individual is changing his or her State of residence and the individual's application to transport the NFA firearm cannot be approved, what options does a lawful possessor have?

NFA firearms may be left in a safe deposit box in his or her former State of residence. Also, the firearm could be left or stored in the former State of residence at the house of a friend or relative in a locked room or container to which only the registered owner has a key. The friend or relative should be supplied with a copy of the registration forms and a letter from the owner authorizing storage of the firearm at that location. The NFA Branch must be notified of the location at which the firearms are stored.

The firearms may also be transferred under the procedures referred to in Question **M15** or abandoned to ATF.

(M24) May a transferor submit an application to transfer an NFA firearm prior to the date on which the transferor receives the weapon?

No.

(M25) If a person has a pistol and an attachable shoulder stock, does this constitute possession of an NFA firearm?

Yes, unless the barrel of the pistol is at least 16 inches in length (and the overall length of the firearm with stock attached is at least 26 inches). However, certain stocked handguns, such as original semiautomatic Mauser "Broomhandles" and Lugers, have been removed from the purview of the NFA as collectors' items.

[27 CFR 179.11]

(M26) Does the owner of a registered NFA firearm have to have any evidence to show it is registered lawfully to him or her?

Yes. The approved application received from ATF serves as evidence of registration of the NFA firearm in the owner's name. This document must be kept available for inspection by ATF officers. It is suggested that a photocopy of the approved application be carried by the owner when the weapon is being transported.

(M27) What is the status of deactivated, unloaded or dummy grenades, artillery shell casings and similar devices?

Such devices would merely be ornaments and not within the purview of the NFA. However, such devices would have to be cut or drilled in such a manner as to preclude possible use as ammunition components for destructive devices.

(M28) Are muzzleloading cannons classified as destructive devices?

Generally, no. Muzzleloading cannons not capable of firing fixed ammunition and manufactured in or before 1898 and replicas thereof are antiques and not subject to the provisions of either the GCA or the NFA.

[26 U.S.C. 5845, 27 CFR 179.11]

(M29) Are grenade and rocket launcher attachments destructive devices?

Grenade and rocket launcher attachments for use on military type rifles generally do not come within the definition of destructive devices. However, the grenades and rockets used in these devices are generally within the definition.

[26 U.S.C. 5845, 27 CFR 179.11]

(M30) What is a "conversion kit?"

A conversion kit is any part or combination of parts designed and intended for use in converting a weapon into a machinegun. A conversion kit is a machinegun for purposes of the NFA.

N. MACHINEGUNS - NATIONAL FIREARMS ACT (NFA)

(N1) May an unlicensed person make a machinegun?

Generally, no. But, in the event that documentation can be provided, along with the Application to Make a Machinegun, which establishes that the weapon is being made for distribution to a Federal or State agency, an individual may be permitted to make the machinegun.

[18 U.S.C. 922(o)(2), 27 CFR 179.105(e)]

(N2) May machineguns be transferred from one registered owner to another?

Yes. If the machinegun was lawfully registered and possessed before May 19, 1986, it may be transferred pursuant to an approved ATF Form 4.

[18 U.S.C. 922(o)(2)]

O. SEMIAUTOMATIC ASSAULT WEAPONS AND LARGE CAPACITY AMMUNITION FEEDING DEVICES

(O1) What restrictions does Federal law impose on semiautomatic assault weapons?

It is generally unlawful for a person to manufacture, transfer, or possess semiautomatic assault weapons after September 13, 1994, the effective date of the law. See the exceptions listed in Question O5.

[18 U.S.C. 922(v)(1)]

(O2) How does the law define the term "semiautomatic assault weapon?"

The term "semiautomatic assault weapon" is defined to include 19 named models of firearms and semiautomatic rifles, semiautomatic pistols, and semiautomatic shotguns that have at least 2 of the features specified in the law. Frames or receivers for firearms are not regulated as semiautomatic assault weapons, since they could be assembled as a firearm other than the 19 named models of firearms. Likewise, frames or receivers are not semiautomatic assault weapons under the "features" test of the law because they do not yet have the features necessary to bring them within the definition.

Semiautomatic assault weapons in knockdown (disassembled) condition consisting of a receiver and all parts needed to assemble a complete semiautomatic assault weapon are subject to regulation if the parts are segregated or packaged together and held by a person as the parts for the assembly of a particular firearm.

[18 U.S.C. 921(a)(30)]

(O3) What restrictions does Federal law impose on large capacity ammunition feeding devices?

It is generally unlawful for a person to transfer or possess a large capacity ammunition feeding device manufactured after September 13, 1994, the effective date of the law. See the exceptions listed in Question O5.

[18 U.S.C. 922(w)(1)]

(04) How does the law define the term "large capacity ammunition feeding device?"

The term "large capacity ammunition feeding device" is defined as a magazine, belt, drum, feed strip, or similar device manufactured after September 13, 1994, that has a capacity of, or that can be readily restored or converted to accept, more than 10 rounds of ammunition. Large capacity ammunition feeding devices in knockdown (disassembled) condition consisting of all parts needed to assemble a complete large capacity ammunition feeding device are subject to regulation if the parts are segregated or packaged together and held by a person as the parts for the assembly of a particular device.

[18 U.S.C. 921(a)(31)]

(05) What exceptions from the prohibitions on semiautomatic assault weapons and large capacity ammunition feeding devices are provided in the law?

Exceptions are provided for semiautomatic assault weapons and large capacity ammunition feeding devices -

(1) lawfully possessed on or before the date of enactment;

(2) manufactured for, transferred to, or possessed by governmental entities or law enforcement officers employed by governmental entities for official use;

(3) transferred to licensees maintaining on-site security at a nuclear power plant required by Federal law, or possession by an employee or contractor of such licensee on-site for such purposes or off-site for purposes of licensee-authorized training or transportation of nuclear materials;

(4) transferred to law enforcement officers by the officer's agency upon the officer's retirement; and

(5) manufactured, transferred, or possessed by licensed manufacturers or licensed importers for the purposes of testing or experimentation as authorized by ATF.

Ammunition feeding devices having a capacity of more than 10 rounds of ammunition that were manufactured on or before September 13, 1994, are excluded from the definition of "large capacity ammunition feeding device" and, therefore, are not subject to the prohibitions.

[18 U.S.C. 922(v)(2), (v)(4), (w)(2) and (w)(3)]

(06) If an NFA firearm has 2 or more of the features specified in the law for semiautomatic assault weapons, will the firearm be regulated under both statutes?

Any firearm that falls within the definition of "semiautomatic assault weapon" and the NFA definition of "firearm" is subject to both laws.

(07) Are replacement parts for grandfathered semiautomatic assault weapons and large capacity ammunition feeding devices subject to regulation under the law?

No. Parts may be replaced in grandfathered semiautomatic assault weapons and grandfathered feeding devices without violating the law. However, if the frame or a receiver for a semiautomatic assault weapon is defective, the replacement must be made by the weapon's manufacturer or importer. The replacement receiver must be marked with the same serial number as the original receiver, and the original receiver must be destroyed. However, a manufacturer or importer who is unable to mark the replacement receiver with the same serial number as the original receiver may seek a marking variance in accordance with 27 CFR 178.92. In addition, the permanent records of the manufacturer or importer should indicate that the receiver for the weapon has been replaced.

(O8) May law enforcement officers purchase and possess semiautomatic assault weapons and high capacity ammunition feeding devices?

Yes. The law provides exceptions for law enforcement officers purchasing assault weapons and magazines for official use. However, assault weapons may not be lawfully distributed to, or received or possessed by, an officer having been convicted of a misdemeanor crime of domestic violence. A licensee may lawfully transfer these items to a law enforcement officer and the officer may lawfully receive and possess them if:

(1) the officer is a "peace officer" having the authority to arrest persons for violations of the law and to obtain and execute search warrants;

(2) the officer is employed by a government agency; and

(3) in the case of a semiautomatic assault weapon, the officer has not been convicted of a misdemeanor crime of domestic violence.

(O9) May law enforcement officers keep their semiautomatic assault weapons and large capacity ammunition feeding devices when they retire or leave their employment with a law enforcement agency?

No. They may not lawfully keep semiautomatic assault weapons and large capacity ammunition feeding devices that they purchased or acquired as their own property. However, the law provides an exception for items that belong to a law enforcement agency and are transferred by the agency to an officer upon the officer's retirement from, or termination of his or her employment with, the agency. Neither this exception nor the exception for official use permits officers to retain their own weapons or feeding devices after retiring or leaving the agency or to acquire additional items. Officers who retire or leave their employment with a law enforcement agency should transfer assault weapons and large capacity ammunition feeding devices that are their own property to a Federal firearms licensee or another qualified officer.

[18 U.S.C. 922(v)(4), (w)(3)]

(O10) If a person is in possession of a frame or receiver for a semiautomatic assault weapon on the date of enactment, may the person acquire the rest of the parts and assemble a complete semiautomatic assault weapon?

No. It is unlawful to make such weapon after the law's effective date.

[18 U.S.C. 922(v)(1)]

(O11) What documentation must a manufacturer, importer, or dealer obtain from law enforcement officers who purchase semiautomatic assault weapons and large capacity ammunition feeding devices for official use?

Licensees may transfer semiautomatic assault weapons and large capacity ammunition feeding devices to law enforcement officers with the following documentation:

(1) a written statement from the purchasing officer, under penalty of perjury, stating that the weapon or device is being purchased for use in performing official duties and that the weapon or device is not being acquired

for personal use or for purposes of transfer or resale; and

(2) a written statement from a supervisor of the purchasing officer, under penalty of perjury, stating that the purchasing officer is acquiring the weapon or feeding device for use in official duties, that the weapon or device is suitable for use in performing official duties, and that the weapon or device is not being acquired for personal use or for purposes of transfer or resale. In the case of a transfer of a semiautomatic assault weapon, the supervisor's written statement must also state that a records check reveals that the purchasing officer has no convictions for misdemeanor crimes of domestic violence.

In the case of semiautomatic assault weapons, licensees are required to retain the above statements in their permanent records for a period of 5 years.

[27 C.F.R. 178.129, 178.132, 178.134]

(O12) May licensed manufacturers, licensed importers, and licensed dealers stockpile semiautomatic assault weapons for future sales to law enforcement agencies and law enforcement officers employed by such agencies?

Yes. Semiautomatic assault weapons may be transferred directly to law enforcement agencies with a purchase order. Licensed manufacturers and licensed dealers may transfer semiautomatic assault weapons to any Federal firearms licensee upon obtaining evidence that the weapons will only be disposed of to law enforcement agencies and law enforcement officers for official use. Examples of acceptable evidence include the following:

1. Contracts between the manufacturer and dealers stating that the weapons may only be sold to law enforcement agencies and law enforcement officers.

2. Copies of purchase orders submitted to the licensee by law enforcement agencies.

3. Copies of letters submitted to the licensee by government agencies or law enforcement officers expressing an interest in purchasing the weapons.

4. Letters from dealers to the manufacturer stating that sales will only be made to law enforcement agencies or law enforcement officers.

5. Letters from law enforcement officers as described in Question **O11**.

The above evidence must be maintained in the records of Federal firearms licensees for a period of 5 years.

ATF Form 6 applications for the importation of nonsporting weapons, including semiautomatic assault weapons, are approved only if the importer submits a purchase order from a governmental entity. Therefore, importers and dealers may not maintain an inventory of imported assault weapons.

[18 U.S.C. 922(v)(4), 27 C.F.R. 178.40, 178.129(e); ATF Rul. 80-8]

(O13) Will manufacturers, importers, and dealers in large capacity ammunition feeding devices be permitted to stockpile devices for sale to governmental entities?

Yes. Possession and transfer of these devices by manufacturers, importers, or dealers will be presumed to be lawful if they maintain evidence that the devices are possessed and transferred for sale to government agencies and law enforcement officers employed by such agencies. Examples of acceptable evidence are the same as those set forth in Question **011**, relating to semiautomatic assault weapons.

The importation of ammunition feeding devices requires an approved ATF Form 6 issued by ATF. ATF will approve an ATF Form 6 application to import such devices when submitted with a purchase order from a law enforcement agency or evidence that the device is being imported for sale to a government agency or law enforcement officer employed by such agency. A Form 6 will also be approved when an application is submitted with a statement by the importer that the devices are being acquired for resale to law enforcement agencies and/or law enforcement officers for official use. ATF will stamp or type a restriction on the Form 6 stating that the devices are approved for importation and sale only to law enforcement agencies or law enforcement officers in accordance with 18 U.S.C. 922(w)(3). A Form 6 will also be approved if there is physical or documentary evidence establishing that the magazines were manufactured on or before September 13, 1994. Examples of such evidence are listed in 27 CFR 178.119(c)(7). Imported magazines manufactured on or before September 13, 1994, may be sold without restriction.

[18 U.S.C. 922(w)(3), 27 CFR 178.40a and 178.119]

(O14) What markings must appear on semiautomatic assault weapons manufactured after September 13, 1994?

In addition to the markings required of all firearms pursuant to 27 C.F.R. 178.92(a)(1), the frames or receivers for semiautomatic assault weapons must be marked "RESTRICTED LAW ENFORCEMENT/GOVERNMENT USE ONLY" or, in the case of weapons manufactured for export, "FOR EXPORT ONLY."

[18 U.S.C. 923(i), 27 CFR 178.92(a)(2)]

(O15) What markings must appear on large capacity ammunition feeding devices manufactured after September 13, 1994?

Persons who import large capacity ammunition feeding devices manufactured after September 13, 1994, or manufacture large capacity ammunition feeding devices must legibly identify each device imported or manufactured by serial number and other prescribed markings. The same serial number may be used for all devices manufactured or imported. Such devices must also be marked "RESTRICTED LAW ENFORCEMENT/GOVERNMENT USE ONLY" or, in the case of devices manufactured for export, "FOR EXPORT ONLY." Domestically made devices must also be marked with the name, city and State of the manufacturer. Imported devices must be marked with the name of the manufacturer, country of origin, and name, city and State of the importer. Persons who manufacture or import metallic links for use in the assembly of belted ammunition are only required to place the prescribed identification marks on the containers used for packaging the links.

[18 U.S.C. 923(i), 27 CFR 178.92(c)]

(O16) Are fixed magazines for weapons specified in APPENDIX A to 18 U.S.C. 922 and fixed magazines for manually operated firearms which hold more than 10 rounds of ammunition "large capacity ammunition feeding devices?"

The law specifically provides that the prohibition on semiautomatic assault weapons shall not apply to any of the firearms specified in Appendix A or any firearm that is manually operated by bolt, pump, lever, or slide action. Accordingly, weapons listed in Appendix A with fixed magazines and manually operated firearms with fixed magazines are exempt from both the assault weapon and feeding device provisions of the law.

[18 U.S.C. 922(v)(3), 922(w)]

(017) What evidence is sufficient for Federal firearms licensees to be sure that particular semiautomatic assault weapons are "grandfathered" weapons which are not subject to the restrictions on possession and transfer?

The requirement that semiautomatic assault weapons be marked "RESTRICTED LAW ENFORCEMENT/GOVERNMENT USE ONLY" was not effective until July 5, 1995. Thus, semiautomatic assault weapons manufactured from September 13, 1994-July 4, 1995, may not be marked with the restrictive markings. Additionally, sporting weapons may have been modified after September 13, 1994, so that they are semiautomatic assault weapons, e.g., a pistol grip and magazine extension are installed on a sporting shotgun. Licensees obtaining semiautomatic assault weapons which do not have the restrictive marking should obtain from the seller an invoice, bill of sale, or other documentation indicating that the weapon in its present configuration was lawfully possessed on or before September 13, 1994.

[27 CFR 178.92(a)(2)]

(018) May semiautomatic assault weapons which have been classified as curios or relics be imported?

Not unless they are being imported for sale to a government agency or law enforcement officer employed by such agency for official use. Since ATF will not approve an importation which would place the importer in violation of the law, ATF would not authorize the importation of semiautomatic assault weapons, even if classified as curios or relics, unless the importer provided evidence that the weapons were being imported for sale to a governmental entity or other exempt purchaser.

[18 U.S.C. 922(v)]

(019) May a licensed dealer lawfully acquire semiautomatic assault weapons manufactured after September 13, 1994, remove the features that bring them within the definition of such weapons in the law, and sell the weapons to the public?

No. The law prohibits possession of semiautomatic assault weapons manufactured after September 13, 1994. However, an exception is provided for licensed dealers possessing and dealing in such weapons that have been manufactured for and are possessed for transfer to government agencies. A dealer's possession of the weapons for sale or transfer to the public is not among the exceptions to the prohibition on their possession. Thus, a dealer who acquires such weapons for the purpose of stripping them of their assault weapon features and selling the modified weapons to the public violates the law. This is true even if the dealer strips the assault weapon features from the weapons so that they no longer meet the definition of semiautomatic assault weapon. The dealer may also have violated the law by making false statements to a supplier that the weapons were being acquired under an exception to the prohibition, e.g., for sales to law enforcement agencies or law enforcement officers.

[18 U.S.C. 922(v), 27 CFR 178.40(c)]

P. BRADY LAW

Editor's Note

Unless otherwise noted, these questions and answers relate to the permanent provisions of the Brady law found in section 922(t) of the Gun Control Act. These provisions, including the requirement for licensees to initiate background checks of individuals to whom all types of fireams are transferred by contacting the National Instant Criminal Background Check System (NICS), became effective on November 30, 1998. They replace the interim provisions of the Brady law that imposed a Federal 5-day waiting period on licensees' sales of handguns and required the sending of Brady forms to State or local officials.

(P1) Does the Brady law require a 5-day waiting period before a firearm can be transferred?

No. Instead of waiting 5 days for a background check through a local law enforcement officer, licensees will initiate a background check through the National Instant Criminal Background Check System (NICS). For the most part, these background checks will be immediate. In some circumstances, however, licensees will have to wait up to 3 business days before getting a response from NICS.

(P2) Who must comply with the requirements of the Brady law?

Federally licensed firearms importers, manufacturers, and dealers must comply with the Brady law prior to the transfer of any firearm to a nonlicensed individual.

(P3) When did the provisions of the permanent Brady law take effect?

The permanent Brady law went into effect on November 30, 1998. Accordingly, any transfer occurring on or after November 30, 1998, is subject to the requirements of this law.

(P4) Is NICS operated by ATF?

No. NICS is operated by the Federal Bureau of Investigation (FBI).

(P5) Do all NICS checks go through the FBI's NICS Operations Center?

No. In many States, licensees initiate NICS checks through the State point of contact (POC). In some cases, the State POC is also the agency that does background checks for firearms transactions under State law.

(P6) If the State is acting as a point of contact (POC), does that mean that all NICS checks go through the POC rather than the FBI?

That depends on the State. In some States, the POC conducts background checks for all firearms transactions. In other States, licensees must contact the POC for handgun transactions and the FBI for longgun transactions. In some States, NICS checks for pawn redemptions are handled by the FBI.

(P7) How does a licensee know whether to contact the FBI or a State point of contact (POC) in order to initiate a NICS check?

Prior to November 30, 1998, ATF sent an open letter to licensees in each State, providing the licensees with instructions as to how to initiate a NICS check in their State. Your local ATF office can also advise you on the appropriate point of contact for NICS checks.

(P8) Is there a charge for NICS checks?

The FBI does not charge a fee for conducting NICS checks. However, States that act as points of contact for NICS checks may charge a fee consistent with State law.

(P9) Must licensees enroll with the FBI to get access to NICS?

Licensees must be enrolled with the FBI before they can initiate NICS checks through the FBI's NICS Operations Center. Licensees who have not received an enrollment package from the FBI should call the FBI NICS Operations Center at 1-877-444-6427 and ask that an enrollment package be sent to them. Licensees in States where a State agency is acting as a point of contact for NICS checks should contact the State for enrollment information.

(P10) Does the Brady law apply to the transfer of long guns as well as handguns?

Yes.

(P11) Does the Brady law apply to the transfer of antique firearms?

No. Licensees need not comply with the Brady law when transferring a weapon that meets the Gun Control Act's definition of an "antique firearm."

(P12) Does the Brady law apply to the transfer of firearms between two licensees?

No. The Brady law only applies when a licensed importer, manufacturer, or dealer is transferring a firearm to a nonlicensee.

(P13) Must licensed collectors comply with the Brady law prior to transferring a curio or relic firearm?

No. Transfers of curio or relic firearms by licensed collectors are not subject to the requirements of the Brady law.

(P14) Is the transfer of a firearm by a licensed dealer to a licensed collector subject to the Brady law?

The Brady law does not apply to the transfer of a curio or relic firearm to a licensed collector. However, a licensed collector who acquires a firearm other than a curio or relic from a licensee would be treated like a nonlicensee, and the transfer would be subject to Brady requirements.

(P15) Must a licensed importer, manufacturer or dealer comply with the Brady law when selling firearms from his or her own personal collection?

No, provided the licensee has maintained the firearm as part of his or her personal collection for at least 1 year from the date the firearm was transferred from the business inventory into the personal collection or otherwise acquired as a personal firearm **and** the licensee complies with the recordkeeping requirements in 27 CFR 178.125a.

(P16) Do the provisions of the Brady law apply to a licensee's loan or rental of a firearm to a nonlicensee?

If the firearm is loaned or rented for use on the licensee's premises, the transaction is not subject to the Brady law. However, if the firearm is loaned or rented for use off the premises, the licensee must comply with the Brady law.

(P17) Must licensees conduct NICS checks for sales of firearms to nonlicensees at gun shows?

Yes. A licensed importer, manufacturer, or dealeer may not transfer a firearm to a nonlicensee at a gun show without first complying with the requirements of the Brady law.

(P18) Is the redemption of a pawned firearm subject to the Brady law?

Yes. Unlike the interim Brady law, the permanent Brady law that went into effect on November 30, 1998 does not contain an exemption for the return of a firearm to the individual from whom it was received. Accordingly, the redemption of a pawned firearm is considered a transfer subject to the permanent Brady law.

(P19) What should a licensed pawnbroker do with a firearm he or she has in pawn when the NICS check results in a "denied" transaction?

The licensee cannot transfer the firearm to the transferee without violating the law and placing the transferee in violation of the law. Licensees with additional questions should contact their local ATF office.

(P20) If an individual repeatedly pawns the same firearm, is the FFL required to do a NICS check each time the firearm is redeemed?

Yes. The fact that the transferee has redeemed the firearm before does not excuse the pawnbroker from complying with the Brady law.

(P21) Can licensed pawnbrokers conduct NICS checks prior to accepting a firearm in pawn?

Yes. The law provides that a NICS check may be done, on an optional basis, prior to accepting a firearm in pawn. If the check results in a "denied" response, the licensee is required to notify local law enforcement officials within 48 hours after receipt of the "denied" response.

(P22) If a pawnbroker conducts an optional NICS check prior to receiving a firearm in pawn, should the owner of the firearm complete a Form 4473?

ATF is developing a new form to cover these transactions. Until this form is distributed, ATF suggests that licensees have the owner complete Section A of the Form 4473, record the results of the NICS check on the form, and retain the form in their records for at least 5 years. These issues were addressed in greater detail in an open letter ATF sent to pawnbrokers.

(P23) What should a licensed pawnbroker do when he or she gets a "denied" response from an optional NICS check conducted prior to the receipt of a firearm in pawn?

The licensee is required to notify local law enforcement officials within 48 hours after receipt of the "denied" response. If the licensee has taken possession of the firearm, he or she may not return it to the individual who offered it for pawn.

(P24) If a licensed pawnbroker conducts a NICS check prior to accepting a firearm in pawn, and gets an "approved" response from NICS, must the pawnbroker conduct another NICS check if the firearm is redeemed from pawn? What if it is redeemed from pawn the same day?

The law provides that another NICS check must be done at the time of redemption, regardless of how recently the pre-pawn NICS check was conducted. Even if the firearm is redeemed the same day, a separate NICS check must be conducted at the time of redemption.

(P25) A firearm is delivered to a licensee by an unlicensed individual for the purpose of repair. Is the return of the repaired firearm subject to the requirements of the Brady law? Would the transfer of a replacement firearm from the licensee to the owner of the damaged firearm be subject to the requirements of the Brady law?

Neither the transfer of a repaired firearm nor the transfer of a replacement firearm would be subject to the requirements of the Brady law. Furthermore, the regulations provide that a Form 4473 is not required to cover these transactions. However, the licensee's permanent acquisition and disposition records should reflect the return of the firearm or the transfer of a replacement firearm.

(P26) Is a licensee's return of a consigned firearm to an unlicensed individual subject to permanent Brady?

Yes.

(P27) Do the requirements of the Brady law apply to sales of firearms to law enforcement officials for official use?

Transfers of firearms to law enforcement officials for their official use are exempt from the provisions of the Brady law when the transaction complies with the conditions set forth in the regulations at 27 CFR 178.134. In general, the purchaser must provide a certification on agency letterhead, signed by a person in authority within the agency (other than the officer purchasing the firearm), stating that the officer will use the firearm in official duties, and that a records check reveals that the purchasing officer has no convictions for misdemeanor crimes of domestic violence. If these conditions are met, the purchasing officer is not required to complete a Form 4473 or undergo a NICS check. However, the licensee must record the transaction in his or her permanent records and retain a copy of the certification letter.

(P28) Do the requirements of the Brady law apply to the sale of a firearm to a law enforcement official for his or her personal use?

Yes. In such transactions, the law enforcement official is treated no differently from any other unlicensed transferee, and a NICS check must be conducted.

(P29) Are there exceptions to the Brady law's requirement for a NICS check prior to a licensee's transfer of a firearm to an unlicensed individual?

Firearm transfers are exempt from the requirement for a NICS check in 3 situations. These include transfers (a) to buyers having a State permit that has been recognized by ATF as an alternative to a NICS check; (b) of National Firearms Act weapons approved by ATF; and (c) certified by ATF as exempt because compliance with the NICS check requirement is impracticable.

(P30) What steps must be followed by an FFL prior to transferring a firearm subject to the requirements of the Brady law?

The following steps must be followed prior to transferring a firearm:

1. The licensee must have the transferee complete and sign ATF Form 4473, Firearms Transaction Record.

2. The licensee must verify the identity of the transferee through a government-issued photo identification.

3. The licensee must contact NICS (through either the FBI or a State point of contact). The licensee will get either a "proceed", "denied" or "delayed" response from the system. If the licensee gets a "delayed" response and there is no additional response from the system, the licensee may transfer the firearm after 3 business days have elapsed. Of course, the licensee must still comply with any waiting period requirements under State law.

(P31) What form of identification must a dealer obtain from a purchaser under the Brady law?

The identification document presented by the purchaser must have a photograph of the purchaser, as well as the purchaser's name, address, and date of birth. The identification document must also have been issued by a governmental entity for the purpose of identification of individuals. An example of an acceptable identification document is a driver's license.

(P32) Under the Brady law, may a licensee transfer a firearm to a nonlicensed individual who does not appear in person at the licensed premises?

In any transaction that is subject to the requirement for a NICS check, the firearm may only be sold over-the-counter. Unless the purchaser appears in person at the licensed premises, the licensee cannot comply with the requirement in the Brady law that the identity of the purchaser be verified by means of a government-issued photo identification document.

(P33) If no NICS check is required, may a licensee transfer a firearm to a nonlicensed individual who does not appear in person at the licensed premises?

Yes, assuming the transaction otherwise complies with Federal law. For example, a licensee may still ship firearms to out-of-state law enforcement officials for official use, as long as the transaction complies with the alternate procedure set forth in the regulations. Furthermore, licensees may ship firearms intrastate to State residents who have valid permits that have been recognized as alternatives to the NICS check requirements, in compliance with the procedures set forth in 27 CFR 178.96(b).

(P34) Do gun purchasers have to fill out a Brady Form under the permanent Brady law?

No. The Brady Form and ATF Form 4473 have been combined into one form, a revised Form 4473.

(P35) Can licensees continue to use old versions of ATF Form 4473 on or after November 30, 1998?

Beginning November 30, 1998, licensees should use the forms dated 10/98. The old forms are obsolete. Licensees who have not yet received their new forms should contact their local ATF office to get copies of the form faxed to them. In addition, licensees may order copies of ATF Form 4473 from the ATF Distribution Center.

(P36) Is the transferee required to provide his or her social security number on the ATF Form 4473?

No. This information is solicited on an optional basis. However, providing this information will help ensure the lawfulness of the sale and avoid the possibility that the transferee will be incorrectly identified as a felon or other prohibited person.

(P37) Will NICS provide an instant response?

NICS will not always provide an instant response. Licensees will receive either a "proceed", "denied" or "delayed" response from NICS. If a "proceed" response is received, the transfer may proceed. If a "denied" response is received, then the transfer may not proceed. If a "delayed" response is received, then the transfer must be delayed until a final response is received from NICS or until after the lapse of 3 business days.

(P38) For purposes of the Brady law, what is meant by a "business day"?

A business day is defined as any day on which State offices are open.

(P39) If a licensee contacts NICS on Thursday, December 3, 1998, and gets a "delayed" response, when may the licensee transfer the firearm if no further response is received from NICS?

The firearm may be transferred on Wednesday, December 9, 1998. Assuming State offices are open on Friday, Monday, and Tuesday, and closed on Saturday and Sunday, 3 business days would have elapsed at the end of Tuesday, December 8, 1998. Therefore, the licensee may transfer the firearm at the start of business on Wednesday, December 9, 1998. (The 3 business days do not include the day the NICS check is initiated.)

(P40) What should a licensee do if he or she gets a "denied" response from NICS or a State point of contact after 3 business days have elapsed, but prior to the transfer of the firearm?

If the licensee receives a "denied" response at any time prior to the transfer of the firearm, he or she may not transfer the firearm.

(P41) What should an FFL tell a transferee who is denied by NICS?

The FFL should inform the transferee that the NICS check indicates that the transfer of the firearm should not be made, but that it does not provide a reason for the denial. The FFL should provide the transferee with the NICS or State transaction number and an appeals brochure. The FBI will provide FFLs with brochures that outline the transferee's appeal rights and responsibilities.

(P42) If a transferee receives a "denied" response from NICS, can the transferee find out why he or she was denied?

Yes. Although the FFL will not know the reason for the denial, the transferee may contact the FBI or the State point of contact in writing to request the reason for denial.

(P43) What information do FFLs have to record on ATF Form 4473 once they hear back from NICS or the State point of contact?

FFLs must record any "proceed", "delayed", or "denied" response received, as well as any transaction number provided.

(P44) What should licensees do if no transaction number is provided?

The FBI's NICS Operations Center will always provide a transaction number at the time of the initial inquiry. Some State points of contact (POC) may not provide a transaction number for denied transactions. If a State POC does not provide a transaction number for a denied transaction, then the licensee should just record the response without a transaction number. Licensees should note, however, that a transaction number is required if the firearm is being transferred within 3 business days of the initiation of a NICS check.

(P45) Do FFLs have to keep a copy of ATF Form 4473 if the transaction is denied or for some other reason is not completed?

FFLs must keep a copy of each ATF Form 4473 for which a NICS check has been initiated, regardless of whether the transfer of the firearm was made. If the transfer is not made, the FFL must keep the Form 4473 for 5 years after the date of the NICS inquiry. If the transfer is made, the FFL must keep the Form 4473 for 20 years after the date of the sale or disposition. Forms 4473 with respect to a transfer that did not take place must be separately maintained.

(P46) When should FFLs contact NICS?

FFLs should contact NICS after the transferee has completed Section A of the ATF Form 4473.

(P47) For what period of time is a NICS check valid?

A NICS check is valid for 30 calendar days, as long as it applies to a single transaction. An FFL may not rely on a NICS check that was conducted more than 30 calendar days prior to the transfer of the firearm.

Example: A NICS check is initiated on December 15, 1998. The FFL receives a "proceed" from NICS. The purchaser does not return to pick up the firearm until January 22, 1999. The FFL must conduct another NICS check before transferring the firearm to the purchaser, and must record the results of the check on the Form 4473.

(P48) Is a NICS check valid for 30 days from when the check was initiated, or from when a "proceed" is issued?

The NICS check is valid for 30 days from when the check was initiated.

Example: A NICS check is initiated on December 15, 1998. The FFL receives a "proceed" response from NICS on December 17, 1998. The purchaser does not return to pick up the firearm until January 16, 1999. The FFL must conduct another NICS check before transferring the firearm to the purchaser.

Example: A NICS check is initiated on December 15, 1998. The FFL receives a "delayed" response from NICS; no further response is received. The purchaser does not return to pick up the firearm until January 16, 1999. The FFL must conduct another NICS check before transferring the firearm to the purchaser.

(P49) A purchaser places an order for a custom-made firearm, which will not be ready for at least 60 days. Should the NICS check be initiated on the date the order is placed or the day the firearm is ready for delivery?

If the licensee knows that it will be more than 30 days before the firearm is ready, the licensee may choose to wait until the firearm is ready for delivery to have the purchaser complete Section A of the Form 4473 and contact NICS for a background check. If the purchaser completes Section A of the Form 4473 on the date the order is placed, and a NICS check is initiated on that day, the licensee will have to conduct a second NICS check prior to transferring the firearm.

(P50) Can one NICS check cover two or more separate firearms transactions?

No. An FFL must initiate separate NICS checks for separate transactions. However, an individual may purchase several firearms in one transaction.

Example: A purchaser completes an ATF Form 4473 for a single firearm on February 15, 1999. The FFL receives a "proceed" from NICS that day. The FFL signs the form, and the firearm is transferred. On February 20, 1999, the purchaser returns to the FFL's premises and wishes to purchase a second firearm. The purchase of the second firearm is a separate transaction. Therefore, a new NICS check must be initiated by the FFL.

Example: A purchaser completes ATF Form 4473 for a single firearm on

February 15, 1999. The FFL receives a "proceed" from NICS that day. The purchaser does not return to pick up the firearm until February 20, 1999. Before the FFL signs the Form 4473 for the first firearm, the purchaser decides to purchase an additional firearm. The second firearm may be recorded on the same Form 4473, in which case the purchase of the 2 firearms is considered a single transaction. Therefore, the licensee is not required to conduct a new NICS check prior to transferring the second firearm.

Example: A purchaser wishes to purchase 1 rifle and 1 handgun. State law requires that a background check be conducted on the sales of all handguns. The State is acting as a point of contact (POC) for NICS checks for handgun sales, while the FBI is conducting NICS checks for longgun transactions in that State. To comply with State law, the dealer initiates a background check through the State POC. There is no need to initiate a separate background check through the FBI for the sale of the rifle, since the 2 firearms are being transferred in one transaction.

(P51) If no exceptions to the NICS check apply, must an FFL always wait 3 business days before transferring a firearm to a transferee?

No. An FFL may transfer a firearm to a transferee as soon as he or she receives a "proceed" from NICS (assuming that the transaction would be in compliance with State law). However, if the FFL does not receive a final "proceed" or "denied" response from NICS, he or she must wait until 3 business days have elapsed prior to transferring the firearm.

(P52) Will a State "instant check" or "point of sale check" system qualify as an alternative to a NICS check?

No. However, it should be noted that many States with their own background check requirements are also acting as points of contact for NICS checks.

(P53) What happens if the transferee successfully appeals the NICS denial but more than 30 calendar days have elapsed since the initial background check was initiated?

The FFL must initiate another NICS check before the firearm may be transferred.

(P54) Are permits to carry concealed firearms included under the permit exception to the NICS check (assuming they meet the other requirements of the exception)?

Yes. The exception includes permits to carry concealed firearms and permits specifically authorizing the purchase of a firearm.

(P55) How does a licensee know whether a permit may be accepted as an alternative to a NICS check?

Prior to November 30, 1998, ATF sent an open letter to licensees in each State, advising them of what permits in that State qualified as alternatives to a NICS check at the time of transfer. Licensees having questions whether particular State permits are acceptable alternatives to NICS checks should contact their local ATF office.

(P56) Does a permit qualify as an alternative to a NICS check if the purchaser is using it to purchase a type of firearm that is not covered by the permit?

Yes, assuming the transaction complies with State law.

Example: ATF recognizes the permit to purchase a handgun and the concealed weapons permit as alternatives to a NICS check in State A. Any purchaser who displays a permit to purchase a handgun or a concealed weapons permit is not required to undergo a NICS check prior to purchasing a rifle, assuming the transaction complies with State law.

Example: In that same State, a person with a concealed weapons permit wants to purchase a handgun. State law prohibits the sale of any handgun to a person unless he or she has a permit to purchase. Accordingly, the licensee cannot lawfully sell the handgun unless the purchaser has a permit to purchase a handgun.

(P57) ATF has recognized the concealed weapons permits in State A and State B as valid alternatives to a NICS check. Can a resident of State A use a concealed weapons permit issued by State A to purchase a longgun in State B without undergoing a NICS check?

No. A permit qualifies as a NICS alternative only if it was issued by the State in which the transfer is to take place.

(P58) ATF's open letter to licensees in State C states that while concealed weapons permits issued by State C were recognized as alternatives to the background check required by the interim Brady law, concealed weapons permits issued by State C on or after November 30, 1998, will not be recognized as alternatives to the NICS check required by the permanent Brady law. Why will these permits no longer be recognized as Brady alternatives?

The change in the status of the permits is probably due to one of the following circumstances: (1) the State does not conduct NICS checks prior to issuing permits; or (2) the State does not disqualify all permit applicants prohibited from possessing firearms under Federal law.

(P59) On December 1, 1998, can a licensee accept as a NICS alternative a "grandfathered" concealed weapons permit that was issued by State C on January 1, 1993?

No. The permit must have been issued within the past 5 years.

(P60) If State law provides that permits are only valid for 2 years, can a licensee accept as a NICS alternative a permit that was issued 4 years ago?

No. The permit must be still valid under State law in order to qualify as an alternative to a NICS check.

(P61) If the State has its own instant check system, must the licensee comply with State requirements for a background check as well as the Brady law?

Yes. If the State is not acting as a point of contact for NICS checks, the licensee may have to initiate 2 background checks by contacting (1) the FBI's NICS Operations Center, and (2) the State system.

(P62) If a licensee gets a "proceed" response from NICS, does he or she still have to wait until the expiration of the State waiting period before transferring the firearm?

Yes. Compliance with the Brady law does not excuse a licensee from compliance with State law.

(P63) If a State is acting as a NICS point of contact (POC), and State law has separate requirements regarding the amount of time that a licensee must wait before transferring a firearm after contacting the State, should the licensee comply with the State requirements, the Federal requirements, or

both?

The licensee must comply with both State and Federal requirements.

Example: State D is acting as a POC for NICS checks. State law requires a background check prior to the transfer of any firearm. State law also requires the licensee to wait 10 days to get a response from the State. The licensee must contact the State POC for a NICS check and a State background check. The licensee must comply with both Federal and State law by waiting 10 days for a response prior to transferring the firearm. If the licensee has not received a response from the State after 10 days, he or she may transfer the firearm.

Example: State E is acting as a POC for NICS checks. State law requires a background check prior to the transfer of any firearm. Under State law, the licensee may transfer the firearm if he or she gets no final response from the State by the next day. The licensee contacts the State POC for a NICS check, and gets a "delayed" response. Assuming that the licensee gets no further response from the State POC, the licensee must comply with both Federal and State law by waiting until 3 business days have elapsed prior to transferring the firearm.

(P64) Does an individual licensee have to conduct a NICS check on himself or herself prior to transferring a firearm to his or her own personal collection?

No. The regulations do not require a licensee to complete a Form 4473 prior to transferring a firearm to his or her own personal collection; accordingly, a NICS check is not required either. Such transfers must be recorded in the manner prescribed by the regulations at 27 CFR 178.125a.

(P65) Is a NICS check required for the sale of firearms registered under the National Firearms Act (NFA)?

No, assuming all NFA requirements have been satisfied.

(P66) An organization without a firearms license wishes to acquire a firearm from a licensee for the purpose of raffling the firearm at an event. How does the licensee comply with the Brady law?

The licensee must comply with the Brady law by conducting a NICS check on the transferee. If the licensee wishes to transfer the firearm to the organization, a representative of the organization must complete a Form 4473, and a NICS check must be conducted on that representative prior to the transfer of the firearm. Alternatively, if the licensee transfers the firearm directly to the winner of the raffle, the winner must complete a Form 4473, and a NICS check must be conducted on the raffle winner prior to the transfer.

Example: A licensee transfers a firearm to the organization sponsoring the raffle. The licensee must comply with the Brady law by requiring a representative of the organization to complete the Form 4473 and undergo a NICS check. Once the firearm has been transferred to the organization, the organization can subsequently transfer the firearm to the raffle winner without a Form 4473 being completed or a NICS check being conducted. This is because the organization is not an FFL; therefore, its transfer of a firearm is not subject to the Brady law.

Example: The licensee or his or her representative brings a firearm to the raffle, so that the firearm can be displayed. After the raffle, the firearm is returned to the licensee's premises. The licensee must comply with the Brady law prior to transferring the firearm directly to the winner of the raffle by requiring the winner to complete the Form 4473 and undergo a NICS check at the licensee's premises prior to the transfer of the firearm.

Example: The raffle meets the definition of an "event" at which a licensee is allowed to conduct business under 27 CFR 178.100(b). The licensee attends the event and transfers the firearm at the event to the winner. The licensee must comply with the Brady law prior to transferring the firearm to the winner by requiring the winner to complete a Form 4473 and undergo a NICS check.

Q. MISDEMEANOR CRIME OF DOMESTIC VIOLENCE

(Q1) What is a "misdemeanor crime of domestic violence"?

A "misdemeanor crime of domestic violence" means an offense that:

(1) is a misdemeanor under Federal or State law;

(2) has, as an element, the use or attempted use of physical force, or the threatened us of a deadly weapon; and

(3) was committed by a current or former spouse, parent, or guardian of the victim, by a person with whom the victim shares a child in common, by a person who is cohabiting with or has cohabited with the victim as a spouse, parent, or guardian, or by a person similarly situated to a spouse, parent, or guardian of the victim.

However, a person is not considered to have been convicted of a misdemeanor crime of domestic violence unless:

(1) the person was represented by counsel in the case, or knowingly and intelligently waived the right of counsel in the case; and

(2) in the case of a prosecution for which a person was entitled to a jury trial in the jurisdiction in which the case was tried, either -

    (a) the case was tried by a jury, or

    (b) the person knowingly and intelligently waived the right to have the case tried by a jury, by guilty plea or otherwise.

In addition, a conviction would not be disabling if it has been expunged or set aside, or is an offense for which the person has been pardoned or has had civil rights restored (if the law of the jurisdiction in which the proceedings were held provides for the loss of civil rights upon conviction for such an offense) unless the pardon, expunction, or restoration of civil rights expressly provides that the person may not ship, transport, possess, or receive firearms, and the person is not otherwise prohibited by the law of the jurisdiction in which the proceedings were held from receiving or possessing firearms.

[18 U.S.C. 921(a)(33), 27 CFR 178.11]

(Q2) What is the effective date of this disability?

The law was effective September 30, 1996. However, the prohibition applies to persons convicted of such misdemeanors at any time, even if the conviction occurred prior to the law's effective date.

[18 U.S.C. 922(g)(9), 27 CFR 178.32(a)(9)]

(Q3) Does application of the law to persons convicted prior to the law's effective date violate constitutional rights?

No. This provision is not being applied retroactively or in violation of the ex post facto clause of the Constitution. This is because the law does not impose additional punishment upon persons convicted prior to the effective date, but merely regulates the future possession of firearms on or after the effective date. The provision is not retroactive merely because the person's conviction occurred prior to the effective date.

[18 U.S.C. 922(g)(9), 27 CFR 178.32(a)(9)]

(Q4) X was convicted of misdemeanor assault on October 10, 1996 for beating his wife. The crime of assault does not make specific mention of domestic violence. May X still lawfully possess firearms or ammunition?

No. X may no longer legally possess firearms or ammunition.

[18 U.S.C. 922(g)(9), 27 CFR 178.32(a)(9)]

(Q5) X was convicted of a misdemeanor crime of domestic violence on September 20, 1996, 10 days before the effective date of the new statute. He possesses a firearm on October 10, 1996. Does X lawfully possess the firearm?

No. If a person was convicted of a misdemeanor crime of domestic violence at any time, he or she may not lawfully possess firearms or ammunition on or after September 30, 1996.

[18 U.S.C. 922(g)(9), 27 CFR 178.32(a)(9)]

(Q6) In determining whether a conviction in a State court is a "conviction" of a misdemeanor crime of domestic violence, does Federal or State law apply?

State law applies. If a conviction of a disqualifying misdemeanor does not occur under State law, the person has not been "convicted" of a misdemeanor crime of domestic violence. Therefore, if the State does not consider the person to be convicted, the person would not have the disability.

[18 U.S.C. 921(a)(33), 27 CFR 178.11]

(Q7) What State and local offenses are "misdemeanors" for purposes of 18 U.S.C. 922(d)(9) and (g)(9)?

The definition of misdemeanor crime of domestic violence in the GCA includes any offense classified as a "misdemeanor" under Federal or State law. In States that do not classify offenses as misdemeanors, the definition includes any State or local offense punishable by imprisonment for a term of 1 year or less. For example, if State A has an offense classified as a "domestic violence misdemeanor" that is punishable by up to 5 years imprisonment, it would be a misdemeanor crime of domestic violence as defined. If State B does not characterize offenses as misdemeanors, but has a domestic violence offense that is punishable by no more than 1 year imprisonment, this offense would be a misdemeanor crime of domestic violence as defined.

[18 U.S.C. 921(a)(33), 27 CFR 178.11]

(Q8) Are local criminal ordinances "misdemeanors under State law" for purposes of sections 922(d)(9) and (g)(9)?

Yes, assuming a violation of the ordinance meets the definition of "misdemeanor crime of domestic violence" in all other respects.

[18 U.S.C. 921(a)(33), 27 CFR 178.11]

(Q9) Does the definition of "misdemeanor crime of domestic violence" include offenses that are punishable only by a fine?

Yes, assuming the offense meets the definition of "misdemeanor crime of domestic violence" in all other respects. Nothing in the language of the definition limits the term to offenses punishable by imprisonment. The legislative history of the definition illustrates that the prohibition on firearm possession by persons convicted of such offenses was to be as broad as possible, for example, covering individuals who plead guilty to minor offenses.

[18 U.S.C. 921(a)(33), 27 CFR 178.11]

(Q10) What is meant by the "similarly situated to a spouse" provision in the definition of a misdemeanor crime of domestic violence?

The phrase means 2 persons who share the same domicile in an intimate relationship, with "domicile" being one's fixed place of dwelling where one intends to reside more or less permanently. Such persons do not have to be married or in a "common law" marriage relationship. However, they must be in more than a "dating" relationship. Furthermore, the "similarly situated to a spouse" provision applies to anyone who was domiciled in an intimate relationship with the victim of the offense either at the time of, or at any time prior to, the offense.

[18 U.S.C. 921(a)(33), 27 CFR 178.11]

(Q11) In order for an offense to qualify as a "misdemeanor crime of domestic violence", does it have to have as an element the relationship part of the definition (e.g., committed by a spouse, parent, or guardian)?

No. The "as an element" language in the definition of "misdemeanor crime of domestic violence" only applies to the use of force provision of the statute and not the relationship provision. However, to be disabling, the offense must have been committed by one of the defined parties.

[18 U.S.C. 921(a)(33), 27 CFR 178.11]

(Q12) Is a person who received "probation before judgment" or some other type of deferred adjudication subject to the disability?

What is a conviction is determined by the law of the jurisdiction in which the proceedings were held. If the State law where the proceedings were held does not consider probation before judgment or deferred adjudication to be a conviction, the person would not be subject to the disability.

[18 U.S.C. 921(a)(33), 27 CFR 178.11]

(Q13) What should a licensee do if he or she has been convicted of a misdemeanor crime of domestic violence?

Federal firearms licensees should verify that they are disabled under the prohibition. A licensee convicted of a disqualifying misdemeanor may not lawfully possess firearms or ammunition. In addition, a licensee who incurs firearms disabilities during the term of a license by reason of such a misdemeanor conviction may not continue operations under the license for more than 30 days after incurring the disability unless the licensee applies for relief from Federal firearms disabilities.

[18 U.S.C. 922(g)(9) and 925(c), 27 CFR 178.144(c)(8) and (i)]

(Q14) What should an individual do if he or she has been convicted of a

misdemeanor crime of domestic violence?

Individuals subject to this disability should immediately dispose of their firearms and ammunition. ATF recommends that such persons transfer their firearms and ammunition to a third party who may lawfully receive and possess them, such as their attorney, a local police agency, or a Federal firearms dealer. The continued possession of firearms and ammunition by persons under this disability is a violation of law and may subject the possessor to criminal penalties. In addition, such firearms and ammunition are subject to seizure and forfeiture.

[18 U.S.C. 922(g)(9) and 924(d)(1), 27 CFR 178.152]

(Q15) Does the disability apply to law enforcement officers?

Yes. The Gun Control Act was amended so that employees of government agencies convicted of misdemeanor crimes of domestic violence would **not** be exempt from disabilities with respect to their receipt or possession of firearms or ammunition. Thus, law enforcement officers and other government officials who have been convicted of a disqualifying misdemeanor may not lawfully possess or receive firearms or ammunition for any purpose, including performance of their official duties. The disability applies to firearms and ammunition issued by government agencies, purchased by government employees for use in performing their official duties, and personal firearms and ammunition possessed by such employees.

[18 U.S.C. 922(g)(9) and 925(a)(1), 27 CFR 178.32(a)(9) and 178.141(a)]

INDEX

SUBJECT	LAW	REGULATIONS	RULINGS, PROCEDURES, INDUSTRY CIRCULARS	ITEMS OF INTEREST	QUESTIONS AND ANSWERS
Address					
Change in business location		178.52, 179.46			
Adjudicated mental defective					
Defined		178.11			
Possession, receipt of firearm/ammunition	922(g)(4)	178.32(a)(4)			
Sale of firearm/ammunition to	922(d)(4)	178.32(d)(4)			
Alien					B
Illegal alien					
Defined		178.11			
Possession, receipt of firearm/ammunition	922(g)(5)	178.32(a)(5)			
Sale to alien	922(d)(5)	178.32(d)(5)			
Legally in U.S.					
Establishing residency to purchase firearm	922(b)(3)	178.11("State of residence"), .124	Rul. 79-7	Item 5	
Receipt of firearm by nonresident alien	922(a)(9)			Item 5	
Sale of firearm to alien for export				Item 5	
State of residence defined		178.11			
Under nonimmigrant visa	922(d)(5), (g)(5)			Item 5	
Waiver by Attorney General	922(y)(3)				
Alternate methods, procedures		178.22(a), 179.26(a)			
Ammunition					F
Armor piercing (see "Armor piercing ammunition")					
Defined	921(a)(17)(A)	178.11			
License					
Importer	923(a)(2)	178.41(a)			
Manufacturer	923(a)(1)	178.41(a)			
Possession, receipt by prohibited person	922(g)	178.32(a)			
Possession by juvenile	922(x)(2)				
Sale to juvenile	922(x)(1)				
Sale to prohibited person	922(d)	178.32(d)			
Sale to underage person	922(b)(1)	178.99(b)		Item 3	
Antique firearm					A
Defined					
Arms Export Control Act		47.11		Item 8	
GCA	921(a)(16)	178.11		Item 8	
Muzzle loading weapon	921(a)(16)			Item 8	
NFA	5845(g)	179.11		Item 8	
Replica	921(a)(16)			Item 8	
Exempt from firearm definition	921(a)(3), 5845(a)	47.11, 178.11, 179.11			
Any other weapon (see also "NFA controls")					
Defined	5845(e)	179.11			
Taser			Rul. 76-6		
Armor piercing ammunition					E
Defined	921(a)(17)(B)	178.11		Item 17	
Exempted				Item 17	
For export	922(a)(7)(B), (8)(B)	178.37(b)			
For government use	922(a)(7)(A), (8)(A)	178.37(a)			
For sporting, industrial purpose	921(a)(17)(C)	178.148			
For testing, experimentation	922(a)(7)(C), (8)(C)	178.37(c), .149			
Identification	923(k)	178.92(b)(1)			
Labeling of packages	923(k)	178.92(b)(2)			
License revocation for transfer	923(e)	178.99(e)			
Importation	922(a)(7)	178.99(d)			
Manufacture	922(a)(7)	178.99(d)			
Record of sale	922(b)(5)	178.125(a)-(d)			
Sale	922(a)(8)	178.99(d)			
Assembly					
Semiautomatic rifle, shotgun from imported parts	922(r)	178.39		Item 18	
For government	922(r)(1)	178.39(b)(1)			
For testing, experimentation	922(r)(2)	178.39(b)(2), .151			
Repair		178.39(b)(3)			
Assault weapon (see "Semiautomatic assault weapon")					
Auctioneer			Rul. 96-2		K
Auto sear			Rul. 81-4		
Bequest					
Acquiring firearm interstate by	922(a)(3)	178.29(a)			
Transfer of firearm interstate to carry out	922(a)(5)	178.30(a)			
Bound book (see "Records")					
Brady Law (National Instant Criminal Background Check System, "NICS")					G, J, P
Accessing NICS	922(t)(1)(A)	178.102(a)			
For purpose other than NICS check		25 6(j)			

INDEX

SUBJECT	LAW	REGULATIONS	RULINGS, PROCEDURES, INDUSTRY CIRCULARS	ITEMS OF INTEREST	QUESTIONS AND ANSWERS
Through FBI		25 6(a), (b)			
Through State point of contact		25 6(d)			
Civil fine	922(t)(5)				
Hearing		178.74			
Notice of		178.73			
Correction of erroneous information	925A	25.10			
Definitions					
NICS		178.11, 25.2			
Other terms		25.2			
Enrollment with FBI					P
Establishment of NICS	922(t)(1)	25.3			
Exceptions					
Certain geographical areas	922(t)(3)(C)	178.102(d)(3), .150			
NFA transfer	922(t)(3)(B)	178.102(d)(2)			
Permit	922(t)(3)(A)	178.102(d)(1)			
Identification					
Document	922(s)(3)(A)	178.11, .102(a)(3), .124(c)(3)(i)			
Transaction identification number	922(t)(2)(A), (4)	178.102(b), .124(c)(3)(iii), 25.6(c)			
Verifying identity of transferee	922(t)(1)(C)	178.102(a)(3), .124(c)(3)(i)			
Liability in civil action	922(t)(6)				
Loan or rental of firearm		178.97			
Out-of-State, mail-order sales		178.96(b)			
Pawnbroker					
Optional NICS check before pawn					J
Penalties					
Prohibited activities		25.11			
Violation of Brady Law	924(a)(5)				
Prohibited transferees	922(g), (n), (t)				
Records					
Destruction	922(t)(2)(C)	25 9			
NICS checks where transfer not made		178.129(b)			
Response to NICS check		178.124(c)(3)(iii)			
Retention periods		178.102(e), .129(b), .131			
Transaction identification number	922(t)(4)	178.124(c)(3)(iii)			
Transactions not subject to NICS		178.131			
Transfer of firearm		178.124(c)			
Remedy for erroneous denial	925A	25.10			
Requirement to contact NICS	922(t)(1)	178.102(a)			
Exceptions	922(t)(3)	178.102(d)			
Responses to NICS checks					
"Delayed"		25 6(c)(1)(B)			
"Denied"		25 6(c)(1)(C)			
"Proceed"		25 6(c)(1)(A)			
Revocation of license	922(t)(5)				
Hearing		178.74			
Notice of		178.73			
Suspension of license	922(t)(5)				
Hearing		178.74			
Notice of		178.73			
System safeguards		25 8			
Time limitation on NICS check		178.102(c)			
Transfer of firearm	922(t)(1)	178.102(a)-(c), 25.6(a)			
Validation of records in system		25 5			
Business					
Changes					
Bankruptcy		179.43			
Death of owner		179.42			
In corporation		179.45			
In partnership		179.44			
New license required for change of address		179.52		Item 10	
Notification of change of control		178.54			
Notification of new trade name		178.53		Item 10	
Continuance by certain successors		178.56			
Continuing operations after indictment, conviction	925(b), (c)	178.143			
Delivery to business with expired license		178.94			
Discontinuance of	923(g)(4)	178.57, .127, 179.105(f)		Item 10	
Hours					
Gunsmith			Rul. 73-13		
Inspection during	923(g)(1)(A)	178.23			
License to engage in	922(a)(1), 923(a)	178.41(a)			
Premises					

INDEX

INDEX

SUBJECT	LAW	REGULATIONS	RULINGS, PROCEDURES, INDUSTRY CIRCULARS	ITEMS OF INTEREST	QUESTIONS AND ANSWERS
Posting sign on juvenile handgun possession		178.103(d)-(f)			
Records (see "Records")					
Registration under NFA	5802	179.34			
Reports (see "Reports")					
Sale (see "Sale or delivery by licensee")					
Defense article					
Importation	22 USC 2778	Part 47			
U.S. Munitions Import List		47 21			
Definitions of terms	921, 5845	47.11, 178.11, 179.11			
Destructive device (see also "NFA controls")					
Defined	921(a)(4) 5845(f)	178.11 179.11			
Gas/flare gun			Rul. 95-3		
Striker-12/Streetsweeper shotgun			Rul. 94-2		
Removal from definition	5845(f)	179.24			
Sale to nonlicensee	922(b)(4)	178.98			
Sale to research organization		178.145			
Transportation	922(a)(4)	178.28		Item 11	
USAS-12 shotgun			Rul. 94-1		
Discharged under dishonorable conditions					B
Defined		178.11			
Possession, receipt of firearm/ammunition by discharged person	922(g)(6)	178.32(a)(6)			
Sale of firearm, ammunition to discharged person	922(d)(6)	178.32(d)(6)			
District of Columbia					
State	921(a)(2)	47.11, 178.11, 179.11			
Drug user, addict (see "unlawful drug user, addict")					
Dwelling as business premises		178.11			
Engaged in the business					
Defined					
Dealer in firearms	921(a)(21)(C)	178.11			
Gunsmith	921(a)(21)(D)	178.11			
Importer of firearms/ammunition	921(a)(21)(E), (F)	178.11			
Manufacturer of firearms/ammunition	921(a)(21)(A), (B)	178.11			
Emergency variations from requirements		178.22(b), 179.26(b)			
Estate, NFA firearm in				Item 14	
Excise tax				Item 16	
Exportation (see also "NFA controls")					L
Application to export NFA firearm	5854	179.114			
Guidelines				Item 5	
Proof of	5854	179.118			
Records	5843	178.171, 179.131			
Shotgun				Item 5	
To U.S. Armed Forces member	925(a)(3)				
Firearm					
Defined					
Antique	921(a)(16), 5845(g)	178.11, 179.11		Item 8	
Arms Export Control Act		47.11			
GCA	921(a)(3)	178.11			
NFA	5845(a)	179.11			
Muzzle loading firearm	921(a)(16)			Item 8	
Taser			Rul. 76-6, 80-20		
Tear gas weapons			Rul. 75-7		
Undetectable firearm	922(p)				
False statement, records					
Arms Export Control Act	22 USC 2778(c)	47 62			
GCA	922(a)(6), (m), 924(a)(1)(A), (3)	178.128			
NFA	5861(l)	179.181			
Frame or receiver					
Firearm	921(a)(3)(C)	178.11			
Machinegun	5845(b)	179.11			
Fugitive from justice					B
Defined		178.11			
Possession, receipt of firearm/ammunition by	922(g)(2)	178.32(a)(2)			
Sale of firearm/ammunition to	922(d)(2)	178.32(d)(2)			
Gun free school zone					
Defined	921(a)(25), (26)				
Discharging firearm in	922(q)(3)(A)				
Exceptions to prohibitions	922(q)(2)(B), (3)(B)				
Possession of firearm in	922(q)(2)(A)				
Gun show					F
Inspection	923(j)	178.23			

INDEX

SUBJECT	LAW	REGULATIONS	RULINGS, PROCEDURES, INDUSTRY CIRCULARS	ITEMS OF INTEREST	QUESTIONS AND ANSWERS
NFA firearm	925(d)(2), (3), 5844	178.111(a), 179.111(a)			
Nonsporting firearm/ammunition	925(d)(3)	178.111(a)	Rul. 80-8		
Permit		47.41-46, 178.111-119, 179.111			
Denial, revocation, suspension		47.44			
Sales sample of NFA firearm	5844(3)	179.111(a)(3)	Rul. 85-2		
Semiautomatic assault rifle	925(d)(3)			Item 20	
Surplus military firearm	925(d)(3)	178.111	Rul. 80-8, 85-10		
Unserviceable firearm	925(d)(1)				
U.S. military firearm/ammunition	22 USC 2778(b)	47 57			
U.S. Munitions Import List		47 21			
War souvenir	925(a)(4)	178.114(c)		Item 9	
Imported parts (see also "Assembly")					
Assembly of semiautomatic rifle, shotgun	922(r)	178.39			
Defined		178.39(c)			
Importer of firearms/ammunition (see also "Brady law", "License", "NFA controls")					A, C
Application for license	923(a)	178.41(b), .44(a)			
Civil fine	922(t)(5)	178.73(a)			
Defined	921(a)(9)	178.11, 179.11			
Engaged in business as importer					
Defined	921(a)(21)(E), (F)	178.11			
License denial	923(d)(2), (f)(1), (2)	178.71, .72			
License requirement	922(a)(1), 923(a)	178.41(a)			
License revocation	923(e)	178.73			
License suspension	922(t)(5)	178.73(a)			
Licensing standards	923(d)(1)	178.47(b)			
Notice to handgun purchaser		178.103(a)-(c)			
Posting of sign on juvenile handgun possession		178.103(d)-(f)			
Records (see "Records")					
Registration					
Arms Export Control Act	22 USC 2778(b)	47 31			
NFA	5802	179.34			
Reports (see "Reports")					
Sales (see Sale or delivery by licensee)					
Indictment					B
Continuing licensed operations after	925(b), (c)				
Defined		178.11			
Receipt of firearm, ammunition by person under	922(n)	178.32(b)			
Sale of firearm, ammunition to person under	922(d)(1)	178.32(d)(1)			
Inspection					
Gun show	923(j)	178.23			
Importers under Arms Export Control Act		47 34(b)			
Licensees under GCA					
Importers, manufacturers, dealers	923(g)(1)(A), (B)	178.23, .121(b)			
Collectors	923(g)(1)(C), (D)	178.23, .121(b)			
Special (occupational) taxpayers under NFA		179.22			
Insular possessions					
State	921(a)(2)				
Interstate or foreign commerce					
Defined	921(a)(2)	178.11			
Interstate transactions					
Licensees					F
Loan, rental of firearm for sporting purposes	922(b)(3)(B)	178.97, .99(a)(2)			
Return, replacement of firearm	922(a)(2)(A)	178.147			
Sale of curio or relic	923(j)	178.100(a)(2)			
Sale of firearm to nonresident	922(b)(3)	178.99(a)			
Sale of rifle or shotgun	922(b)(3)(A)	178.99(a)(1)			
Transfer of firearm to writer, consultant			IC 72-23		
Transfer of firearm to agent, employee			IC 72-23		
Nonlicensees					B
Acquisition of firearm generally	922(a)(3)	178.29			
Acquisition of rifle/shotgun from licensee	922(a)(3)(B)	178.29(a)			
Interstate shipment of firearm to licensee		178.147			
Loan, rental of firearm	922(a)(5)(B)	178.30(b)			
Transfer of firearm to nonlicensee generally	922(a)(5), (b)(3)	178.30			
Prohibited persons	922(g)	178.32(a)-(c)			
Transfer, receipt of firearm to carry out					
Bequest	922(a)(3)(A), (5)(A)	178.29(a), 30(a)			
Intestate succession	922(a)(3)(A), (5)(A)	178.29(b), 30(a)			
Transportation	922(a)(4), 926A	178.28			
Intestate succession					
Acquiring firearm interstate by	922(a)(3)(A)	178.29(a)			

INDEX

INDEX

INDEX

INDEX

SUBJECT	LAW	REGULATIONS	RULINGS, PROCEDURES, INDUSTRY CIRCULARS	ITEMS OF INTEREST	QUESTIONS AND ANSWERS
Sale to nonlicensee	922(b)(4)	178.98			
Sale to research organization		178.145			
Weapon made from	5845(a)(2)				
Silencer (Muffler) (see also "NFA controls")	921(a)(3)(C), 5845(a)(7)				
Defined	921(a)(24)	179.11			
Skeet, trap, similar activity		178.35			
State of residence					B
Alien legally in U.S.		178.11			
College student			Rul. 80-21		
Defined		178.11			
Member of Armed Forces	921(b)	178.11			
Stolen firearm					F
Possession, receipt	922(j)	178.33			
Report of theft, loss	923(g)(6)	178.39a		Item 19	
Theft from licensee	922(u), 924(m)	178.33a			
Theft from any person	924(l)				
Transportation, shipment	922(i)	178.33			
Storage, guns	921(a)(34), 923(d)(1)(G)				
Straw purchase				Item 15	
Theft of firearm (see "Stolen firearm")					
Transportation					
By prohibited person	922(g)	178.32(a)-(c)			
Interstate transport of					
Destructive device	922(a)(4)	178.28		Item 11	
Machinegun	922(a)(4)	178.28		Item 11	
Short-barreled rifle	922(a)(4)	178.28		Item 11	
Short-barreled shotgun	922(a)(4)	178.28		Item 11	
Personal firearm	926A	178.38		Item 11	
Undetectable firearm	922(p)				
Unlawful drug user, addict					B
Defined		178.11			
Possession, receipt of firearm/ammunition by	922(g)(3)	178.32(a)(3)			
Sale of firearm/ammunition to	922(d)(3)	178.32(d)(3)			
United States					
Defined	921(a)(2)	47.11, 179.11			
U.S. Munitions Import List		47 21			
Verification of identity of licensee			IC 74-13		F
Certified copy of license	926(a)(1)	178.95			
Certified list of licenses		178.94			
Waiver by Attorney General	922(y)(3)				
Weapon made from rifle, shotgun (see "NFA controls")					

Form Approval: OMB No. 1512-0525 (05/31/2001)

DEPARTMENT OF THE TREASURY
BUREAU OF ALCOHOL, TOBACCO AND FIREARMS
APPLICATION FOR AN AMENDED FEDERAL FIREARMS LICENSE
UNDER 18 U.S.C. CHAPTER 44, FIREARMS

All entries must be in ink. Please read attached instructions carefully before completing this form.

SECTION I

1. NAME OF OWNER OR CORPORATION *(If partnership, include name of each partner)*	2. SOCIAL SECURITY NUMBER 3. DATE OF BIRTH
4. TRADE OR BUSINESS NAME *(If any)*	5. FEDERAL FIREARMS LICENSE NUMBER *(Include expiration date)*
6. CURRENT BUSINESS ADDRESS *(Include county)*	7. NEW BUSINESS ADDRESS *(Include county)*
8. CURRENT MAILING ADDRESS	9. NEW MAILING ADDRESS
10. CURRENT HOME ADDRESS	11. NEW HOME ADDRESS
12. PRESENT TELEPHONE NUMBER *(With area code)* BUSINESS _____ RESIDENCE _____	13. NEW TELEPHONE NUMBER *(With area code)* BUSINESS _____ RESIDENCE _____

IF BUSINESS IS OBTAINED FROM SOMEONE ELSE, GIVE:

14. NAME	15. LICENSE NUMBER

16. HOURS OF OPERATION OF LICENSEE'S BUSINESS

TIME	SUNDAY	MONDAY	TUESDAY	WEDNESDAY	THURSDAY	FRIDAY	SATURDAY
OPEN							
CLOSE							

17. ARE THE LICENSEE'S BUSINESS PREMISES OPEN TO THE GENERAL PUBLIC DURING THESE HOURS?

☐ YES   ☐ NO

18. WILL ANY OTHER BUSINESS, OTHER THAN THAT FOR WHICH THE LICENSE WAS ISSUED, BE CONDUCTED ON THE NEW PREMISES?

☐ YES   ☐ NO

19. LICENSEE'S NEW PREMISES ARE

☐ OWNED   ☐ LEASED   ☐ RENTED

20. IF THE NEW PREMISES ARE RENTED OR LEASED, PLEASE SPECIFY THE DATE THE LEASE OR RENTAL AGREEMENT BEGINS, THE DATE THE AGREEMENT TERMINATES, THE TERM OF THE LEASE OR RENTAL, AND THE NAME, ADDRESS AND TELEPHONE NUMBER OF THE OWNER OF THE PROPERTY, ATTACH A COPY OF THE LEASE OR RENTAL AGREEMENT TO THIS APPLICATION.

21. INDICATE TYPE OF BUSINESS PREMISES AT THE NEW LOCATION:

COMMERCIAL

☐ STOREFRONT

☐ OFFICE

☐ ROD & GUN CLUB

☐ MILITARY BASE

☐ OTHER (Specify) _____

RESIDENTIAL

☐ SINGLE/FAMILY DWELLING

☐ CONDOMINIUM

☐ APARTMENT

☐ HOTEL/MOTEL

☐ PUBLIC HOUSING

☐ OTHER (Specify) _____

IF THE NEW PREMISES ARE LOCATED ON A MILITARY INSTALLATION, ATTACH A WRITTEN AUTHORIZATION FROM THE BASE COMMANDER TO CONDUCT THE FIREARMS BUSINESS ON THE MILITARY INSTALLATION.

22. IF THE NEW PREMISES ARE OWNED BY THE APPLICANT, ARE THERE ANY RESTRICTIONS OR COVENANTS WHICH PROHIBIT THE OWNER FROM OPERATING A BUSINESS ON THE PREMISES? IF YES, ON A SEPARATE SHEET, DESCRIBE SUCH RESTRICTIONS OR COVENANTS.

☐ YES   ☐ NO

23. DO ANY ZONING ORDINANCES PROHIBIT THE CONDUCT OF A FIREARMS BUSINESS ON THE NEW PREMISES? ☐ YES ☐ NO	24. IS A LICENSE, PERMIT, OR PAYMENT OF OCCUPATIONAL OR BUSINESS TAX REQUIRED UNDER STATE OR LOCAL LAW TO ENGAGE IN A FIREARMS BUSINESS FROM THE PREMISES? IF SO, PROVIDE COPIES. ☐ YES ☐ NO

SIGN HERE	TITLE	DATE

SECTION II - CERTIFICATION OF COMPLIANCE WITH STATE AND LOCAL LAW

AS REQUIRED BY 18 U.S.C. 923(d)(1), I CERTIFY THAT:

1. THE BUSINESS TO BE CONDUCTED UNDER THE FEDERAL FIREARMS LICENSE IS NOT PROHIBITED BY STATE OR LOCAL LAW AT THE PREMISES INDICATED IN QUESTION 5.

2. WITHIN 30 DAYS AFTER THE APPLICATION IS APPROVED THE BUSINESS TO BE CONDUCTED AT THE LOCATION INDICATED IN QUESTION 5 WILL COMPLY WITH THE REQUIREMENTS OF STATE AND LOCAL LAW APPLICABLE TO THE CONDUCT OF BUSINESS.

3. BUSINESS WILL NOT BE CONDUCTED UNDER THE AMENDED LICENSE UNTIL THE REQUIREMENTS OF STATE AND LOCAL LAW APPLICABLE TO THE BUSINESS HAVE BEEN MET.

4. NOTIFICATION OF THIS APPLICATION HAS BEEN PROVIDED TO THE CHIEF LAW ENFORCEMENT OFFICER OF THE LOCALITY IN WHICH THE PREMISES INDICATED IN QUESTION 5 ARE LOCATED BY MAILING COPY 2 OF THIS FORM TO SUCH OFFICER.

25. CERTIFICATION: UNDER THE PENALTIES IMPOSED BY 18 U.S.C. 924, I DECLARE THAT I HAVE EXAMINED THIS APPLICATION AND THE DOCUMENTS SUBMITTED IN SUPPORT HEREOF, AND TO THE BEST OF MY KNOWLEDGE AND BELIEF, THEY ARE TRUE, CORRECT AND COMPLETE.

SIGN HERE	DATE

ATF F 5300.38 (10-96)

www.ingramcontent.com/pod-product-compliance
Lightning Source LLC
Chambersburg PA
CBHW081430170526
45166CB00008B/2156